Crop Development, Growth and Yield

Crop Development, Growth and Yield

Edited by Lancaster Mason

SYRAWOOD
PUBLISHING HOUSE

New York

Published by Syrawood Publishing House,
750 Third Avenue, 9th Floor,
New York, NY 10017, USA
www.syrawoodpublishinghouse.com

Crop Development, Growth and Yield
Edited by Lancaster Mason

International Standard Book Number: 978-1-68286-704-4 (Hardback)

Cataloging-in-Publication Data

Crop development, growth and yield / edited by Lancaster Mason.
   p. cm.
Includes bibliographical references and index.
ISBN 978-1-68286-704-4
1. Crops--Development. 2. Crops--Growth. 3. Crop yields. 4. Agronomy.
I. Mason, Lancaster.
SB112.5 .C76 2019
630--dc23

TABLE OF CONTENTS

PREFACE

This book has been an outcome of determined endeavour from a group of educationists in the field. The primary objective was to involve a broad spectrum of professionals from diverse cultural background involved in the field for developing new researches. The book not only targets students but also scholars pursuing higher research for further enhancement of the theoretical and practical applications of the subject.

The study of crop development, growth and yield falls under the domain of agricultural science. It focuses on ways to increase crop yield through the right use of fertilizers and pesticides. A good crop production management system addresses both crop production optimization and environmental sustainability. Research in crop science involves developing strategies for improving cultivars, crop yield, disease resistance and climate resilience. The field also uses tools of biotechnology and genetic engineering in order to alter the genetic traits of crops and to produce a better quality yield. This book is a compilation of chapters that discuss the most vital concepts and emerging trends in the field of crop development, growth and yield. It provides a comprehensive understanding of modern theories and advanced concepts of this field. This book will serve as a reference to a broad spectrum of readers.

It was an honour to edit such a profound book and also a challenging task to compile and examine all the relevant data for accuracy and originality. I wish to acknowledge the efforts of the contributors for submitting such brilliant and diverse chapters in the field and for endlessly working for the completion of the book. Last, but not the least; I thank my family for being a constant source of support in all my research endeavours.

Editor

Improvement of Faba Bean Yield Using *Rhizobium/Agrobacterium* Inoculant in Low-Fertility Sandy Soil

Sameh H. Youseif *, Fayrouz H. Abd El-Megeed and Saleh A. Saleh

National Gene Bank and Genetic Resources, Agricultural Research Center, Giza 12619, Egypt;
fayrouz_1983@hotmail.com (F.H.A.E.-M.); sasaleh2006@hotmail.com (S.A.S.)
* Correspondence: samehheikal@hotmail.com

Academic Editor: Peter Langridge

Abstract: Soil fertility is one of the major limiting factors for crop's productivity in Egypt and the world in general. Biological nitrogen fixation (BNF) has a great importance as a non-polluting and a cost-effective way to improve soil fertility through supplying N to different agricultural systems. Faba bean (*Vicia faba* L.) is one of the most efficient nitrogen-fixing legumes that can meet all of their N needs through BNF. Therefore, understanding the impact of rhizobial inoculation and contrasting soil rhizobia on nodulation and N_2 fixation in faba bean is crucial to optimize the crop yield, particularly under low fertility soil conditions. This study investigated the symbiotic effectiveness of 17 *Rhizobium/Agrobacterium* strains previously isolated from different Egyptian governorates in improving the nodulation and N_2 fixation in faba bean cv. *Giza 843* under controlled greenhouse conditions. Five strains that had a high nitrogen-fixing capacity under greenhouse conditions were subsequently tested in field trials as faba bean inoculants at Ismaillia Governorate in northeast Egypt in comparison with the chemical N-fertilization treatment (96 kg N·ha^{-1}). A starter N-dose (48 kg N·ha^{-1}) was applied in combination with different *Rhizobium* inoculants. The field experiments were established at sites without a background of inoculation under low fertility sandy soil conditions over two successive winter growing seasons, 2012/2013 and 2013/2014. Under greenhouse conditions, inoculated plants produced significantly higher nodules dry weight, plant biomass, and shoot N-uptake than non-inoculated ones. In the first season (2012/2013), inoculation of field-grown faba bean showed significant improvements in seed yield (3.73–4.36 ton·ha^{-1}) and seed N-yield (138–153 Kg N·ha^{-1}), which were higher than the uninoculated control (48 kg N·ha^{-1}) that produced 2.97 Kg·ha^{-1} and 95 kg N·ha^{-1}, respectively. Similarly, in the second season (2013/2014), inoculation significantly improved seed yield (3.16–4.68 ton·ha^{-1}) and seed N-yield (98–155 Kg N·ha^{-1}) relative to the uninoculated control (48 kg N·ha^{-1}), which recorded 2.58 Kg·ha^{-1} and 80 kg N·ha^{-1}, respectively. Interestingly, faba bean inoculated with strain *Rlv* NGB-FR 126 showed significant increments in seed yield (35%–48%) and seed N-yield (34%–49%) compared to the inorganic N fertilizers treatment (96 kg N·ha^{-1}) over the two cropping seasons, respectively. These results indicate that inoculation of faba bean with effective rhizobial strains can reduce the need for inorganic N fertilization to achieve higher crop yield under low fertility soil conditions.

Keywords: *Rhizobium*; *Agrobacterium*; inoculation; soil fertility; faba bean

1. Introduction

Faba bean (*Vicia faba* L., broad bean, horse bean) is a major grain legume widely cultivated in many countries for food and feed purposes [1]. Due to its multiple uses, high nutritional value, and ability to grow over a wide range of climatic and soil conditions, cultivation of faba bean is suitable for sustainable agriculture in many marginal areas [2].

Faba bean is one of the oldest legume crops grown in Egypt [3]. However, production has declined considerably from 523,000 tonnes in 1998 to 158,000 tonnes in 2014 [4]—often a result of susceptibility to foliar diseases, the effects of parasites [5], and/or competition with other crops. Egypt now is the world's largest importer of faba bean; its annual requirement of half million tonnes accounts for over half of global imports [6]. Therefore, increasing faba bean production and improving yield quality is a major target to meet the demand of the increasing Egyptian population, since faba bean constitutes a major part of the diet of Egyptian people [7].

Arid land with low nutrient availability, like most of the Egyptian land available for agriculture expansion [8], covers around 30% of the world's land area [9]. In such poor ecosystems, application of high levels of inorganic fertilizers is a common practice to compensate for nitrogen deficiency which is very costly and is a crucial obstruction toward increasing production of food crops including legumes [10]. In addition, more than 50% of the applied nitrogen fertilizers are somehow lost through different processes which not only represent a cash loss to the farmers, but also a source of pollution for the environment [11]. Consequently, there has been a growing interest in environmental friendly sustainable agricultural practices [12].

Biological nitrogen fixation, especially rhizobia-legumes symbiosis, is one of the alternative solutions and the promising technologies which play an important role in reducing the consumption of chemical N-fertilizers, increasing soil fertility, decreasing the production cost, and eliminating the undesirable pollution impact of chemical fertilizers in the environment [13]. Worldwide, N_2 fixed by nodulated legumes (pulses and oilseeds legumes) is estimated to contribute 21.45 Tg N annually to global agricultural systems [14].

Like other legumes, faba bean contributes to sustainable agriculture by fixing atmospheric nitrogen in symbiosis with soil rhizobia [15]. Faba bean commonly establishes effective nitrogen fixation symbiosis with fast-growing rhizobia of the species *Rhizobium leguminosarum sv. viciae* (*Rlv*) [16]. Later, *R. fabae* [17], *R. laguerrereae* [18], *R. etli* [19,20], and *Agrobacterium tumefaciens* [20] were also identified as faba bean-nodulating microsymbionts.

Faba bean is one of the most efficient nitrogen-fixing legumes and faba bean plants can meet all of their N needs through biological nitrogen fixation (BNF) [14,21]. Globally, the amounts of N_2-fixed by faba bean were estimated in the range from 45 to 300 kg N·ha^{-1} [22]. Under different Egyptian field conditions, the amount of N_2-fixed by this legume ranged between 121 and 171 kg N·ha^{-1} [23,24].

Populations of soil rhizobia often vary considerably in their abundance and effectiveness in nodulating and fixing atmospheric nitrogen (N_2) symbiotically with their legume hosts [25,26]. Low fertile soils, particularly sandy soils, contain insufficient numbers of indigenous rhizobia to form efficient symbiotic relationships with their appropriate legumes. In such cases, the reliance on soil rhizobia as the sole source of inoculants can restrict legume yields [27,28]. Therefore, research on the impact of legume inoculation with efficient rhizobial strains can assist in defining the potential of inoculation to improve legume yields and increase the contribution of legume fixed N to the agriculture system [29].

In a previous study, the taxonomic diversity of 42 rhizobial strains that had been isolated from nodules of faba bean grown under different agro-ecological conditions in Egypt was studied using multilocus sequence analyses (MLSA) [20]. Interestingly, only 17 strains were identified as *Rlv*, while 24 strains were identified as *A. tumefaciens*, and one strain was classified as *R. etli*. All isolated strains formed effective symbioses with faba bean plants in Leonard jar assemblies. The present study is complementary to the previous work, and is intended to investigate the potential of highly efficient strains as faba bean inoculants to enhance the crop yield and productivity under low fertility sandy soils compared to the recommended inorganic N-fertilization (96 kg N·ha^{-1}).

2. Material and Method

2.1. Bacterial Strains

Seventeen Egyptian strains of faba bean nodulating rhizobia including twelve *A. tumefaciens* and five *R. leguminosarum sv. viciae* [20] were used in this study. All strains were grown in yeast extract-mannitol (YEM) medium [30].

2.2. Symbiotic Effectiveness under Greenhouse Conditions

Studies of the symbiotic properties (nodulation and nitrogen uptake) of 17 rhizobial strains were conducted with faba bean cultivar *Giza 843* in a pot experiment. Plastic pots (30 cm diameter) were filled with 10 kg of sandy soil and arranged in a complete randomized block design with three replicates. Low fertility sandy soil samples with no history of inoculation were collected from Agricultural Research Station, Ismaillia Governorate (30°37'00.10" N and 32°14'38.57" E). Soil characteristics used in the pot experiment were determined according to [31], and are shown in Table 1. Four seeds were planted in each pot. Each seed was inoculated with 1 mL of a log phase rhizobial culture (10^9 cells mL^{-1}). Growth conditions of faba bean plants were 12–25 °C (night/day), a relative humidity of 50%–60%, and a photoperiod of 10 hr. At flowering stage, 50 days after sowing, plants were uprooted and assayed for dry weight of nodules, shoots and roots dry weight, as well as shoot N-uptake by faba bean plants.

Table 1. Physical and chemical properties of different sandy soils used in this study.

Property	Value		
	Greenhouse Experiment	Field Winter Growing Season 2012/2013	Field Winter Growing Season 2013/2014
Texture grade	Sandy	Sandy	Sandy
$CaCo_3$ (%)	2.85	2.80	1.95
Saturation percent S.P (%)	19.60	27.90	27.60
pH	7.75	8.20	7.94
Electrical conductivity ($dS \cdot m^{-1}$ at 25 °C)	0.67	0.90	0.81
Soluble cations (meq/L)			
Ca^{+2}	2.12	3.40	3.00
Mg^{+2}	1.10	1.30	1.18
Na^+	2.57	5.30	3.61
K^+	1.10	0.40	0.35
Soluble anions (meq/L)			
CO_3^{-2}	0.00	0.00	0.00
HCO_3^-	1.95	1.20	1.10
Cl^-	3.00	3.40	3.10
SO_4^{-2}	1.94	5.80	3.94
Total N (%)	0.017	0.024	0.021
Total Soluble-N ($mg \cdot Kg^{-1}$)	8.50	20.00	19.50
Available-P ($mg \cdot Kg^{-1}$)	3.10	4.90	4.45
Available-K ($mg \cdot Kg^{-1}$)	179.00	252.00	238.50
Organic matter (%)	0.31	0.31	0.30
Available micronutrients ($mg \cdot Kg^{-1}$)			
Fe	1.30	4.30	3.92
Mn	0.90	2.10	1.95
Zn	1.00	1.20	1.35
Cu	0.03	0.02	0.05

2.3. Field Experiments

Field experiments were carried out at El Wasfeya village, Ismaillia Governorate, Egypt (30°34′27.30″ N, 32°10′26.21″ E) in the two successive winter growing seasons, 2012/2013 and 2013/2014. Soil characteristics from experimental sites were determined according to [31]. The main physical and chemical properties of the soil are shown in Table 1. Faba bean variety *Giza 843* was used due to its tolerance to drought stress conditions. Faba bean seeds were sown in the rate of 100 kg seeds ha^{-1} and were cultivated in strips. Each strip (4 m × 12 m) consisted of four plots. Each plot area was 12 m^2 and consisted of four rows, spaced 0.6 m apart. An additional fifth row was placed in each plot and served as a border, and was not involved in calculations. Each strip was spaced apart by 1 m apart to prevent bacterial migrations. Weeds, insects, and fungal pathogens were controlled by chemical spray applications, as required, at rates according to manufacturers' recommendations. At flowering stage, 50 days after sowing, plant samples from each plot were randomly selected from as uniform of an area as possible (in the middle of the second and third lines), in order to avoid heterogeneous conditions or disturbed sites, for estimating nodules dry weight, plant dry matters, and shoot N-content. At harvest, biological yield was determined by the mechanical harvesting of the entire plot using a plot harvester.

2.4. Fertilization

All treatments received the recommended dose of phosphate and potassium fertilization in the rate of 75 kg P$_2$O$_5$ ha^{-1} and 115 kg K$_2$O ha^{-1}, respectively. All bacterial treatments received 48 kg N·ha^{-1} as a starter N-dose. In addition, three un-inoculated controls were involved; (T$_0$) the uninoculated non-N fertilized control; (T) the uninoculated with starter N-dose (48 kg N·ha^{-1}); and (TN) the uninoculated with full N-fertilizers (96 kg N·ha^{-1}).

2.5. Inocula Preparation and Seed Inoculation

Vermiculite supplemented with 10% peat was used as a powder carrier [32], packed in polyethylene bags (300 g carrier per bag), sealed and sterilized by gamma irradiation (2.5 × 10^6 rads). Rhizobial strains were grown in YEM medium [30]. Cultures of (1 × 10^9 colony-forming unit mL^{-1}) were injected into the carrier to satisfy 60% of water holding capacity. At sowing, faba bean seeds were coated with different rhizobial inoculants at a rate of 10 g of inoculant/1 kg seeds, using Arabic gum solution (16%) as the adhesive agent for seed coating [33].

2.6. Statistical Analysis

Data was analyzed for variance using the MSTAT analysis software [34].

3. Results

3.1. Symbiotic Effectiveness under Greenhouse Conditions

The symbiotic efficiency of 17 rhizobial strains related to *Rlv* and *A. tumefaciens* was assessed with faba bean cv. *Giza 843* in a pot experiment under greenhouse conditions. The effect of inoculation on dry weight of nodules, plant dry weight accumulation, and shoot N-content is shown in Table 2. All strains successfully nodulated faba bean and showed different nodulation patterns which ranged from 82 to 366 mg nodules/plant (Table 2). *Rlv* strains NGB-FR 126 and NGB-FR 128 resulted in the highest dry weight of nodules with 366 and 295 mg nodules/plant, respectively. In case of *A. tumefaciens* strains, the maximum dry weight of nodules (230 mg nodules/plant) was produced by strain NGB-FR 39. Nevertheless, the uninoculated controls T$_0$, T, and TN resulted in 48, 63 and 52 mg nodules/plant, respectively. Shoot dry weight of faba bean plants increased significantly in response to effective inoculation, however, no significant variations were observed for root dry weight. Out of the tested strains, eight strains (NGB-FR 39, 62, 65, 70, 107, 126, 128, and 142), resulted in significant increment in shoot dry matter, which ranged from 4.01 to 4.27 g/plant relative to the uninoculated control (T) with 48 kg N·ha^{-1}. In the same trend, shoot N-content was clearly affected according to the type of inoculated strains (Table 2). All inoculated strains, except for NGB-FR 26 and 51, produced shoot N content significantly higher than that obtained with non-inoculated control (T). Faba bean plants inoculated by *Rlv* strain NGB-FR 126 showed the highest shoot N-content (152 mg N/plant) which was significant greater than all other treatments, including the full N-fertilized treatment (TN), which recorded 142 mg N/plant (Table 2). On the other hand, all inoculated strains except for strain NGB-FR 26 significantly increased shoot dry weight and accumulated higher N in plants than the uninoculated non-fertilized control (T$_0$).

Table 2. List of faba bean nodulating rhizobia used in this study, their identity *, and symbiotic properties under greenhouse condition.

Treatment	Identity *	Dry Weight of Nodules (mg)		Dry wt. (g/Plant)				Shoot N Content (mg N/Plant)	
				Root		Shoot			
NGB-FR 10	*A. tumefaciens*	103	g	1.68	a	3.62	cdefg	122	hi
NGB-FR 25	*A. tumefaciens*	82	hi	1.57	a	3.61	defg	123	h
NGB-FR 26	*A. tumefaciens*	82	hi	1.49	a	3.40	fgh	109	l
NGB-FR 27	*A. tumefaciens*	96	gh	1.54	a	3.83	abcdefg	130	g
NGB-FR 39	*A. tumefaciens*	230	d	1.80	a	4.18	abcde	142	de
NGB-FR 51	*A. tumefaciens*	94	gh	1.47	a	3.62	cdefg	114	k
NGB-FR 62	*A. tumefaciens*	130	f	1.68	a	4.27	ab	145	bc
NGB-FR 65	*R. leguminosarum sv. viciae*	172	e	1.78	a	4.16	abcde	141	de
NGB-FR 70	*R. leguminosarum sv. viciae*	191	e	1.79	a	4.22	abcd	139	ef
NGB-FR 99	*A. tumefaciens*	104	g	1.49	a	3.67	bcdefg	118	j
NGB-FR 107	*A. tumefaciens*	100	gh	1.78	a	4.01	abcdef	137	f
NGB-FR 122	*A. tumefaciens*	182	e	1.62	a	3.82	abcdefg	130	g
NGB-FR 126	*R. leguminosarum sv. viciae*	366	a	1.94	a	4.26	ab	152	a
NGB-FR 128	*R. leguminosarum sv. viciae*	295	b	1.89	a	4.24	abc	144	cd
NGB-FR 132	*A. tumefaciens*	98	gh	1.60	a	3.60	efg	119	ij
NGB-FR 140	*R. leguminosarum sv. viciae*	146	f	1.87	a	3.82	abcdefg	122	hi
NGB-FR 142	*A. tumefaciens*	130	f	1.60	a	4.13	abcde	136	f
T$_0$		48	j	1.32	a	2.87	h	89	m
T		63	ij	1.46	a	3.34	gh	114	k
TN		52	j	1.91	a	4.32	a	142	cde

* Bacterial identification based on 16S rRNA sequencing and multilocus sequence typing [20]. Data per plant are means of four replicates (three plants per replicate). Values followed by the same letter within each column are not significantly different at $p < 0.05$. T$_0$: uninoculated seeds and non-chemical N-fertilizers. T: uninoculated seeds plus starter N-fertilizer (48 kg N·ha^{-1}). TN: uninoculated seeds and full N-fertilizer (96 kg N·ha^{-1}).

3.2. Evaluation of Faba Bean Inoculation under Field Trials

Two field experiments were conducted over two successive growing seasons (2012/2013 and 2013/2014), in low fertility sandy soils at Ismaillia Governorate in order to investigate the symbiotic properties of the rhizobial strains that had a high nitrogen-fixing capacity under controlled conditions in the greenhouse. The results showed that all selected strains were able to nodulate faba bean cv. *Giza 843* under field-grown conditions (Tables 3–6).

In the first season (2012/2013), at flowering stage, the highest nodules dry mass per plant was achieved by *Rlv* NGB-FR 128 (367 mg/plant) followed by *Rlv* NGB-FR 126 (322 mg/plant), which was significantly higher than that obtained by other tested strains (Table 3). On the other hand, the uninoculated controls (T$_0$, T, and TN) showed nodulation status ranged from 11–83 mg nodules/plant. Highly significant differences were observed in the dry matter of faba bean plants according to different rhizobial inoculations (Table 3). Faba bean plants inoculated by *Rlv* NGB-FR 126 and *A. tumefaciens* NGB-FR 62 showed the highest shoot dry weights (6.2 and 6.1 g/plant, respectively), which were significantly higher relative to the full N-uninoculated control (TN) that resulted in 5.4 g/plant. Similarly, *Rlv* NGB-FR 126 resulted in the maximum shoot N-content (281 mg N/plant), which was significantly greater as compared to the uninoculated controls where the accumulated N in shoots ranged from 128–258 mg N/plant.

Table 3. Effect of different rhizobial strains on nodulation, growth parameters, and shoot N content of faba bean plants after 50 days of sowing under field conditions (winter growing season 2012/2013).

Treatment	Dry wt. of Nodules (mg/Plant)		Dry wt. (g/Plant)				Shoot N Content (mg N/Plant)	
			Root		Shoot			
NGB-FR-39	85	e	0.84	d	4.3	e	172	d
NGB-FR-62	287	c	1.21	a	6.1	ab	256	b
NGB-FR-70	143	d	1.08	b	4.7	e	181	d
NGB-FR-126	322	b	1.20	a	6.2	a	281	a
NGB-FR-128	367	a	1.14	ab	5.5	bc	251	b
T_0	28	f	0.69	e	3.1	f	128	e
T	83	e	0.96	c	4.8	de	211	c
TN	11	f	1.19	a	5.4	cd	258	b

Data per plant are means of four replicates (three plants per replicate). Values followed by the same letter within each column are not significantly different at $p < 0.05$. T_0: uninoculated seeds and non-chemical N-fertilizers. T: uninoculated seeds plus starter N-fertilizer (48 kg N·ha^{-1}). TN: uninoculated seeds and full N-fertilizer (96 kg N·ha^{-1}).

Table 4. Effect of different rhizobial strains on different growth parameters and yield of faba bean plants under field conditions (winter growing season 2012/2013).

Treatment	Plant Height (cm)		No of Branches/ Plant		No of Pods/Plant		No of Seeds/Plant		Seed Index		Yield (ton/ha)				Seed N Yield kgN/ha	
											Straw		Seed			
NGB-FR-39	89	d	3.0	a	18	cd	48	de	75.7	c	3.6	bcd	2.83	d	85	e
NGB-FR-62	127	b	3.3	a	22	ab	65	ab	79.2	a	4.6	ab	4.29	a	150	a
NGB-FR-70	90	d	3.3	a	15	de	43	e	72.3	d	2.9	cd	2.78	d	83	e
NGB-FR-126	134	a	3.7	a	24	a	72	a	79.0	a	5.9	a	4.36	a	153	a
NGB-FR-128	128	b	3.3	a	20	bc	61	bc	78.3	ab	4.4	b	3.73	b	138	b
T_0	72	e	3.0	a	13	e	31	f	69.2	e	2.4	d	1.76	e	53	f
T	92	d	3.3	a	19	bc	53	cd	76.1	bc	3.8	bc	2.97	d	95	d
TN	101	c	3.7	a	22	ab	62	bc	78.1	abc	4.5	ab	3.24	c	114	c

Values followed by the same letter within each column are not significantly different at $p < 0.05$. T_0: uninoculated seeds and non-chemical N-fertilizers. T: uninoculated seeds plus starter N-fertilizer (48 kg N·ha^{-1}). TN: uninoculated seeds and full N-fertilizer (96 kg N·ha^{-1}).

Table 5. Effect of different rhizobial strains on nodulation, growth parameters, and shoot N content of faba bean plants after 50 days of sowing under field conditions (winter growing season 2013/2014).

Treatment	Dry wt. of Nodules (mg/Plant)		Dry wt. (g/plant)				Shoot N Content (mgN/Plant)	
			Root		Shoot			
NGB-FR-39	309	e	1.86	c	10.4	c	312	c
NGB-FR-62	421	d	1.86	c	10.2	c	307	c
NGB-FR-70	711	b	2.32	b	12.9	b	402	b
NGB-FR-126	814	a	2.53	a	14.9	a	483	a
NGB-FR-128	581	c	1.73	c	9.4	d	283	d
T_0	19	f	1.10	e	5.8	f	151	f
T	59	f	1.51	d	8.3	e	233	e
TN	14	f	1.80	c	9.5	d	285	d

Data per plant are means of four replicates (three plants per replicate). Values followed by the same letter within each column are not significantly different at $p < 0.05$. T_0: uninoculated seeds and non-chemical N-fertilizers. T: uninoculated seeds plus starter N-fertilizer (48 kg N·ha^{-1}). TN: uninoculated seeds and full N-fertilizer (96 kg N·ha^{-1}).

Table 6. Effect of different rhizobial strains on different growth parameters and yield of faba bean plants under field conditions (winter growing season 2013/2014).

Treatment	Plant Height (cm)		No of Branches/ Plant		No of Pods/Plant		No of Seeds/Plant		Seed Index		Yield (ton/ha)				Seed N Yield kgN/ha	
											Straw		Seed			
NGB-FR-39	125	c	4.5	a	18	b	52	b	90.5	bcd	6.6	c	3.51	b	109	b
NGB-FR-62	123	cd	4.3	ab	17	bc	45	bc	92.3	abc	6.5	c	3.43	bc	107	b
NGB-FR-70	150	b	4.5	a	20	a	61	a	93.0	ab	8.4	b	4.35	a	148	a
NGB-FR-126	173	a	4.5	a	21	a	64	a	95.3	a	8.9	a	4.68	a	155	a
NGB-FR-128	116	d	4.0	ab	14	d	40	cd	89.0	cd	6.1	d	3.16	c	98	b
T_0	80	f	3.0	c	8	e	25	e	81.3	f	3.4	f	1.58	e	47	d
T	102	e	3.5	bc	13	d	32	de	84.8	e	5.3	e	2.58	d	80	c
TN	117	cd	4.0	ab	15	cd	38	cd	88.5	d	6.0	d	3.17	c	104	b

Values followed by the same letter within each column are not significantly different at $p < 0.05$. T_0: uninoculated seeds and non-chemical N-fertilizers. T: uninoculated seeds plus starter N-fertilizer (48 kg $N \cdot ha^{-1}$). TN: uninoculated seeds and full N-fertilizer (96 kg $N \cdot ha^{-1}$).

At harvest, in the first season (2012/2013), rhizobial inoculation induced significant increases in plant height, number of pods/plant, number of seeds/plant, and seed index of faba beans according to the type of inoculated strain (Table 4). Inoculated plants with strains NGB-FR 62, 126, and 128 enhanced plant height (127–134 cm), which was significantly higher than the uninoculated control plants (T). Likewise, inoculation with strain NGB-FR 126 produced the highest number of pods/plant (24 pods), which was significantly greater than the uninoculated control (T), that produced 19 pods/plant. Similarly, strains NGB-FR 62 and 126 gave the maximum number of seeds/plant and seed index, which were significantly higher than the uninoculated control (T). On the other hand, number of branches had no significant variations among tested strains and the uninoculated controls (T_0, T, and TN). Faba bean inoculated with *Rlv* NGB-FR 126 and *A. tumefaciens* NGB-FR 62 produced the maximum seed yield (4.36 and 4.29 $ton \cdot ha^{-1}$, respectively) and the maximum seed N-yield (153 and 150 kg $N \cdot ha^{-1}$, respectively), which were significantly higher compared to the full-N fertilized uninoculated control (TN).

In the second season 2013/2014, at flowering stage, the significant effect of rhizobial inoculations on nodulation and plant growth parameters was obvious (Table 5). Strain *Rlv* NGB-FR 126 produced the highest dry weight of nodules (814 mg nodules/plant), while the uninoculated controls (T_0, T, and TN) gave nodulations with a range of 14–59 mg nodules/plant. Faba bean plants inoculated with *Rlv* NGB-FR 126 recorded the maximum root dry weight (2.53 g/plant), shoot dry weight (14.9 g/plant), and shoot N-content (483 mg N/plant) with significant increases higher than the full N-fertilizer uninoculated control (TN), which resulted in 1.8 g/plant, 9.5 g/plant, and 285 mg N/plant, respectively (Table 5).

At harvest, plant height was significantly increased upon inoculation by all tested strains, which was greater than the uninoculated control (T). In the same trend, inoculation by all rhizobial strains, except in case of NGB-FR 128, produced significant increments in number of pods/plant (17–21 pods) and number of seeds/plant (45–64 seeds) relative to the uninoculated control (T) which gave 13 pods/plant and 32 seeds/plant, respectively. Crop yield and seed N-yield of the inoculated faba bean in the second season (2013/2014) surpassed those obtained by the un-inoculated controls (Table 6). All over again, the capacity of *Rlv* NGB-FR 126 to produce the uppermost seed yield (4.68 $ton \cdot ha^{-1}$) and seed N-yield (155 kg $N \cdot ha^{-1}$) was confirmed and was significantly greater than the full-N fertilizers treatment (TN), which recorded 3.17 $ton \cdot ha^{-1}$ and 104 kg $N \cdot ha^{-1}$, respectively (Table 6).

Over the two experimental seasons, seed inoculation and nitrogen fertilization treatments (48 and 96 kg $N \cdot ha^{-1}$) produced significantly higher seed yield and seed N-yield compared to non-inoculated non-fertilized control (T_0), indicating that N availability under such low fertility soil is a major constraint for crop productivity.

4. Discussion

Low fertility of soil is one of the major constraints limiting crop productivity [8]. The success of legume grain crops is dependent on their capacity to form effective nitrogen-fixing symbioses with root-nodule bacteria. However, many soils may do not have adequate amounts of native rhizobia in terms of number, quality, or effectiveness to enhance biological nitrogen fixation [29]. *Rhizobium*-legume association can be manipulated, through inoculation under N-limiting field conditions, to improve crop production easily and inexpensively [35]. Where natural N_2 fixation is not optimal, inoculation is essential, ensuring that a high and effective rhizobial population is available in the rhizosphere of the plant [36]. The use of *Rhizobium* inoculants in legumes is the oldest agro-biotechnological application [37]. Several reports demonstrated significant improvement of yield and yield components in faba bean with *Rhizobium* inoculation [29,38–40].

Generally, the common practice of faba bean cultivation in Egypt is planting the seeds without inoculation. Therefore, most farmers depend on application of high levels of chemical fertilizers to supply N to plants, particularly under sandy soil conditions with low fertility nature. Since biological N_2 fixation is not active at early stages of plant growth, especially under low fertility soils, a starter N-dose (48 Kg $N \cdot ha^{-1}$) was applied in this study to enhance plant growth and eventually improve the grain yield production. The application of a starter N-dose with the rate of 48 Kg $N \cdot ha^{-1}$ was previously reported to increase nodulation and nitrogen fixation of faba bean under Egyptian soil conditions [41]. In another study, an amount of 40 kg $N \cdot ha^{-1}$ was used as a starter N-dose by [42], when they measured the field performance of rhizobial inoculants for some important legumes (lentils, soybeans, faba beans, and peanuts) in Egypt under both clay loam Nile Delta soils and virgin sandy soils. Our results are consistent with previous studies which have reported that the application of an amount of N fertilizer enhances nodulation of different legume crops [33,43,44].

In the present study, we reported the potential use of *Rhizobium/Agrobacterium* inoculants as a powerful alternate source of N in low nutrient ecosystems. Under greenhouse conditions (Table 2), all strains nodulated faba bean cultivar *Giza 843*. Out of the tested strains, eight strains (NGB-FR 39, 62, 65, 70, 107, 126, 128, and 140) could establish an effective nitrogen fixation association with this cultivar, producing a dry weight and shoot N content significantly higher than those obtained by the uninoculated control (T) with 48 Kg $N \cdot ha^{-1}$. Previous studies have identified that there are often strong relationships between shoot dry matter and the amount of N_2 fixed [45,46].

Under field conditions, growth and grain yield of faba bean increased significantly in response to inoculation with the most effective rhizobial strains (Tables 3–6). Increases in N_2 fixation translated to greater grain N concentration, and therefore resulted in increased N export from the field at harvest. In the first season (2012/2013), faba bean inoculated with strains *A. tumefaciens* NGB-FR 62 and *Rlv* NGB-FR 126 showed significant increases in seed yield (44%–47%) and seed N-yield (58%–61%), respectively, relative to the uninoculated control (T). While, in the second growing season (2013/2014), inoculation with strains *Rlv* NGB-FR 70 and *Rlv* NGB-FR 126 produced significant increases in seed yield (69%–81%) and seed N-yield (85%–94%), respectively, over the uninoculated control (T). These results are in line with previous report that was published by [29]. They found that in Australia, at sites without soil rhizobia, faba bean grain yield and total grain N increased by 59% and 132%, respectively, due to different inoculation rates.

Unexpectedly, in the first season (2012/2013), faba bean plants inoculated by *A. tumefaciens* strain NGB-FR 39 and *Rlv* strain NGB-FR 70 showed significantly less N uptake compared to the uninoculated control (T) with 48 kg $N \cdot ha^{-1}$ (Table 3). This trend was also observed in regards to the final seed N-yield parameter (Table 4). This could be due to the presence of effective indigenous rhizobia or highly competitive but ineffective indigenous strains [47]. Our results are consistent with those published by [48], who reported that N uptake and N_2 fixation response to indigenous soil rhizobia in regards to uninoculated cowpea plants surpassed those of inoculated treatments.

A. tumefaciens were previously isolated from the root nodules of several tropical legumes [49]; *Phaseolus vulgaris* [50], *Sesbania* spp. [51], and *Vicia faba* [20,52]. The ability of *A. tumefaciens* to nodulate legumes roots may be attributed to the possession of a transferred *Sym* plasmid which enabled them to form root nodules and fix nitrogen symbiotically [53]. However, many *Agrobacterium* strains isolated from root nodules failed to re-nodulate their original hosts [50,54], which makes *Agrobacterium* a poor choice for legume inoculation [55]. On the contrary, our results revealed the highly symbiotic stability of tested local *A. tumefaciens* strains to nodulate faba bean roots under both greenhouse and field experiments. The stability of nodulating machinery of *Agrobacterium* strains with soybean was recently reported [33].

Data presented in this study showed that the increase in seed yield in response to rhizobial inoculation was variable depending upon the strain type and climatic conditions of the cropping year. Similar findings were previously reported on soybean by [33,56]. The increments in seed yields in the full N-fertilized plots (TN) and/or inoculated plots, in relation to the uninoculated non-N fertilized plots (T_0) controls indicate that, in these soils, nitrogen is a limiting factor, and that crop yields could be strongly improved by means of inoculation or fertilization. However, we found that response to inoculation with the best rhizobial strains was greater than the full N fertilization (96 Kg N·ha^{-1}). This study demonstrated the highest potential of rhizobial inoculation as successful alternates of chemical N fertilizers, where effective inoculation with *Rlv* NGB-FR 126 showed significant increments in the final grain yield (35%–48%) and grain N-yield (34%–49%) compared to the inorganic N-fertilized treatments (TN) over the two cropping seasons, respectively. Our results showed that faba bean inoculation could effectively reduce the need of applied inorganic N-fertilizers while achieving higher grain yield. These findings are in line with those published by [36], who reported that, in a field experiment, inoculation of lentil by *Rhizobium* strains Lt29 increased seed yield by 59% while N fertilizer (50 kg urea ha^{-1}) enhanced yields by 40% over the uninoculated non-fertilized control. Our results are also in agreement with another study [56] which indicated that inoculated soybean under field conditions produced higher or not significantly different seed yields and seed N-yield than the fertilized uninoculated control with 200 kg N·ha^{-1}.

5. Conclusions

Field experiments conducted through the two successive growing seasons have demonstrated that nodulation, total N uptake, and faba bean yield and yield components could be significantly improved through the combined use of *Rhizobium/Agrobacterium* inoculations and starter N application (48 kg N·ha^{-1}) under low fertility sandy soil conditions. Effective inoculation with strain *Rlv* NGB-FR 126 reduced 50% of the applied chemical N-fertilizers, while maintaining faba bean productivity at levels significantly higher than those that resulted from having added inorganic N inputs (96 kg N·ha^{-1}). The results of this study indicate the possibility of using this strain for the development of commercial faba bean inoculants and for achieving better crop yields with reduced usage of N fertilization.

Acknowledgments: This work has been financed by Science and Technology Development Fund (STDF), Egypt, project ID: STDF 901.

Author Contributions: S.H.Y. designed, conducted the field experiments, analyzed the data and wrote the manuscript. F.H.A. and S.H.Y. prepared the bacterial formulations. S.A.S. conceived the research and revised the manuscript.

Conflicts of Interest: The authors declare no conflict of interest.

References

1. Sillero, J.C.; Villegas-Fernández, A.M.; Thomas, J.; Rojas-Molina, M.M.; Emeran, A.A.; Fernández-Aparicio, M.; Rubiales, D. Faba bean breeding for disease resistance. *Field Crops Res.* **2010**, *115*, 297–307. [CrossRef]
2. Nadal, S.; Suso, M.; Moreno, M. Management of *Vicia faba* genetic resources: Changes associated to the selfing process in the major, equina and minor groups. *Genet. Resour. Crop Evol.* **2003**, *50*, 183–192. [CrossRef]

3.   Nassib, A.M.; Khalil, S.A.; Hussein, A.H.A. Faba bean production and consumption in Egypt. In *Present Status and Future Prospects of Faba Bean Production and Improvement in the Mediterranean Countries*; Cunero, J.I., Saxena, M.C., Eds.; CIHEAM: Zaragoza, Ethiopia, 1991; pp. 127–131.

4.   FAOSTAT. Food and Agriculture Organizations of the United Nations: Statistics Division. 2015. Available online: http://faostat3.fao.org/browse/Q/QC/E (accessed on 21 May 2016).

5.   Abdel-Monaim, M.F. Improvement of biocontrol of damping-off and root rot/wilt of faba bean by salicylic acid and hydrogen peroxide. *Mycobiology* **2013**, *41*, 47–55. [CrossRef] [PubMed]

6.   GRDC, Grains Research and Development Corporation. 2014. Available online: https://grdc.com.au/Research-and-Development/GRDC-UpdatePapers/2014/07/Faba-bean-marketing-and-the-Middle-East (accessed on 2 June 2016).

7.   Hegab, A.S.A.; Fayed, M.T.B.; Hamada, M.M.A.; Abdrabbo, M.A.A. Productivity and irrigation requirements of faba-bean in North Delta of Egyptin relation to planting dates. *Ann. Agric. Sci.* **2014**, *59*, 185–193.

8.   Zahran, H.H. Rhizobium-legume symbiosis and nitrogen fixation under severe conditions and in an arid climate. *Microbiol. Mol. Biol. Rev.* **1999**, *63*, 968–989. [PubMed]

9.   Malagnoux, M. Arid land forest of the world: Global environmental perspectives. Forestry Department (FAO), 2007. Available online: http://www.fao.org/3/a-ah836e.pdf (accessed on 28 April 2016).

10.   Mmbaga, G.W.; Mtei, K.M.; Ndakidemi, P.A. Extrapolations on the use of *Rhizobium* inoculants supplemented with phosphorus (P) and potassium (K) on growth and nutrition of legumes. *Agric. Sci.* **2014**, *5*, 1207–1226. [CrossRef]

11.   Ladha, J.K.; Padre, A.T.; Punzalan, G.C.; Castillo, E.; Singh, U.; Reddy, C.K. Nondestructive estimation of shoot nitrogen in different rice genotypes. *Agron. J.* **1998**, *90*, 33–40. [CrossRef]

12.   Lowe, P.; Baldock, D. Integration of environmental objectives into agricultural policy making. In *CAP Regimes and the European Countryside: Prospects for Integration between Agricultural, Regional and Environmental Policies*; Brouwer, F., Lowe, P., Eds.; CABI Publishing: Wallingford, UK, 2000; pp. 31–54.

13.   Peoples, M.B.; Herridge, D.F.; Ladha, J.K. Biological nitrogen fixation: An efficient source of nitrogen for sustainable agricultural production? *Plant Soil* **1995**, *174*, 3–28. [CrossRef]

14.   Herridge, D.F.; Peoples, M.B.; Boddey, R.M. Global inputs of biological nitrogen fixation in agricultural systems. *Plant Soil* **2008**, *311*, 1–18. [CrossRef]

15.   Van Berkum, P.; Beyene, D.; Vera, F.T.; Keyser, H.H. Variability among *Rhizobium* strains originating from nodules of *Vicia faba*. *Appl. Environ. Microbiol.* **1995**, *61*, 2649–2653. [PubMed]

16.   Allen, O.N.; Allen, E.K. *The Leguminosae—A Source Book of Characteristics, Uses and Nodulation*; Macmillan Publishers Ltd.: London, UK, 1981.

17.   Tian, C.F.; Wang, E.T.; Wu, L.J.; Han, T.X.; Chen, W.F.; Gu, C.T.; Gu, J.G.; Chen, W.X. *Rhizobium fabae* sp. nov., a bacterium that nodulates *Vicia faba*. *Int. J. Syst. Evol. Microbiol.* **2008**, *58*, 2871–2875. [CrossRef] [PubMed]

18.   Saïdi, S.; Ramírez-Bahena, M.H.; Santillana, N.; Zúñiga, D.; Álvarez-Martínez, E.; Peix, A.; Mhamdi, R.; Velázquez, E. *Rhizobium laguerreae* sp. nov. nodulates *Vicia faba* on several continents. *Int. J. Syst. Evol. Microbiol.* **2014**, *64*, 242–247. [CrossRef] [PubMed]

19.   Tian, C.F.; Wang, E.T.; Han, T.X.; Sui, X.H.; Chen, W.X. Genetic diversity of rhizobia associated with Vicia faba in three ecological regions of China. *Arch. Microbiol.* **2007**, *188*, 273–282. [CrossRef] [PubMed]

20.   Youseif, S.H.; Abd El-Megeed, F.H.; Ageez, A.; Cocking, E.; Saleh, S.A. Phylogenetic multilocus sequence analysis of native rhizobia nodulating faba bean (*Vicia faba* L.) in Egypt. *Syst. Appl. Microbiol.* **2014**, *37*, 560–569. [CrossRef] [PubMed]

21.   Lindemann, W.C.; Glover, C.R. *Nitrogen Fixation by Legumes. Guide A-129*; Cooperative Extension Service, College of Agriculture and Home Economics New Mexico State University: New Mexico, NM, USA, 2003; pp. 1–4. Available online: http://aces.nmsu.edu/pubs/_a/a-129.pdf (accessed on 25 May 2016).

22.   Smil, V. Nitrogen in crop production: An account of global flows. *Glob. Biogeochem. Cycles* **1999**, *13*, 647–662. [CrossRef]

23.   Abdel-Daiem, M.; Hassan, M.F.; Hamdi, Y.A.; Abdel-Ghaffar, A.S. Nitrogen fixation and yield of faba bean, lentil and chickpea in response to selected agricultural practices in Egypt. In *World Crops: Cool Season Food Legumes*; Summerfield, R.J., Ed.; Kluwer Academic Publishers: London, UK, 1988; pp. 189–204.

24.   Abdel-Ghaffar, A.S. Effect of edaphic factors on biological nitrogen fixation in *Vicia faba* under Egyptian field conditions. In *Nitrogen Fixation by Legumes in Mediterranean Agriculture*; Beck, D.P., Materon, L.A., Eds.; Martinus Nijhoff Publishers: Dordrecht, The Netherlands, 1988; pp. 325–326.

25. Denton, M.D.; Coventry, D.R.; Bellotti, W.D.; Howieson, J.G. Distribution, abundance and symbiotic effectiveness of *Rhizobium leguminosarum bv. trifolii* from alkaline pasture soils in South Australia. *Aust. J. Exp. Agric.* **2000**, *40*, 25–35. [CrossRef]

26. Ballard, R.A.; Charman, N.; McInnes, A.; Davidson, J.A. Size, symbiotic effectiveness and genetic diversity of field pea rhizobia (*Rhizobium leguminosarum bv. viciae*) populations in South Australian soils. *Soil. Biol. Biochem.* **2004**, *36*, 1347–1355. [CrossRef]

27. Evans, J. An evaluation of potential *Rhizobium* inoculant strains used for pulse production in acidic soils of southeast Australia. *Aust. J. Exp. Agric.* **2005**, *45*, 257–268. [CrossRef]

28. Hungria, M.; Franchini, J.C.; Campo, R.J.; Crispino, C.C.; Moraes, J.Z.; Sibaldelli, R.N.R.; Mendes, I.C.; Arihara, J. Nitrogen nutrition of soybean in Brazil: Contributions of biological N_2 fixation and N fertilizer to grain yield. *Can. J. Plant Sci.* **2006**, *86*, 927–939. [CrossRef]

29. Denton, M.D.; Pearce, D.J.; Peoples, M.B. Nitrogen contributions from faba bean (*Vicia faba* L.) reliant on soil rhizobia or inoculation. *Plant Soil* **2013**, *365*, 363–374. [CrossRef]

30. Vincent, J.M. *A Manual for the Practical Study of Root-Nodule Bacteria (International Biological Programme, Handbook No. 15)*; Blackwell Scientific: Oxford, UK, 1970; p. 164.

31. Page, A.L.; Miller, R.H.; Keeney, D.R. *Method of Soil Analysis. Part 2—Chemical and Microbiological Properties*; American Society of Agronomy & Soil Science Society of America: Madison, WI, USA, 1982.

32. Saleh, S.A.; Mekhemar, G.A.A.; Abo El-Soud, A.A.; Ragab, A.A.; Mikhaeel, F.T. Survival of *Azorhizobium* and *Azospirillum* in different carrier materials: Inoculation of wheat and *Sesbania rostrata*. *Bull. Fac. Agric. Univ. Cairo* **2001**, *52*, 319–338.

33. Youseif, S.H.; Abd El-Megeed, F.H.; Khalifa, M.A.; Saleh, S.A. Symbiotic effectiveness of *Rhizobium* (*Agrobacterium*) compared to *Ensifer* (*Sinorhizobium*) and *Bradyrhizobium* genera for soybean inoculation under field conditions. *Res. J. Microbiol.* **2014**, *9*, 151–162. [CrossRef]

34. Snedecor, G.W.; Cochran, W.G. *Statistical Methods*, 7th ed.; Iowa State University Press: Ames, IA, USA, 1980; pp. 255–269.

35. Freiberg, C.; Fellay, R.; Bairoch, A.; Broughton, W.J.; Rosenthal, A.; Perret, X. Molecular basis of symbiosis between *Rhizobium* and legumes. *Nature* **1997**, *387*, 394–401. [CrossRef] [PubMed]

36. Tena, W.; Wolde-Meskel, E.; Walley, F. Symbiotic efficiency of native and exotic Rhizobium strains nodulating Lentil (*Lens culinaris* Medik.) in soils of southern Ethiopia. *Agronomy* **2016**, *6*. [CrossRef]

37. Lindström, K.; Murwira, M.; Willems, A.; Altier, N. The biodiversity of beneficial microbe host mutualism: The case of rhizobia. *Res. Microbiol.* **2010**, *161*, 453–463. [CrossRef] [PubMed]

38. Elsheikh, E.A.E.; Elzidany, A.A. Effects of Rhizobium inoculation, organic and chemical fertilizers on yield and physical properties of faba beanseeds. *Plant Foods Hum. Nutr.* **1997**, *51*, 137–144. [CrossRef] [PubMed]

39. Amanuel, G.; Kühne, R.F.; Tanner, D.G.; Vlek, P.L.G. Biological nitrogen fixation in faba bean (*Vicia faba* L.) in the Ethiopian highlands as affected by P fertilization and inoculation. *Biol. Fertil. Soils* **2000**, *32*, 353–359. [CrossRef]

40. Argaw, A. Characterization of symbiotic effectiveness of rhizobia nodulating faba bean (*Vicia faba* L.) isolated from central Ethiopia. *Res. J. Microbiol.* **2012**, *7*, 280–296. [CrossRef]

41. Abdel-Ghaffar, A.S. Nodulation problems and response to inoculation. In First OAU/STRC International African Conference on Bio-Fertilizers, Cairo, Egypt, 22–26 March 1982; 1982.

42. Moawad, H.; El-Din, S.B.; Khalfallah, M.A. Field performance of rhizobial inoculants for some important legumes in Egypt. In *Nitrogen Fixation by Legumes in Mediterranean Agriculture*; Beck, D.P., Materon, L.A., Eds.; ICARDA: Aleppo, Syria, 1989; pp. 235–244.

43. Daba, S.; Haile, M. Effects of rhizobial inoculant and nitrogen fertilizer on yield and nodulation of common bean. *J. Plant Nutr.* **2000**, *23*, 581–591. [CrossRef]

44. Argaw, A.; Tsigie, A. Indigenous rhizobia population influences the effectiveness of *Rhizobium* inoculation and need of inorganic N for common bean (*Phaseolus vulgaris* L.) production in eastern Ethiopia. *Chem. Biol. Technol. Agric.* **2015**, *2*. [CrossRef]

45. Peoples, M.B.; Brockwell, J.; Herridge, D.F.; Rochester, I.J.; Alves, B.J.R.; Urquiaga, S.; Boddey, R.M.; Dakora, F.D.; Bhattarai, S.; Maskey, S.L.; et al. The contributions of nitrogen fixing crop legumes to the productivity of agricultural systems. *Symbiosis* **2009**, *48*, 1–17. [CrossRef]

46. Unkovich, M.J.; Baldock, J.; Peoples, M.B. Prospects and problems of simple linear models for estimating symbiotic N_2 fixation by crop and pasture legumes. *Plant Soil* **2010**, *329*, 75–89. [CrossRef]

47. Samuel, M.; Herrmann, L.; Pypers, P.; Matiru, V.; Mwirichia, R.; Lesueur, D.D. Potential of indigenous bradyrhizobia versus commercial inoculants to improve cowpea (*Vigna unguiculata* L. *walp.*) and green gram (*Vigna radiata* L. *wilczek.*) yields in Kenya. *Soil Sci. Plant Nutr.* **2012**, *58*, 750–763.

48. Aliyu, I.A.; Yusuf, A.A.; Abaidoo, R.C. Response of grain legumes to rhizobial inoculation in two Savanna soils of Nigeria. *Afr. J. Microbiol. Res.* **2013**, *7*, 1332–1342.

49. De Lajudie, P.; Willems, A.; Nick, G.; Mohamed, T.S.; Torck, U.; Filai-Maltouf, A.; Kersters, K.; Dreyfus, B.; Lindstrom, K.; Gillis, M. *Agrobacterium bv.* 1 strains isolated from nodules of tropical legumes. *Syst. Appl. Microbiol.* **1999**, *22*, 119–132. [CrossRef]

50. Mhamdi, R.; Mrabet, M.; Laguerre, G.; Tiwari, R.; Aouani, M.E. Colonization of *Phaseolus vulgaris* nodules by *Agrobacterium*-like strains. *Can. J. Microbiol.* **2005**, *51*, 105–111. [CrossRef] [PubMed]

51. Cummings, S.P.; Gyaneshwar, P.; Vinuesa, P.; Farruggia, F.T.; Andrews, M.; Humphry, D.; Elliott, G.N.; Nelson, A.; Orr, C.; Pettitt, D.; et al. Nodulation of *Sesbania* species by *Rhizobium* (*Agrobacterium*) strain IRBG74 and other rhizobia. *Environ. Microbiol.* **2009**, *11*, 2510–2525. [CrossRef] [PubMed]

52. Tiwary, B.N.; Prasad, B.; Ghosh, A.; Kumar, S.; Jain, R.K. Characterization of two novel biovar of *Agrobacterium tumefaciens* isolated from root nodules of *Vicia faba*. *Curr. Microbiol.* **2007**, *55*, 328–333. [CrossRef] [PubMed]

53. Sawada, H.; Kuykendall, L.D.; Young, J.M. Changing concepts in the systematics of bacterial nitrogen-fixing legume symbionts. *J. Gen. Appl. Microbiol.* **2003**, *49*, 155–179. [CrossRef] [PubMed]

54. Wang, L.L.; Wang, E.T.; Liu, J.; Li, Y.; Chen, W.X. Endophytic occupation of root nodules and roots of *Melilotus dentatus* by *Agrobacterium tumefaciens*. *Microb. Ecol.* **2006**, *52*, 436–443. [CrossRef] [PubMed]

55. Shamseldin, A.A.; Vinuesa, Y.P.; Thierfelder, H.; Werner, D. *Rhizobium etli* and *Rhizobium gallicum* nodulate *Phaseolus vulgaris* in Egyptian soils and display cultivar-dependent symbiotic efficiency. *Symbiosis* **2005**, *38*, 145–161.

56. Albareda, M.; Rodríguez-Navarro, D.N.; Temprano, F.J. Use of *Sinorhizobium* (*Ensifer*) *fredii* for soybean inoculants in South Spain. *Europ. J. Agron.* **2009**, *30*, 205–211. [CrossRef]

2

Round-Bale Silage Harvesting and Processing Effects on Overwintering Ability, Dry Matter Yield, Fermentation Quality, and Palatability of Dwarf Napiergrass (*Pennisetum purpureum* Schumach)

Satoru Fukagawa [1], Kenichi Kataoka [1] and Yasuyuki Ishii [2,*]

[1]   Nagasaki Agricultural and Forestry Technical Development Center, Shimabara, Nagasaki 859-1404, Japan; s.fukagawa-123@pref.nagasaki.lg.jp (S.F.); jazz@pref.nagasaki.lg.jp (K.K.)
[2]   Faculty of Agriculture, University of Miyazaki, Miyazaki 889-2192, Japan
*   Correspondence: yishii@cc.miyazaki-u.ac.jp

Academic Editor: Jerry H. Cherney

Abstract: Round-bale silage harvesting and processing methods were assessed to evaluate overwintering ability and dry matter (DM) yield, fermentation quality and palatability of overwintered dwarf Napiergrass (*Pennisetum purpureum* Schumach) in the two years following establishment in Nagasaki, Japan, in May 2013 using rooted tillers with a density of 2 plants/m^2. In 2014, harvesting methods under no-wilting treatment were compared for flail-type harvesting with a round-baler (Flail/baler plot) and mower conditioning with a round-baler (Mower/baler plot), which is common for beef-calf–producing farmers in the region. In 2015, the effect of ensilage with wilting was investigated only in the Mower/baler plot. Dwarf Napiergrass was cut twice, in early August (summer) and late November (late autumn), each year. The winter survival rate was greater than 96% in May both years. The DM yield in the Mower/baler plot did not differ significantly for the first summer cutting or the annual total from the Flail/baler plot, but did show inferior yield for the second cutting. The fermentation quality of the second-cut plants, estimated using the V2-score, was higher in the Flail/baler plot than in the Mower/baler plot, possibly because of higher air-tightness, and the second-cut silage tended to have better fermentation quality than the first-cut silage in both harvesting plots. Wilting improved the fermentation quality of dwarf Napiergrass silage in summer, but not in autumn. The palatability of the silage, as estimated by alternative and voluntary intake trials using Japanese Black beef cattle, did not differ significantly between plots. The results suggest that dwarf Napiergrass can be better harvested using a mower conditioner with processing by a round-baler, an approach common to beef-calf–producing farmers, than with the flail/baler system, without reducing the persistence, yield, or palatability of the silage. Moreover, wilting treatment improved the fermentation quality of the dwarf Napiergrass silage when processed in summer.

Keywords: harvesting; silage processing; wintering ability; fermentation quality; dwarf Napiergrass

1. Introduction

Relative to the average for tropical grasses, the dwarf variety of late-heading-type Napiergrass (*Pennisetum purpureum* Schumach) has a high nutritive value as estimated by its crude protein (CP) concentration and in vitro dry matter (DM) digestibility (IVDMD) [1,2]. Dwarf Napiergrass shows high winter survival when receiving a mid-November to late November closing cut in the northern Kyushu including Nagasaki Prefecture, where the average daily temperature drops below 15 °C [2], and thus, the grass can be used as a perennial.

Dwarf Napiergrass has better silage quality in terms of palatability than sorghum (*Sorghum bicolor* Moench) and Sudangrass (*Sorghum sudanense* (Piper) Stapf.) [3]. Dwarf Napiergrass silage was able to be substituted for corn silage with only a small decrease in the milk yield of dairy cows [4]. Furthermore, the total digestible nutrient (TDN) concentration in the grass species is high at 550–600 mg/g, as estimated by digestion trials with Japanese Black breeding cattle [3]. Therefore, dwarf Napiergrass has been verified as a promising forage species for beef cow production in the region [2,3].

The fermentation quality of round-bale silage in dwarf Napiergrass was evaluated by cutting with a flail-type harvester equipped with a round-baler in our previous studies [2,5,6]. However, this is not the usual machinery for the silage processing of grasses by beef cow producers, who most often use a mower conditioner and round-baler. In the present study, we investigated ensiling methods, comparing the effects of the previous flail-type harvester with the current mower conditioner on DM yield, overwintering ability, palatability, and fermentation quality of dwarf Napiergrass.

Since the fermentation quality of the round-bale silage of dwarf Napiergrass produced in summer is considered inconsistent [3,5], we have had success in improving the amount of satisfactory quality silage by adding lactic acid bacteria with *Acremonium* cellulose or fermented juice of epiphytic lactic acid bacteria [6]. Wilting is an effective method to improve the fermentation quality of tropical grass silage [7], though in the previous study [6], a no-wilting ensilage method was examined, and only for dwarf Napiergrass. Thus, we also investigated the effect of wilting on the fermentation quality of dwarf Napiergrass silage under processing by a mower conditioner and round-baler.

2. Materials and Methods

2.1. Cultivation of the Plants

The Napiergrass (*Pennisetum purpureum* Schumach) genotype was a dwarf variety of a late heading type [8,9]. Experiments were conducted at the Livestock Research Division, Nagasaki Agricultural and Forestry Technical Development Center in Shimabara, Nagasaki (32°14′ N, 130°20′ E) in 2014–2015. The dwarf Napiergrass pasture (500 m^2) examined was established by transplanting rooted tillers at 2 plants/m^2 with 1 m inter-row and 0.5 m intra-row spacing on 24 May 2013, as described in our previous study [6]. Additional fertilizer at 10 g/m^2 of N and K$_2$O were applied immediately after harvest in mid-June and in each experiment of both years.

2.2. Experimental Design

2.2.1. Ensiling Methods in 2014 (Experiment 1)

Overwintering plants, established in May 2013, were used after cutting on 11 June 2014. Experimental plots were set at two ensiling methods without wilting before ensilage, specifically, harvested using a flail-type harvester equipped with a round-baler (JCB1420 Combination Baler; IHI Corp., Sapporo, Japan), designated as the Flail/baler plot, and harvested using a mower conditioner (John Deere 1340 Mower Conditioner; Deere & Company, Moline, IL, USA) with subsequent processing using a round-baler (TCR0930WT; IHI Corp., Sapporo, Japan), as for the Mower/baler plot, which is common in beef cattle production in the region. The bales were wrapped with six layers of polyethylene film (KS Wrap, Kaneko Seeds Co. Ltd., Maebashi, Japan). Dwarf Napiergrass was harvested in the first summer cut on 11 August at 61 days' regrowth and in the second autumn cut on 21 November at 102 days' regrowth.

2.2.2. Wilting Effect in 2015 (Experiment 2)

The experiment used overwintering plants after cutting on 10 June 2015. Plants were harvested by a mower conditioner, followed by wilting treatments (with wilting for one day under sunshine or without wilting) and processed with a round-baler. The machinery was the same as in Experiment 1.

Dwarf Napiergrass was harvested twice, with the first cut in summer on 3 August at 54 days' regrowth and the second cut in autumn on 19 November at 108 days' regrowth.

2.3. Measurements

Plant growth was evaluated by plant length at harvest. Plant samples cut at 10 cm above the ground surface were divided into leaf blade, stem (inclusive of leaf sheath), and dead parts. The DM weight of each plant part was determined after oven-drying at 70 °C for 72 h. Crude protein (CP) concentrations and in vitro DM digestibility (IVDMD) of the samples in Experiment 1 were analyzed using a Kjeltec analyzer (FOSS, Hillerod, Denmark) and pepsin-cellulase assay [10].

The palatability of the silages in Experiment 1 was evaluated using alternative and voluntary intake methods for the first-cut and second-cut samples, respectively. Relative consumption of the silages in both Mower/baler plot and Flail/baler plot was measured at a four-day interval by an alternative method where two materials were fed to two paired Japanese Black breeding beef cows, or four head in total (live weight averaged at 475 kg and 419 kg for the high and low groups, respectively) over 2 h. The four-day term for this measurement consisted of a one-day preliminary period, followed by a three-day examination period to evaluate the consumption rate, expressed as a percentage of whole intake of fed materials. The voluntary intake was measured with four Japanese Black breeding beef cows (live weight averaged at 485.5 kg) penned individually in a two-treatment changeover design. Each trial consisted of a two-day preliminary period followed by a three-day examination period. The silages without concentrates were fed to beef cows ad libitum in sufficient quantities to evaluate dry matter intake per day in each term. Fukagawa et al. [3] reported that the palatability of Napiergrass silages was successfully evaluated by the alternative and voluntary intake methods for the short and long periods, respectively.

Silage extraction methods were the same as reported in previous research [6]. Samples (25 g) of silage, which were taken from three round bales per plot 60 days after ensiling, were mixed with 200 mL distilled water and stored in a refrigerator at 5 °C overnight. The pH values of the extracts of the silage samples were measured using a pH meter (S20 SevenEasy pH, Mettler-Toledo AG, Schwerzenbach, Switzerland), ammoniacal nitrogen (NH_3-N) was measured using a Kjeltec analyzer (Kjeltec system 1035, FOSS A/S Co. Ltd., Hillerod, Denmark), and organic acids were measured with a bromothymol blue (**BTB**) post-labeling method using an high performance liquid chromatography (HPLC) system with an RI-2031 Plus refractive index detector (JASCO Corporation, Tokyo, Japan) and a Shodex RS Pak KC-811 column (Showa Denko K.K., Tokyo, Japan), as described previously [11]. The V2-score values for assessing the silage fermentation quality were determined from the concentrations of acetic, propionic, butyric, caproic, and valeric acids, and NH_3-N [12], based on the calculation method listed in Table 1. CP concentration and IVDMD of the silage samples in Experiment 1 were analyzed by a Kjeltec analyzer and pepsin-cellulase assay, respectively.

Table 1. Calculation of V2-score evaluation (Y) for silage (% Fresh Weight, FW).

NH_3-N [1]	Evaluation	$C_2 + C_3$ [2]	Evaluation	$C_4 + C_5 + C_6$ [3]	Evaluation	V2-Score Evaluation [4]
Xa	Ya	Xb	Yb	Xc	Yc	Y
≤ 20	Ya = 50	≤ 0.2	Yb = 10	0	Yc = 40	
20–200	Ya = (1000 − 5Xa)/18	0.2–1.5	Yb = (150 − 100Xa)/13	0–0.5	Yc = 40 − 80Xc	Y = Ya + Yb + Yc
>200	Ya = 0	>1.5	Yb = 0	>0.5	Yc = 0	

[1] ammoniacal nitrogen, [2] Sum of acetic and propionic acids; [3] Sum of butyric, caproic and valeric acids containing each isomer; [4] V2-score indicates fermentation quality of silages with judging by the score of good (above 80), fair (60–80) and poor (below 60).

2.4. Statistical Analysis

Independent Student's *t*-tests were conducted using StatView for Windows software ver. 5.0 (SAS Institute Inc., Cary, NC, USA) for the differences between two pairs at the 5% probability level.

3. Results

Overwintering ability and growth, yield, and quality attributes of dwarf Napiergrass plants before ensiling are presented in Table 2 for Experiment 1. The winter survival rate was above 96% in both plots for the two-year experiment. No significant differences were detected between plots in the attributes of plant length or DM ratio of leaf blade per stem with leaf sheath (LB/ST) in the first or second cut, or in the CP concentration or IVDMD of the second-cut samples. Neither the DM yield of the first-cut samples nor the annual total differed significantly in the Mower/baler plot from the Flail/baler plot, though the Flail/baler plot had a superior second-cut yield.

Table 2. Plant length, ratio of leaf blade to stem with leaf sheath (LB/ST), dry matter yield, crude protein (CP) concentration and in vitro dry matter (DM) digestibility (IVDMD) at ensiling in the first (1st)- and second (2nd)-cut samples and winter survival rate of dwarf Napiergrass.

Experiment (Exp.)	Plot	Plant Length (cm)		LB/ST		DM Yield (Mg/ha)			CP (% DM)		IVDMD (%)		Winter Survival Rate (%)
		1st	2nd	1st	2nd	1st	2nd	Annual	1st	2nd	1st	2nd	
Exp. 1 [1]	Mower/baler [3]	160.6 ns [6]	157.7 ns	1.18 ns	1.06 ns	9.8 ns	9.6 b	19.4 ns	10.6 ns	9.56 a	66.1 ns	58.0 a	99.5 ns
	Flail/baler [4]	155.9	162.1	1.31	0.93	10.3	12.2 a	22.5	10.6	8.87 b	64.7	56.1 b	97.5
Exp. 2 [2]	Wilting [5]	136.9	146.7	1.66	0.81	6.8	10.6	17.4					96.1 ns
	Unwilting	137.6	145.4	1.4	0.91	7.8	9.8	17.6					97.4

[1] Dates of establishment, the first and second cuttings were 24 May 2013, 11 August 2014 and 21 November 2014, respectively. [2] Dates of establishment, the first and second cutting were 24 May 2013, 3 August 2015 and 19 November 2015, respectively. [3] Cutting and ensiling methods were mower conditoner and round baler, of which bale was sized at 90 cm diameter and 86 cm height. [4] Cutting and ensiling methods were combination-baler, of which bale was sized at 90 cm diameter and 86 cm height. [5] Cutting, ensiling methods and bale size were the same as in the Mower/baler plot in Experiment 1, except for wilting at 1 day under sunshine after cutting (wilting) or unwilting imposed. [6] The values with different small letters (a, b) within the same column are significantly different at 5 % level by student t-test. ns: Non-significant.

The moisture, bale density, organic acid composition, and fermentation quality of the silages are shown in Table 3. Bale density was only examined for the second autumn-cut samples in Experiment 1, where a significantly higher density in the Flail/baler plot was obtained compared with the Mower/baler plot. In Experiment 1, lower pH and higher lactic acid, combined with a tendency for lower acetic and propionic acid concentrations, were obtained in the Flail/baler plot compared to the Mower/baler plot, which corresponded with a significantly higher fermentation quality in the Flail/baler plot based on the V2-score. However, samples in the first summer-cut plants tended to have lower lactic acid and higher pH in both harvesting plots compared with the second-cut samples. The fermentation quality estimated using the V2-score did not differ significantly between harvesting plots for the first summer-cut samples, while it was higher quality in the Flail/baler plot than in the Mower/baler plot for the second-cut samples.

Table 3. Moisture and fermentation quality of silage and density of bale in dwarf Napiergrass, as affected by ensiling methods (Exp.1) and wilting treatment (Exp.2).

Experiment (Exp.)	Time of Cut (Season)	Plot	Moisture (%)	Organic Acid Composition			Fermentation Quality		Density of Bale (g/cm³)
				Lactic [1] (% FW)	$C_2 + C_3$ [1] (% FW)	$C_4 + C_5 + C_6$ [2] (% FW)	pH	V2-Score [3]	
Exp. 1 [1]	1st (Summer)	Mower/baler	84.5 ns	0.02	4.14 a [4]	1.57 ns	4.84 b	33 ns	-
		Flail/baler	84.0	nd	3.81 b	1.49	5.06 a	30 bb	-
	2nd (Autumn)	Mower/baler	83.7 ns	0.60 b	1.15 ns	0.20	5.02 a	77 b	0.443 b
		Flail/baler	83.2	1.72 a	0.86	nd [5]	3.96 b	95 a	0.557 a
Exp. 2 [2]	1st (Summer)	Wilting	55.4 b	2.48 a	0.57 b	nd	5.11	84 a	-
		Unwilting	84.7 a	1.09 b	1.01 a	1.16	4.81	45 b	-
	2nd (Autumn)	Wilting	74.1 b	0.12 ns	0.86 ns	0.15	5.94	75 ns	-
		Unwilting	81.1 a	0.18	0.78	0.24	5.60	70 bb	-

[1] Sum of acetic and propionic acids. [2] Sum of butric, capronic and valeric acids containing each isomer. [3] V2-score indicates fermentation quality of silages with judging by the score of good (above 80), fair (60-80) and poor (below 60). [4] The values with different small letter (a, b) within the same column on each time of cut in the same experiment are significantly different at 5% level by student-t test. ns: Not-significant. [5] nd = Not ditected.

In Experiment 2, the wilting treatment did not affect pH, while it increased the lactic acid concentration significantly, corresponding with no detection of butyric, caproic or valeric acid in the summer samples. On the other hand, the decline of the moisture concentration with wilting occurred to a lesser extent in the second-cut samples in Experiment 2, compared with the greater decline of nearly 30 points in the summer samples.

Figure 1 presents the palatability of the silages, as assessed by the alternative and voluntary intake trials using Japanese Black beef cows. The relative DM intake of the dwarf Napiergrass silage increased with the increase in the tester live weight (LW), which showed a similar tendency for the alternative and voluntary intake trials. Concentrations of structural carbohydrates such as neutral detergent fiber (NDF) and acid detergent fiber (ADF) of the silages did not differ between the plots, as shown in Table 4.

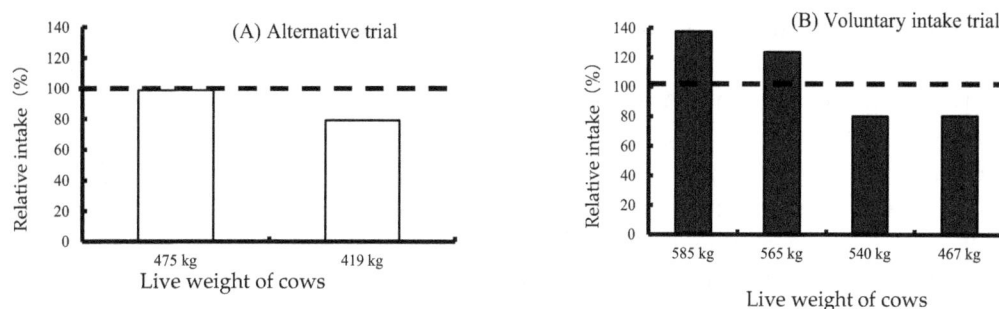

Figure 1. Relative dry matter (DM) intake in Mower/baler plot to Flail/baler plot (as a percentage) in the two palatability tests from dwarf Napiergrass silages in Experiment 2. (**A**) The DM intake over 2 h was determined using two paired Japanese Black breeding cows (mean live weight (LW): high live weight (HLW) plot = 475 kg, low live weight (LLW) plot = 419 kg) with a one-day preliminary period followed by a three-day examination period; (**B**) The DM intake per day, determined using four Japanese Black breeding cows (mean LW = 485.5 kg) with a two-day preliminary period followed by a three-day examination period.

Table 4. Nutritive values of crude protein (CP), neutral detergent fiber (NDF) and acid detergent fiber (ADF) concentrations in dwarf Napiergrass silages of the first (1st)- and second (2nd)-cut samples in Experiment 1.

Experiment (Exp.)	Time of cut (Season)	Plot	CP (% DM [1])	NDF (% DM)	ADF (% DM)
Exp. 1 [2]	1st (Summer)	Mower/baler	7.1 [a 1]	61.1 [ns]	44.8 [ns]
		Flail/baler	6.5 [b]	60.3	41.8
	2nd (Autumn)	Mower/baler	9.0 [ns]	58.4 [ns]	39.7 [ns]
		Flail/baler	9.1	56.4	39.2

[1] Dry matter. [2] The values with different small letter (a, b) within the same column on each time of cut are significantly different at 5 % level by student t-test. ns: Non-significant.

4. Discussion

The persistence of dwarf Napiergrass must be maintained after machine harvest. We found that the machinery ensiling methods did not affect the overwintering ability, which showed a nearly perfect ($\geq$96%) winter survival rate when the closing cut was conducted in mid- to late November.

The bale density was significantly higher in the Flail/baler plot (0.557 g/cm^3) than in the Mower/baler plot (0.443 g/cm^3), possibly due to the operation of the flail-type harvester, an effect also observed by Shao et al. [13]. Therefore, the fermentation quality of Napiergrass, as assessed by the V2-score, was superior in the Flail/baler plot to the Mower/baler plot. The V2-scores of both plots of dwarf Napiergrass tended to be higher in silages processed in autumn than in summer, corresponding to our earlier research findings [5], which were affected by higher concentrations of monosaccharides and oligosaccharides in the plants [5].

Good fermentation quality of five Napiergrass clones including dwarf types was reported to lead to a silage with a pH above 4.2, which was related to the low DM concentration of the plants, below 20% [14]. Since a wilting treatment should increase the DM percentage of Napiergrass by at least 30% DM in order to get a satisfactory fermentation quality of the silage [15], the wilting treatment for the first summer cut in Experiment 2 was successful because it decreased the moisture to 55.4%, leading to an improved silage fermentation quality.

The results of ensiling and the wilting effects suggested that the fermentation quality of dwarf Napiergrass silage in summer can be improved by wilting treatment without additives, due to the faster decline of the moisture concentration that was observed after treatment in summer, compared with no positive effects when treated in autumn.

In our previous research [6], ensilage of dwarf Napiergrass covered by plastic bags without additives after harvest by a flail-type harvester in autumn proved to lead to a satisfactory silage quality of the grass species, which may be applicable to some small-holder beef cow and calf–producing farmers in the region as winter-stored forage. In this research, a satisfactory fermentation quality of the silage was confirmed when adopting the practical apparatus of a mower and round-baler system for ordinary cow-calf farmers.

In our previous research [3], the palatability of the dwarf Napiergrass silage fed to Japanese Black beef cattle, assessed by alternative trials, was superior to that of Sudangrass silage, because of the lower NDF and ADF concentration in Napiergrass. In the present study, no difference in palatability was apparent between the two harvesting and processing plots, as reflected by the absence of any significant ensilage effects on the NDF and ADF concentrations.

5. Conclusions

A flail-type harvester equipped with a round-baler had no significant effects on the overwintering ability, total DM yield or palatability of silage of dwarf Napiergrass compared with conventional harvesting equipment. Thus, we recommend that breeding beef cow producers ensile dwarf Napiergrass using a mower conditioner, followed by a round-baler, using the normal ensiling processes used in the region. As a pretreatment, wilting effectively improves the fermentation quality of silage when dwarf Napiergrass is ensiled in summer.

Author Contributions: Satoru Fukagawa and Yasuyuki Ishii conceived and designed the experiments; Satoru Fukagawa performed the experiments, analyzed the data and Kenichi Kataoka contributed materials and analysis tools; Satoru Fukagawa and Yasuyuki Ishii wrote the paper.

Conflicts of Interest: The authors declare no conflict of interest.

Nomenclature

Napiergrass	*Pennisetum purpureum* Schumach
DM	dry matter
CP	crude protein
IVDMD	in vitro dry matter digestibility
LB/ST	ratio of leaf blade to stem with leaf sheath
NH_3-N	ammoniacal nitrogen
FW	fresh weight
C_2	acetic acid
C_3	propionic acid
C_4	butyric acid
C_5	caproic acid
C_6	valeric acid
NDF	neutral detergent fiber
ADF	acid detergent fiber
HPLC	high performance liquid chromatography

References

1. Ishii, Y.; Sunusi, A.A.; Mukhtar, M.; Idota, S.; Fukuyama, K. Herbage quality of dwarf Napier grass under a rotational cattle grazing system two years after establishment. In Proceedings of the A Satellite Workshop of the XXth International Grassland Congress, Glasgow, Scotland, 3–5 July 2005; p. 150.

2. Fukagawa, S.; Ogasawara, S.; Ishii, Y. Defoliation management aiming at suitable quality herbage production of dwarf napiergrass (*Pennisetum purpureum* Schum.) in northern Kyushu. *J. Jpn. Grassl. Sci.* **2015**, *61*, 59–66.

3. Fukagawa, S.; Hirokawa, J.; Ohkushi, M.; Ishii, Y. Fermentative quality and feed characteristics of dwarf napiergrass (*Pennisetum purpureum* Schumach). *J. Jpn. Grassl. Sci.* **2010**, *56*, 26–33.

4. Ruiz, T.M.; Sanchez, W.K.; Staples, C.R. Comparison of 'Mott' dwarf elephantgrass silage and corn silage for lactating dairy cows. *J. Dairy Sci.* **1992**, *55*, 533–543. [CrossRef]

5. Fukagawa, S.; Maruta, S.; Mine, Y.; Ishii, Y. Season and defoliation frequency effects on dwarf Napiergrass (*Pennisetum purpureum* Schumach) production, utilization, winter survival, and ensilage. *J. Warm Reg. Soc. Anim. Sci.* **2016**, *59*, 105–113.

6. Fukagawa, S.; Ishii, Y.; Hattori, I. Fermentation quality of round-bale silage as affected by additives and ensiling seasons in dwarf napiergrass (*Pennisetum purpureum* Schumach). *Agronomy* **2016**, *4*, 48. [CrossRef]

7. Catchpool, V.R.; Henzell, E.F. Silage and silage-making from tropical herbage species. *Herb. Abst.* **1971**, *41*, 213–221.

8. Ishii, Y.; Tudsri, S.; Ito, K. Potentiality of dry matter production and overwintering ability in dwarf napiergrass introduced from Thailand. *Bull. Fac. Agric. Miyazaki Univ.* **1998**, *45*, 1–10.

9. Ishii, Y.; Hamano, K.; Kang, D.J.; Kannika, R.; Idota, S.; Fukuyama, K.; Nishiwaki, A. C_4-napier grass cultivation for cadmium phytoremediation activity and organic livestock farming in Kyushu, Japan. *J. Agric. Sci. Technol. A* **2013**, *3*, 321–330.

10. Goto, I.; Minson, D.J. Prediction of the dry matter digestibility of tropical grasses using a pepsin-cellulase assay. *Anim. Feed Sci. Technol.* **1977**, *2*, 247–253. [CrossRef]

11. Akiyama, F. A study of sample preparation for determination of oligo-saccharides of forage crops with high performance liquid chromatography (HPLC). *Bull. Natl. Grassl. Res. Inst.* **1998**, *58*, 17–25.

12. Association of Self-supply Feed Evaluation. *Guidebook for Forage Evaluation*; Japanese Grassland Agriculture Forage Seed Association: Tokyo, Japan, 2009; pp. 64–78. (In Japanese)

13. Shao, T.; Wang, T.; Shimojo, M.; Masuda, Y. Effect of ensiling density on fermentation quality of guineagrass (*Panicum maximum* Jacq.) silage during the early stage of ensiling. *Asian-Aust. J. Anim. Sci.* **2005**, *18*, 1273–1278.

14. Dos Santos, R.J.C.; de Lira, M.A.; Guim, A.; dos Santos, M.V.F.; Dubeux, J.C.B., Jr.; de Mello, A.C.d.L. Elephant grass clones for silage production. *Sci. Agric.* **2013**, *70*, 6–11. [CrossRef]

15. Moran, J. Making quality silage. In *Tropical Dairy Farming: Feeding Management for Small Holder Farmers in the Tropics*; Landlinks Press, Department of Primary Industry: Brisbane, Australia, 2005; pp. 83–97.

3

Influence of Irrigation Scheduling Using Thermometry on Peach Tree Water Status and Yield under Different Irrigation Systems

Huihui Zhang [1,2,*], Dong Wang [1] and Jim L. Gartung [1]

[1] USDA-ARS Water Management Research Unit, San Joaquin Valley Agricultural Sciences Center, 9611 S. Riverbend Ave., Parlier, CA 93648, USA; dong.wang@ars.usda.gov (D.W.); jim.gartung@ars.usda.gov (J.L.G.)

[2] Curent Affiliation: USDA-ARS Water Management and System Research Unit, 2150 Centre Avenue, Building D, Suite 320, Fort Collins, CO 80526, USA

* Correspondence: Huihui.zhang@ars.usda.gov

Academic Editor: Rakesh S. Chandran

Abstract: Remotely-sensed canopy temperature from infrared thermometer (IRT) sensors has long been shown to be effective for detecting plant water stress. A field study was conducted to investigate peach tree responses to deficit irrigation which was controlled using canopy to air temperature difference (ΔT) during the postharvest period at the USDA-ARS (U.S. Department of Agriculture, Agricultural Research Service) San Joaquin Valley Agricultural Sciences Center in Parlier, California, USA. The experimental site consisted of a 1.6 ha early maturing peach tree orchard. A total of 18 IRT sensors were used to control six irrigation treatments including furrow, micro-spray, and surface drip irrigation systems with and without postharvest deficit irrigation. During the postharvest period in the 2012–2013 and 2013–2014 growing seasons, ΔT threshold values at mid-day was tested to trigger irrigation in three irrigation systems. The results showed that mid-day stem water potentials (ψ) for well irrigated trees were maintained at a range of -0.5 to -1.2 MPa while ψ of deficit irrigated trees dropped to lower values. Soil water content in deficit surface drip irrigation treatment was higher compared to deficit furrow and micro-spray irrigation treatments in 2012. The number of fruits and fruit weight from peach trees under postharvest deficit irrigation treatment were less than those well-watered trees; however, no statistically significant (at the $p < 0.05$ level) reduction in fruit size or quality was found for trees irrigated by surface drip and micro-spray irrigation systems by deficit irrigation. Beside doubles, we found an increased number of fruits with deep sutures and dimples which may be a long-term (seven-year postharvest regulated deficit irrigation) impact of deficit irrigation on this peach tree variety. Overall, deployment of IRT sensors provided real-time measurement of canopy water status and the information is valuable for making irrigation management decisions.

Keywords: irrigation scheduling; early-maturing peach; canopy-to-air temperature; drip irrigation; furrow irrigation; micro-spray irrigation

1. Introduction

The United States is the third largest peach producer in the world. The total peach production was estimated at 847 thousand tons in 2015 [1]. California was shown to be the largest producer on a state level, accounting for about 70% of the U.S. total in the same year. However, due to the continuous drought situation and warm winter temperatures in California in recent years, water supplies for agricultural irrigation have declined. Water shortages require farmers and producers to improve water management and utilize limited irrigation water effectively to meet crop demands.

Regulated deficit irrigation (RDI) has been applied to peach trees, as an irrigation management strategy, for saving water [2–5]. RDI during the second stage of fruit development and postharvest stages on late maturing peach trees in deep soils could save 23%–35% of irrigation water [6]. Compared with late maturing peach trees, early maturing varieties of peaches usually ripen and harvest in late May or early June and have their highest water demand during the postharvest summer months. For early maturing varieties of peaches, RDI can be applied during postharvest non-fruit bearing periods [7]. A four-year project with different levels of postharvest irrigation on an early season peach in California showed no reduction in yield and fruit size or a progressive decline in tree vigor and health [8]. Falagán et al. [9] also confirmed that RDI on early maturing peaches allowed saving a significant amount of water and provided peaches with overall good quality and vitamin C status. In addition, Bryla et al. [10] has studied furrow, micro-spray, surface and subsurface drip irrigation systems with different irrigation frequencies for early maturing peaches. Surface and subsurface drip irrigation systems had higher irrigation efficiency and produced higher growth and production. A study in the same orchard indicated RDI with furrow and drip irrigation during postharvest period can substantially save water without significantly impacting the yield [11]. Johnson and Phene [12] suggested RDI should be applied in June and July only during postharvest season to reduce double fruits, which had been considered to be a potential negative effect of RDI for peaches.

Traditionally irrigation can be scheduled with measurements of soil water content or water potential, stem or leaf water potential, direct or indirect estimate of evapotranspiration, or the check book type of methods. Irrigation scheduling with different plant- and climate-based water stress indicators has been reported on both late and early-maturing varieties of peaches, such as stem water potential [13], water supply index [14], and trunk diameter [15]. The downsides of these techniques are the need of intensive labor and additional irrigation equipment. Canopy temperature is a direct response to plant water status [16] and has the advantage of less laborious or more timely response than other traditional techniques. It also has the potential for upscaling by using remote sensing such as drones, airplanes, or satellite-borne thermal sensors for making the temperature measurement. With the advance of infrared thermometer technology in the late 70s, continuous measurement of canopy temperature by IRTs has been previously used for monitoring crop water status and controlling irrigation for a variety of crops and in different parts of the world for annual crops [17,18]. Clawson and Blad [19] used canopy temperature variability and average canopy temperature above that of a well-watered reference plot to schedule irrigation in corn (*Zea mays* L.). Wanjura and Upchurch [20] developed a temperature time-threshold model and demonstrated applications in irrigation scheduling in cotton. Infrared thermometry and thermal imaging have also been used on various fruit trees to measure canopy temperature. Glenn et al. [21] examined infrared measurement techniques for evaluating the canopy temperature in peaches. Sepulcre-Cantó et al. [22] investigated the detection of water stress in an olive (*Olea European* L.) orchard with airborne remote sensing imagery for individual trees and found a high correlation between leaf water potential and crown canopy to air temperature differences obtained from the imagery. Most of the peaches in California are irrigated by micro-spray and furrow surface irrigation systems. Surface drip and subsurface drip system have the advantage of reducing evaporation and better control of deep percolation. With less irrigation amount and high efficiency, drip irrigation has been adopted by farmers. There is a need to determine the performance of a canopy temperature based irrigation scheduling technique on peaches under different irrigation types.

The objectives of this study were to: (1) determine the effectiveness of the canopy-to-air temperature difference method to trigger irrigations for early maturing peach trees under different irrigation systems in an arid climate; and (2) to evaluate the peach tree response to deficit irrigation treatments.

2. Materials and Methods

2.1. Study Site and Irrigation System Descriptions

The study was conducted over a period of two growing seasons (2012–2013 and 2013–2014) in a 1.6 ha peach orchard at the USDA-ARS (U.S. Department of Agriculture, Agricultural Research Service) San Joaquin Valley Agricultural Sciences Center near Parlier, CA (36°37′ N; 119°31′ W). The soil is Hanford fine sandy loam characterized as coarse-loamy, mixed, thermic Typic Xerorthents, and low organic matter content (1.38% for 0–20 cm, and 0.24% for 20–100 cm soil depth). The averaged bulk density was 1.55 g·cm^{-3}.

The early maturing peach trees (Crimson Lady) were planted in 1999 at a spacing of 1.8 m × 4.9 m (6 feet by 16 feet) and were trained to a perpendicular-V shape. This peach variety blooms in February–March, were commercially thinned each spring, and were harvested at the end of May or early June every year. Tree canopy was pruned annually in the winter by commercial contractors. For the 2012–2013 growing season, deficit irrigation was applied during the postharvest period in 2012 and fruit yield was evaluated in late May or early June after harvest in 2013. Similarly for the 2013–2014 season, deficit irrigation occurred in 2013 after harvest and fruit yield and quality were determined in 2014.

The experimental design was a randomized block with furrow, micro-spray and surface drip irrigation treatments as the main effect and levels of postharvest deficit irrigation as the sub-effect with six replications. Each treatment plot consisted of three rows of eight trees. Three trees from the center of the middle rows were used for plant IRT measurements while the rest served as guard trees. Furrow treatments were irrigated in 1 m wide, 0.2 m deep, and 9.8 m long V-shaped furrows on both sides of the tree row, running parallel to the row, and located 1 m (furrow center) from the tree trunks. Drip treatments were irrigated with drip tubing containing 0.002 m^3·h^{-1} (1/2 gph) integral turbulent flow embedded emitters spaced 0.91 m (3 feet) apart (GeoFlow, Charlotte, NC, USA). Two tubing laterals were used for each tree row, one on each side at a distance of 1 m from the tree trunks. Micro-spray treatments were applied with one 40 L/h Fan-jet emitter with a 4 m diameter, 230° spray pattern (Bowsmith, Inc., Exeter, CA, USA), and located near the base of each tree. Irrigation amount was measured using turbine water meters (Model SRII and W-120 Invensys Metering Systems, Uniontown, PA, USA). Detailed descriptions of the orchard, soil, and irrigation systems can be found in [10,23].

2.2. Irrigation Treatments and Irrigation Control

Johnson et al. [24] developed a daily crop coefficient (k_c) curve using a weighing lysimeter located in a nearby peach orchard with the same variety, planting density, and training system as trees used in this current study. Crop evapotranspiration requirements (ET_c) were estimated using this k_c and current reference evapotranspiration (ET_o) obtained from a nearby weather station (California Irrigation Management Information Systems or CIMIS, California Department of Water Resources, Sacramento, CA, USA).

All trees received uniform irrigation matching the full ET_c requirement during early growing season in spring. The last full orchard irrigation was applied on 30 May 2012 and 31 May 2013. After that, the experimental treatments during the postharvest irrigation scheduling period were designated as:

FF—Full irrigation treatment by furrow irrigation system, where trees were targeted to irrigate with enough water to replace 100% of ET_c,

FD—Deficit irrigation treatment by furrow irrigation system, where trees were targeted to irrigate with the same amount of water per irrigation event as in FF but with less frequency,

MF—Full irrigation treatment by micro-spray irrigation system, where trees were targeted to irrigate with enough water to replace 100% ET_c,

MD—Deficit irrigation treatment by micro-spray irrigation system, where trees were targeted to irrigate with enough water to replace 25% ET_c,

SF—Full irrigation treatment by surface drip irrigation system, where trees were targeted to irrigate with enough water to replace 100% ET_c,

SD—Deficit irrigation treatment by surface drip irrigation system, where trees were targeted to irrigate with enough water to replace 25% ET_c.

To measure real-time tree canopy temperatures, eighteen IRT sensors (Model SI-100 series, Apogee Instruments, Inc., Logan, UT, USA) were installed in the field on 23–25 April 2012 by mounting them on galvanized metal pipes 5.5 m above the soil surface. The field of view (FOV) of the IRT sensor was 36° with the accuracy at $\pm 0.5°$. The IRT sensors were calibrated by the manufacturer before installation. The sensors were used to measure canopy temperature for the irrigation treatments FF, FD, SF, SD, MF, and MD, for three of the six replications used in the study. The metal pipes were installed on the north side of the middle tree of the center row in each plot. The sensors were mounted on the pipes and pointed southward at approximately 30° from nadir with the center of the FOV aimed at the middle trees of the center row. Canopy temperature was measured at 1 Hz and an average value was recorded at 15-min intervals using a CR3000 datalogger (Campbell Scientific Inc., Logan, UT, USA). An MD9 multi-drop networksystem (Campbell Scientific, Inc., Logan, UT, USA) was used to connect the six sensors in each rep through a coax cable and the data were retrieved at a central station located outside the orchards. Air temperature was measured with a thermistor as part of an air temperature and relative humidity sensor (Vaisala HMP 45C, Campbell Scientific, Inc., Logan, UT, USA) located in the orchard at the top of canopy level. Detailed descriptions regarding the IRT sensor set up and irrigation scheduling management can be found in [11,25].

All irrigation events during the postharvest stages were triggered by the threshold values of canopy temperature to air temperature difference (ΔT) at 14:00 PDT (Pacific Daylight Time) that were determined based on the results in previous studies [11,25]. Wang and Gartung [11] found a linear correlation between mid-day ΔT and stem water potential (SWP) ($\Delta T = -5.3709 \times SWP - 5.3289$), using two-year data in this orchard. Although the emprical linear relationship is site-specific and plant specific, it may be robust enough to be used as a guide for irrigation. From their study, the highest SWP value for full irrigation treatment was about -0.7 MPa during the postharvest season. Therefore, if the goal is to maintain the peach orchard without water stress, instead of measuring SWP, it might be possible to use the ΔT value of -1.5 °C for full irrigation treatment. To maintain a deficit irrigation, we could use a ΔT value of 2.5 °C while SWP is less than -1.5 MPa. Thus, in this study the threshold values of ΔT for FF and FD plots were tested at -1.5 °C and 2.5 °C, respectively. A decision was made daily on whether to irrigate based on if the specific threshold of the treatment was exceeded. Due to the limitation of the capacity of furrow irrigation system, irrigation decisions were made only starting seven days after an irrigation event until a decision to irrigate was made. During each irrigation event, 75 mm water was applied and completed in three days. In the previous studies, ΔT performance had not been investigated for the micro-spray treatment plots. We determined a threshold value for it based on the findings in furrow and surface drip treatment plots. The threshold value of ΔT for MF plots was tested at -0.5 °C initially and changed to -2.5 °C in August 2012. An irrigation decision was made daily beginning five days after an irrigation event until a decision to irrigate was made when the specific threshold of the treatment was exceeded. During each irrigation event, 55 mm water of was applied and completed in two days. MD plots were irrigated at the same time with 14 mm of water. The threshold value of ΔT for SF plots was tested at -1.5 °C. A three-day irrigation cycle was used. An amount of 25 mm and 6.25 mm irrigation water was applied in SF and SD plots, respectively. When an irrigation event was triggered, the plots were irrigated immediately. The irrigation scheduling to control targeted full and deficit irrigation treatments started on 7 June 2012 (Day-of-year, DOY159) and 7 June 2013 (DOY158) to the end of August in each year.

2.3. Stem Water Potential

Stem water potential (ψ) was measured approximately weekly in all treatment plots after harvest at the end of May and continued for the rest of the year for 2012–2013 and 2013–2014 growing season.

Stem water potential was measured using a pressure chamber (Model 3000–1412, Soil Moisture Equipment Corp., Santa Barbara, CA, USA) between 12:00–14:00 PDT following the procedures described in [10]. Six measurements per plot were taken from the middle three trees (two leaves per tree) of the center row. Total hermetic aluminum foil bags were placed on each leaf at least 2 h before taking stem water potential measurements.

2.4. Soil Water Content

Soil water content for the root zone profile was monitored weekly at 15, 45, 75, 105, and 135 cm depths using a neutron probe (Series 4300, Troxler International, LTD., Research Triangle Park, NC, USA) with galvanized steel access tubes located at the middle of the center row within each treatment plot. The neutron probe was calibrated with volumatric soil samples taken from 15–135 cm depths in the study field in 2012 (N = 30, R^2 = 0.98).

2.5. Fruit Yield

Peach fruit yield was measured and fruit quality was assessed for both the 2012–2013 and 2013–2014 seasons. Marketable-sized fruits were picked by a commercial harvesting crew (Sunny Cal, Reedley, CA, USA) following typical farming procedures. A total of two picks, about three days apart, were used during each season. For the experimental plots, the total number of peaches per tree and weight per tree were measured for each treatment plot. Fruit quality was assessed by randomly selecting 120 peaches per plot and counting the number of peaches with doubles, deep sutures, external splits, dimples, deformation, or internal split pits. Ten fruit per plot were also assessed for skin color, flesh firmness, soluble solids, pH, and titratable acidity after one or two weeks of storage at 1 °C. Fruit skin color was measured by a handheld Chroma Meter (CR-400, Konica Minolta, Japan). A slice of skin of each fruit was removed and flesh firmness was measured by a penetrometer and recorded as pounds force and then converted to Newtons (N). Two pieces of each fruit were removed, and juiced using a juice processer to form a composite juice sample for each treatment. A few drops of peach juice were tested for soluble solids concentration with a handheld Brix refractometer (Atago Inc., Bellevue, WA, USA). Titratable acidity (%) and pH values were measured by titrating a 5-g sample of juice diluted with 50 mL of deionized water using 0.1 N NaOH to an endpoint of 8.6 pH using TIM 850 Titration Radiometer analytical workstation (Radiometer Analytical SAS, Lyon, France).

2.6. Statistical Analysis

Production data was analyzed by analysis of variance (ANOVA) test using JMP (SAS Institute, Cary, NC, USA). Means were separated at the 0.05 level using Tukey's HSD (honest significant difference) test.

3. Results

3.1. Climatology

Because of the Mediterranean climate and drought conditions, precipitation was near zero during the 2012 and 2013 summer growing seasons at Parlier, California, which is about the same amount received for the past 10 years during this same period. The maximum air temperatures in June and July of 2013 were higher than temperatures observed in respective months in 2012, and mean wind speed was higher in 2012 than 2013. Average daily ET_0 values were similar between 2013 and 2012 (Table 1).

Table 1. Climatic conditions for the 2012 and 2013 irrigation scheduling periods. Min RH—Average monthly minimum relative humidity; Max RH—Average monthly maximum relative humidity; Min Air—Average monthly minimum air temperature; Max Air—Average monthly maximum air temperature.

Month	Min RH (%)	Max RH (%)	Min Air (°C)	Max Air (°C)	Total Monthly Precipitation (mm)	Wind Speed (m/s)	Average Daily ET_0 (mm·day^{-1})
June 2012	26	79	14.6	31.9	0	2.4	6.95
July 2012	28	83	16.9	34.1	0.1	1.8	6.6
August 2012	25	84	17.9	36.9	0	1.6	5.99
June 2013	32	78	16.7	33.6	1	2.1	6.89
July 2013	24	77	18.8	39.6	0	1.7	6.65
August 2013	24	78	16.2	35.2	0.3	1.7	6.1

3.2. Irrigation Controlling by ΔT

Figure 1 shows the dynamic of ΔT values and irrigation events triggered by ΔT in treatment FF, MF, SF, and FD in 2012 and FD and MF in 2013 as examples of irrigation scheduling by thermometry on tree canopy. Deficit irrigations started right after harvest at the end of May in 2012. In general, most of the irrigation events were triggered by ΔT values. For example, treatment FF was furrow irrigated when ΔT exceeded the threshold value on 18 July 2012; consequently, ΔT dropped and fell below the threshold value. Seven days after the irrigation, ΔT value exceeded the threshold value on 27 July 2012 and triggered another irrigation event. The threshold value for treatment MF was determined initially based on the previous results on furrow and surface drip systems. The results showed only a few ΔT values were greater than the threshold value, so it hardly triggered an irrigation event for this full irrigation treatment. Since the irrigation frequency was low, it did not meet the 100% of ET requirement. We changed the threshold value to $-2.5\,°C$ at the beginning of August. From then on, more irrigation signals were received. The threshold value in treatment SF triggered more irrigation events than those in FF and MF. There were delayed irrigation or missed irrigation events, although irrigation signals were received in those days (i.e., 1 July (DOY183) and 7 July (DOY 190)). Two irrigation decisions were also not based on ΔT signals. Only one irrigation event was triggered by the ΔT signal during the period for treatment FD. In general, ΔT values ranged from -5 to $3\,°C$ for all full irrigation treatments. The maximum ΔT value of $5.9\,°C$ was found in FD in 2013, while the minimum ΔT value was $-6.6\,°C$ in MF in 2013. Due to the system failure, MF and MD plots were over-irrigated on 19–21 July 2013.

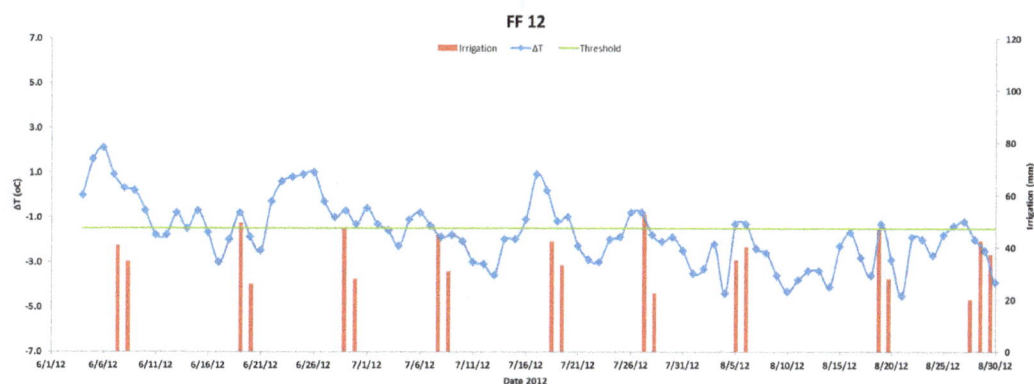

Figure 1. *Cont.*

Figure 1. *Cont.*

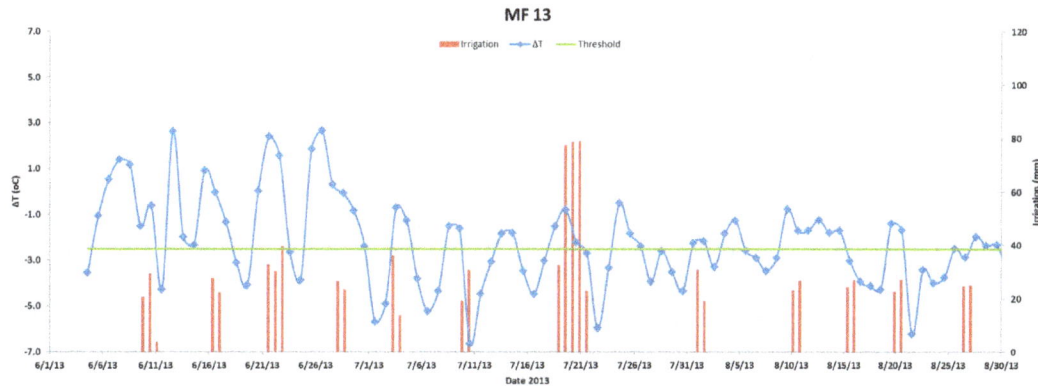

Figure 1. The canopy-to-air temperature difference (ΔT) values plotted with irrigation amounts for the FF, MF, SF, and FD treatments in 2012 and FD and MF treatments in 2013. FF—Full irrigation treatment by furrow irrigation system, where trees were targeted to irrigate with enough water to replace 100% of ET_c; FD—Deficit irrigation treatment by furrow irrigation system, where trees were targeted to irrigate with the same amount of water per irrigation event as in FF but with less frequency; MF—Full irrigation treatment by micro-spray irrigation system, where trees were targeted to irrigate with enough water to replace 100% ET_c ; SF—Full irrigation treatment by surface drip irrigation system, where trees were targeted to irrigate with enough water to replace 100% ET_c.

3.3. Soil Water and Irrigations

The soil water content for all the plots at the beginning of irrigation scheduling period (7 June ~27 August 2012; DOY 159-240) was similar and not statistically different. The mean values were 0.16, 0.20 and 0.27 $m^3 \cdot m^{-3}$ at the 15 cm, 75 cm and 135 cm soil depths, respectively. Over the postharvest period, the soil water profile showed a decreasing trend (Figure 2) but remained above 0.12 $m^3 \cdot m^{-3}$ at the 75 cm depth and above 0.15 $m^3 \cdot m^{-3}$ at the 135 cm depth in three full irrigation treatments. The soil water profile in the deficit plots dropped to as low as 0.05 $m^3 \cdot m^{-3}$ at the end of this period at both 75 cm and 135 cm depths. We also found that for all three soil depths, the deficit irrigation treatment SD had higher soil water content than MD and FD although the irrigation amount was similar. Soil water content responded more to irrigation signals in furrow and micro-spray than for drip irrigation at the surface depth (15 cm). Also the high water content readings on 27 August 2012 (DOY 240) in the FD plots at 15 and 75 cm depths were caused by the residual effect of the large irrigation event on 10 August 2012 (DOY 233).

On 31 May in 2013 (DOY151), the initial soil water content was not significantly different in all depths among treatments. Differential irrigation treatments in 2012 did not result in soil water variability amongst treatments since we fully irrigated all the plots from early season to harvest. Over the irrigation scheduling period, 31 May~1 September 2013 (DOY 151–244), deficit irrigation treatments (FD, MD) received 54% and 59% of the irrigation amount in the full irrigation treatment FF and MF, respectively, and SD received 34% of the irrigation amount in SF (Table 2). The soil profiles in full irrigation treatments remained above 0.15 $m^3 \cdot m^{-3}$ at the 75 cm depth and 0.20 $m^3 \cdot m^{-3}$ at the 135 cm depth except for SF. We observed that soil water content in MD increased suddenly on 26 July 2013 (DOY 207) due to the over irrigation (Figure 2) on 19–21 July 2013 (DOY 200-202).

The ET_c in June–August 2012 were similar to the values in 2013 (Table 1); however, the irrigation amount in 2013 was increased compared to 2012 (Table 2). For other treatments, the major difference of irrigation amount between two years happened in June. For example, irrigation signals were triggered twice in June 2012, but four times in June 2013 for MF. There was no irrigation in June 2012 for FD, but irrigation was triggered twice in June 2013 (Figure 2). Treatment FF received more irrigation in July and August, 2013 than for the comparable months in 2012. Although irrigation amount was higher in

2013 than 2012, the soil profiles in both SF and SD dropped lower in 2013 compared with the values in 2012. The irrigation amount in SF in both years and MF in 2012 did not meet full crop ET_c.

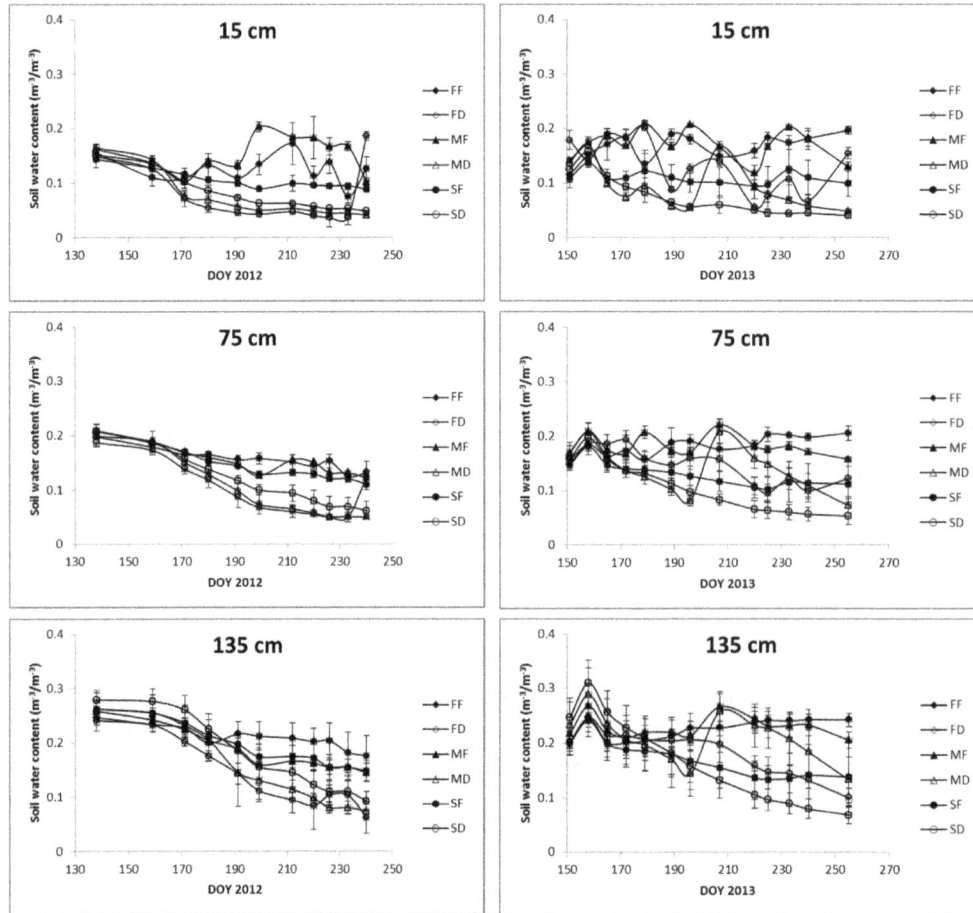

Figure 2. Soil water contents at 15, 75 and 135 cm soil depths in six treatments during the postharvest periods in 2012 and 2013. Vertical bars indicate standard deviation.

Table 2. Reference crop evapotranspiration (ET_0), potential crop evapotranspiration (ET_c), and irrigation amount applied to each treatment during the irrigation scheduling periods in 2012 and 2013. MD—Deficit irrigation treatment by micro-spray irrigation system, where trees were targeted to irrigate with enough water to replace 25% ET_c; SD—Deficit irrigation treatment by surface drip irrigation system, where trees were targeted to irrigate with enough water to replace 25% ET_c. See Figure 1 for previously provided definitions.

Month	ET_0 (mm)	ET_c (mm)	Irrigation Amount (mm)					
			FF	FD	MF	MD	SF	SD
2012								
June	208	207	228	2	106	27	107	27
July	205	240	225	0	200	50	149	37
August	186	227	250	100	210	51	164	41
Total	599	674	703	102	516	128	420	105
2013								
June	207	205	229	225	239	93	189	95
July	206	242	301	150	472	401	175	44
August	189	231	301	75	220	51	175	44
Total	602	678	831	450	931	545	539	183

3.4. Stem Water Potential

Mid-day stem water potential varied widely among treatments (Figure 3). The ψ values were similar among all treatments at the beginning of the irrigation scheduling period. In 2012, ψ deceased progressively with increasing stress in deficit treatments, reaching the lowest values by 20 August 2012 (DOY 233) when the last measurement was taken (-1.29 MPa in FD, -1.31 MPa in SD, and -1.47 MPa in MD). However, ψ values in the full irrigation treatments were maintained above -1.0 MPa by the end of July and then decreased to about -1.2 MPa by 20 August 2012. In 2013, ψ values in FF and MF were above -0.9 MPa for the entire period, and ψ in SF was lower than -1.0 MPa in early July and then increased and remained above -1.0 MPa, which was close to the values in FF and MF. ψ in SD dropped to -1.9 MPa and was significantly more negative compared to those in MD and FD.

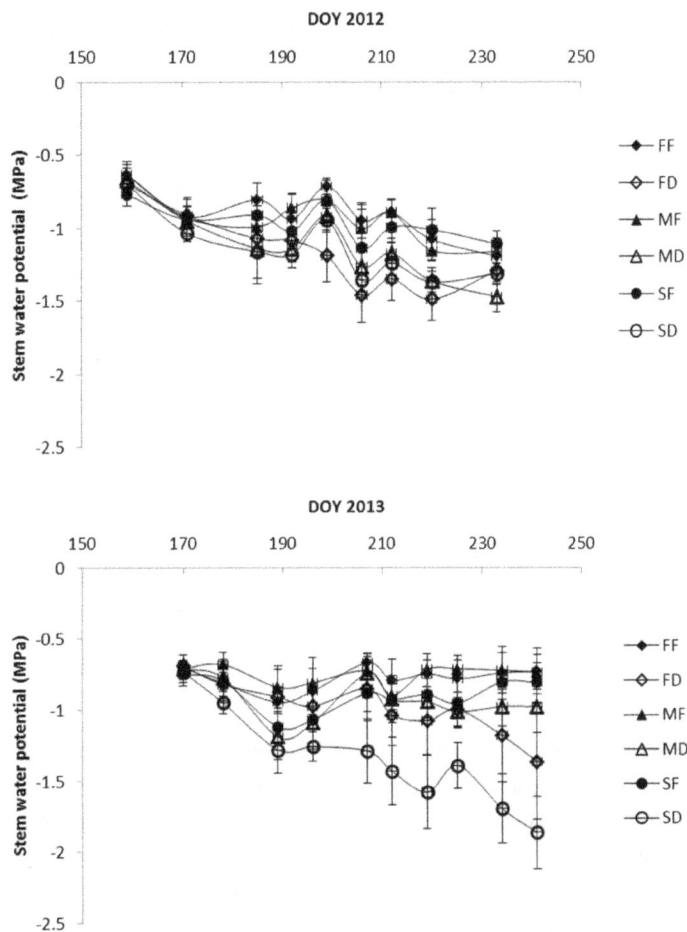

Figure 3. Stem water potential (ψ) in 2012 and 2013. Vertical bars indicate standard deviation.

3.5. Yield and Fruit Quality

Table 3 summarizes the yield data for the 2012–2013 and 2013–2014 growing seasons by irrigation treatment during postharvest. The total fruit number per tree and fruit weight per tree showed statistically significant difference between treatments in both seasons. The reduction on yield in 2013 with respect to 2012 was attributed to differences in pruning and thinning work done by different contractors. Fruit size was not affected by irrigation treatment in the 2012–2013 growing season. In the 2013–2014 growing season, the fruit size in FF was significantly greater than the other five treatments. There was no statistical difference in total soluble solids (%), pH values of the juice, and flesh firmness (N) among all treatments in both years (Table 4). In addition, no statistical difference in fruit skin color and titratable acidity was found among all treatments in both seasons (data not shown). Table 5 shows fruit physical quality data for the 2012–2013 and 2013–2014 growing seasons by irrigation treatment during postharvest. The number of fruits with doubles and external split were found to be statistically different among treatments in 2012–2013 and for doubles only in the 2013–2014 growing season.

Table 3. Peach yield and yield parameters in the 2012–2013 and 2013–2014 growing season res ponding to different irrigation system and full/deficit irrigation treatments [1].

Treatment	2012–2013			2013–2014		
	Fruit number per Tree	Fruit Weight (kg) per Tree	Fruit Weight (g/fruit)	Fruit Number per Tree	Fruit Weight (kg) per Tree	Fruit Weight (g/fruit)
FF	155 ± 5.6 [a]	20.5 ± 0.7 [a]	137 ± 4.4	105 ± 3.2 [a]	13.2 ± 0.4 [a]	125 ± 1.1 [a]
FD	138 ± 4.7 [ab]	18.0 ± 0.7 [ab]	136 ± 4.0	74 ± 3.2 [c]	8.8 ± 0.4 [c]	120 ± 1.4 [b]
MF	114 ± 5.9 [c]	15.2 ± 0.7 [cd]	137 ± 3.4	101 ± 3.4 [a]	12.2 ± 0.4 [a]	119 ± 1.2 [b]
MD	107 ± 4.5 [c]	14.3 ± 0.6 [d]	135 ± 1.7	90 ± 4.1 [b]	10.7 ± 0.5 [b]	119 ± 1.0 [b]
SF	129 ± 5.7 [bc]	17.9 ± 0.7 [abc]	144 ± 3.4	76 ± 3.4 [c]	9.1 ± 0.4 [c]	119 ± 1.2 [b]
SD	122 ± 5.2 [bc]	16.2 ± 0.5 [acd]	136 ± 2.4	58 ± 3.0 [d]	6.9 ± 0.4 [d]	118 ± 1.6 [b]
p-value	<0.0001	<0.0001	0.4423	<0.0001	<0.0001	0.0018

[1] Means (±SD) followed by a different letter (within a column) are significantly different at $p = 0.05$ according to the Tukey's studentized range (HSD) test.

Table 4. Peach flesh quality parameters in the 2012–2013 and 2013–2014 growing season responding to different irrigation system and full/deficit irrigation treatments. TSS = total soluble solids.

Treatment	2012–2013			2013–2014		
	TSS (%)	Firmness (N)	pH	TSS (%)	Firmness (N)	pH
FF	11.0 ± 0.2	37.5 ± 0.7	3.41 ± 0.06	10.3 ± 0.3	39.2 ± 0.8	3.23 ± 0.03
FD	11.2 ± 0.1	38.2 ± 0.6	3.35 ± 0.06	10.8 ± 0.5	41.7 ± 0.8	3.18 ± 0.04
MF	11.0 ± 0.1	36.9 ± 0.7	3.35 ± 0.05	10.2 ± 0.5	42.9 ± 0.8	3.22 ± 0.02
MD	11.4 ± 0.2	36.5 ± 0.6	3.42 ± 0.06	10.3 ± 0.4	41.1 ± 0.9	3.28 ± 0.03
SF	10.8 ± 0.2	37.0 ± 0.7	3.28 ± 0.05	10.2 ± 0.4	39.9 ± 0.5	3.25 ± 0.02
SD	11.4 ± 0.2	34.3 ± 0.7	3.42 ± 0.05	10.4 ± 0.4	41.6 ± 0.9	3.27 ± 0.02
p-value	0.2480	0.0571	0.4030	0.8817	0.8712	0.0931

When comparing full and deficit irrigation treatment for each irrigation system, we found significantly greater fruit number and weight per tree in FF than FD (F value = 6.3, $p = 0.01$; F value = 7.7, $p = 0.07$); larger fruit in SF compared to SD (F value = 4.2, $p = 0.05$); more doubles in MD than MF and in SD than SF (F value = 10.2, $p = 0.02$; F value = 8.2, $p = 0.04$); and more external split fruits in SD compared to SF (F value = 9.0, $p = 0.03$) for the 2012–2013 growing season. For the 2013–2014 growing season, full irrigation treatments produced greater fruits in number and weight per tree than deficit irrigation treatments (FF vs. FD, MF vs. MD, and SF vs. SD). The number of fruit was significantly greater in FF compared to FD (F value = 4.17, $p = 0.05$) and no statistical difference was found in fruit size between MF and MD, and SF and SD. There were more deep suture fruits in FD than FF (F value = 15.9, $p = 0.01$) and doubles in SD than SF (F value = 20.9, $p = 0.006$). No other quality parameters were significant.

For the 2012–2013 growing season, various irrigation systems had a significant effect on fruit number and weight in full irrigation treatments (FF, MF, and SF) (F value = 12.8, $p < 0.0001$; F value = 14.1, $p < 0.0001$) and deficit irrigation treatments (FD, MD, and SD) (F value = 10.5, $p < 0.0001$; F value = 9.9, $p = 0.0001$). The results confirmed that trees irrigated by micro-spray had lower yield than trees irrigated by surface drip irrigation [23]. There were no statistical physical quality differences in both full and deficit irrigation treatments. For the 2013–2014 growing season, the total fruit number per tree and fruit weight per tree also showed significant difference among full irrigation treatments (FF, MF, and SF) (F value = 21.4, $p < 0.0001$; F value = 22.7, $p < 0.0001$) and among deficit irrigation treatments (FD, MD, and SD) (F value = 20.6, $p < 0.0001$; F value = 21.8, $p < 0.0001$). Treatment FF produced larger fruits than the other two full irrigation treatments MF and SF, but no statistically different fruit size was observed among deficit irrigation treatments. Physical quality differences among full irrigation treatments were insignificant, but there were more doubles in SD than the other deficit treatments FD and MD (F value = 5.6, $p = 0.02$).

Table 5. Peach fruit physical quality parameters in the 2012–2013 and 2013–2014 growing season res ponding to different irrigation system and full/deficit irrigation treatments [1].

	Doubles (%)	Deep sutures (%)	External splits (%)	Dimples (%)	Deformed (%)	Split pit (%)
Treatment	Fruit quality parameters 2012–2013					
FF	1.53 [b]	0.83	0.14 [b]	0.41	0.76	
FD	1.94 [b]	0.14	0.55 [ab]	1.68	0.38	
MF	1.25 [b]	0.70	0.69 [ab]	0.92	0.86	
MD	5.83 [a]	1.81	1.25 [ab]	0.43	1.85	
SF	0.69 [b]	0.69	0.28 [b]	2.93	0.84	
SD	3.33 [ab]	0.56	1.53 [a]	2.11	1.55	
p-value	0.0005	0.2846	0.0055	0.1984	0.423	
Treatment	Fruit quality parameters 2013–2014					
FF	0 [b]	0.69	0.14	1.39	1.53	0.14
FD	0.7 [b]	1.81	0.28	5.00	0.70	0.14
MF	0 [b]	1.39	0.14	1.11	1.25	0.28
MD	0.28 [b]	1.94	0.00	2.64	1.95	0.14
SF	0.14 [b]	1.25	0.00	6.81	1.39	0
SD	1.67 [a]	2.36	0.00	8.19	2.78	0.42
p-value	<0.0001	0.4283	0.6032	0.1125	0.3319	0.5795

[1] Means followed by a different letter (within a column) are significantly different at $p = 0.05$ according to the Tukey's studentized range (HSD) test.

4. Discussion

The relationship between canopy-to-air temperature difference (ΔT) and other crop water stress indicators have been confirmed on tree crops [22,26,27]. A well-watered crop will transpire to its potential rate so canopy temperature should be below surrounding air temperature; while a water stressed crop will tend to close stomata, preventing transpiration water loss but at the same time causing a rise in canopy temperature. The potential of using ΔT for irrigation scheduling on early maturing peach trees has been evaluated by [11,25]. Irrigation amount, interval, and ΔT signals for various irrigation systems, furrow, micro-spray, and surface drip, were decided based on the capacity and limitation of each irrigation system and the findings in the previous studies.

Irrigation controlled by ΔT signals in full irrigation treatments to meet full crop ET during the postharvest period was successful in furrow irrigation, but not in micro-spray and surface drip (Table 2). Treatments MF and SF were not treated as targeted full irrigation treatments, but only met crop ET requirement partially (except for MF in 2013 due to an over irrigation event). It may be related to ΔT threshold values and irrigation intervals. For example, we had to adjust the threshold values to increase irrigation frequency (Figure 1) in treatment MF.

Soil water contents showed the decreasing trend with decreasing irrigation application. Although with less irrigation in MF and SF compared to FF in 2012, the soil water content stayed stable after 17 July 2012 (DOY 199) at 75–135 cm soil depths with values close to treatment FF. Bryla et al. [10] has indicated that 90% roots in furrow, micro-spray and drip irrigation system were located at <1 m depth in the soil layer. The soil water content of treatment SF in 2013 became stable at 75–135 cm soil depth after 29 July 2013 (DOY 210); however, the values were much lower than those in FF and MF. Therefore, with irrigation at 34% of ET_c and low soil water content (dropped even below deficit treatments FD and MD) in treatment SF in 2013, we believe that the targeted full irrigation treatment goal for SF was not achieved which likely resulted in some degree of water stress.

The ψ values recorded in the fully irrigated trees were similar to those found in well-irrigated peach trees by other studies [4,6]. Comparing fully irrigated treatments with deficit treatments that received 25% of ET_c (Table 2), small negative ψ values were found in the fully irrigated trees with clear differences from stressed trees in July and August 2012 when seasonal water demand was the highest. Vera et al. [28] found the lowest ψ value was -1.9 MPa during the postharvest when irrigation satisfied 25% of ET_c, which confirms our results in SD (25% of ET_c). The ψ values of treatment SF in 2013 were similar to those in FF and MF. Trees from the treatment SF, did not show the level of water stress that should have been expected taking into account the soil water content profile of that treatment (Figure 2). It might indicate that stem water potential may not be a good indicator of peach tree water stress under surface drip systems at certain stress levels.

Although the fruit sizes were smaller in 2013–2014 than in the 2012–2013 growing season, the results were comparable to previous studies in this orchard [10,11,23,25] or for the same peach variety in a nearby field [12]. Also fruit in 2013–2014 growing season tended to be undersized at harvest because they were picked early for better pricing due to the hot weather. The two pick days in 2014 were on 14 May and 17 May, two weeks earlier than usual. When comparing treatment MF with FF in 2013, more irrigation water was applied to MF than FF, but larger fruit was recorded in FF than MF. The results confirmed that more water is actually needed for the micro-spray system to maximize yield and meet crop water requirement due to higher soil evaporation compared with the furrow system [29]. Deficit irrigation caused fewer numbers of fruits and lower total weight per tree, but fruit size remained relatively unchanged [15,25]. It again confirmed that postharvest deficit irrigation could save about 50% of irrigation water without impacting fruit size [25]. Among all defective fruits, deep sutures and dimples were a much more serious problem compared to doubles in the 2013–2014 growing season. Deep sutures and dimples increased more than doubles and dimples were significantly increased in the most severely stressed SD treatment compared to the 2012–2013 growing season.

5. Conclusions

In this study, full and deficit irrigation treatments were applied by furrow, micro-spray and surface drip systems on an early maturing variety of peach during the postharvest season from June to August in 2012 and 2013. Irrigation was controlled by canopy-to-air temperature difference. Using canopy-to-air temperature difference to signal irrigation worked well in the furrow irrigation system, but needs improved techniques in the micro-spray and surface drip irrigation systems. Both irrigation systems and full/deficit irrigation treatments had no significant effect on peach flesh qualities, but due to the seven-year postharvest RDI study in this orchard, the long-term deficit irrigation treatment did cause more defective fruits.

Acknowledgments: The authors would like to acknowledge Matthew Gonzales and Don Tucker for assistance with field and lab work; and Sharon Peterson and Rick Emershad for helping with fruit quality assessment.

Author Contributions: Huihui Zhang and Dong Wang conceived and designed experiments; Huihui Zhang and Jim Gartung performed the experiments and collected data; Huihui Zhang analyzed data and wrote the paper.

Conflicts of Interest: The authors declare no conflict of interest.

References

1.   National Agricultural Statistical Service. Non Citrus Fruits and Nuts USDA 2015. Available online: http://usda.mannlib.cornell.edu/usda/current/NoncFruiNu/NoncFruiNu-07-06-2016.pdf (accessed on 23 September 2016).

2.   Chalmers, D.J.; Mitchell, P.D.; Heek, L.V. Control of peach tree growth and productivity by regulated water supply, tree density, and summer pruning. *J. Am. Soc. Hortic. Sci.* **1981**, *106*, 307–312.

3. Goldhamer, D.A.; Salinas, M.; Crisosto, C.; Day, K.R.; Soler, M.; Moriana, A. Effects of regulated deficit irrigation and partial root zone drying on late harvest peach tree performance. *Acta Hortic.* **2002**, *592*, 343–350. [CrossRef]

4. Girona, J.; Mata, M.; Arbonès, A.; Alegre, S.; Rufat, J.; Marsal, J. Peach tree response to single and combined regulated deficit irrigation regimes under shallow soils. *J. Am. Soc. Hortic. Sci.* **2003**, *128*, 432–440.

5. Sotiropoulos, T.; Kalfountzos, D.; Aleksiou, I.; Kotsopoulos, S.; Koutinas, N. Response of a clingstone peach cultivar to regulated deficit irrigation. *Sci. Agric.* **2010**, *67*, 164–169. [CrossRef]

6. Girona, J.; Gelly, M.; Mata, M.; Arbonès, A.; Rufat, J.; Marsal, J. Peach tree response to single and combined deficit irrigation regimes in deep soils. *Agric. Water Manag.* **2005**, *72*, 97–108. [CrossRef]

7. Larson, K.; DeJong, T.; Johnson, R. Physiological and growth responses of mature peach trees to postharvest water stress. *J. Am. Soc. Hortic. Sci.* **1988**, *113*, 296–300.

8. Johnson, R.S.; Handley, D.F.; DeJong, T.M. Long-term response of early maturing peach trees to postharvest water deficits. *J. Am. Soc. Hortic. Sci.* **1992**, *117*, 881–886.

9. Falagán, N.; Artés, F.; Gómez, P.A.; Artés-Hernández, F.; Conejero, W.; Aguayo, E. Deficit irrigation strategies enhance health-promoting compounds through the intensification of specific enzymes in early peaches. *J. Sci. Food Agric.* **2016**, *96*, 1803–1813. [CrossRef] [PubMed]

10. Bryla, D.R.; Dickson, E.; Shenk, R.; Johnson, R.S.; Crisosto, C.H.; Trout, T.J. Influence of irrigation method and scheduling on patterns of soil and tree water status and its relation to yield and fruit quality in peach. *Hortscience* **2005**, *40*, 2118–2124.

11. Wang, D.; Gartung, J. Infrared canopy temperature of early-ripening peach trees under postharvest deficit irrigation. *Agric. Water Manag.* **2010**, *97*, 1787–1794. [CrossRef]

12. Johnson, R.S.; Phene, B.C. Fruit quality disorders in an early maturing peach cultivar caused by postharvest water stress. *Int. Soc. Hortic. Sci.* **2008**, 385–390. [CrossRef]

13. Mirás-Avalos, J.M.; Pérez-Sarmiento, F.; Alcobendas, R.; Alarcón, J.J.; Mounzer, O.; Nicolás, E. Using midday stem water potential for scheduling deficit irrigation in mid–late maturing peach trees under Mediterranean conditions. *Irrig. Sci.* **2016**, *34*, 161–173. [CrossRef]

14. Ghrab, M.; Masmoudi, M.M.; Ben Mimoun, M.; Ben Mechlia, N. Plant- and climate-based indicators for irrigation scheduling in mid-season peach cultivar under contrasting watering conditions. *Sci. Hortic.* **2013**, *158*, 59–67. [CrossRef]

15. Conejero, W.; Mellisho, C.D.; Ortuño, M.F.; Moriana, A.; Moreno, F.; Torrecillas, A. Using trunk diameter sensors for regulated deficit irrigation scheduling in early maturing peach trees. *Environ. Exp. Bot.* **2011**, *71*, 409–415. [CrossRef]

16. Jackson, R.D.; Idso, S.B.; Reginato, R.J.; Pinter, P.J. Canopy temperature as a crop water stress indicator. *Water Resour. Res.* **1981**, *17*, 1133–1138. [CrossRef]

17. O'Shaughnessy, S.A.; Evett, S.R.; Colaizzi, P.D.; Howell, T.A. A crop water stress index and time threshold for automatic irrigation scheduling of grain sorghum. *Agric. Water Manag.* **2012**, *107*, 122–132. [CrossRef]

18. Baker, J.T.; Mahan, J.R.; Gitz, D.C.; Lascano, R.J.; Ephrath, J.E. Comparison of deficit irrigation scheduling methods that use canopy temperature measurements. *Plant Biosyst.* **2013**, *147*, 40–49. [CrossRef]

19. Clawson, K.L.; Blad, B.L. Infrared thermometry for scheduling irrigation of corn1. *Agron. J.* **1982**, *74*, 311–316. [CrossRef]

20. Wanjura, D.F.; Upchurch, D.R. Accounting for humidity in canopy-temperature-controlled irrigation scheduling. *Agric. Water Manag.* **1997**, *34*, 217–231. [CrossRef]

21. Glenn, D.; Worthington, J.; Welker, W.; McFarland, M. Estimation of peach tree water use using infrared thermometry. *J. Am. Soc. Hortic. Sci.* **1989**, *114*, 737–741.

22. Sepulcre-Cantó, G.; Zarco-Tejada, P.J.; Jiménez-Muñoz, J.C.; Sobrino, J.A.; Miguel, E.D.; Villalobos, F.J. Detection of water stress in an olive orchard with thermal remote sensing imagery. *Agric. For. Meteorol.* **2006**, *136*, 31–44. [CrossRef]

23. Bryla, D.R.; Trout, T.J.; Ayars, J.E.; Johnson, R.S. Growth and production of young peach trees irrigated by furrow, microjet, surface drip, or subsurface drip systems. *HortScience* **2003**, *38*, 1112–1116.

24. Johnson, R.; Williams, L.; Ayars, J.; Trout, T. Weighing lysimeters aid study of water relations in tree and vine crops. *Calif. Agric.* **2005**, *59*, 133–136. [CrossRef]

25. Zhang, H.; Wang, D. Management of postharvest deficit irrigation of peach trees using infrared canopy temperature. *Vadose Zone J.* **2013**, *12*. [CrossRef]

26. Andrews, P.K.; Chalmers, D.J.; Moremong, M. Canopy-air temperature differences and soil water as predictors of water stress of apple trees grown in a humid, temperate climate. *J. Am. Soc. Hortic. Sci.* **1992**, *117*, 453–458.

27. Gonzalez-Dugo, V.; Zarco-Tejada, P.; Berni, J.A.J.; Suárez, L.; Goldhamer, D.; Fereres, E. Almond tree canopy temperature reveals intra-crown variability that is water stress-dependent. *Agric. For. Meteorol.* **2012**, *154–155*, 156–165. [CrossRef]

28. Vera, J.; Abrisqueta, I.; Abrisqueta, J.M.; Ruiz-Sánchez, M.C. Effect of deficit irrigation on early-maturing peach tree performance. *Irrig. Sci.* **2013**, *31*, 747–757. [CrossRef]

29. Layne, R.E.C.; Tan, C.S.; Hunter, D.M.; Cline, R.A. Irrigation and fertilizer application methods affect performance of high-density peach orchards. *HortScience* **1996**, *31*, 370–375.

4

Is There a Positive Synergistic Effect of Biochar and Compost Soil Amendments on Plant Growth and Physiological Performance?

M. Lukas Seehausen, Nigel V. Gale *, Stefana Dranga, Virginia Hudson, Norman Liu, Jane Michener, Emma Thurston, Charlene Williams, Sandy M. Smith and Sean C. Thomas

Faculty of Forestry, University of Toronto, 33 Willcocks Street, Toronto, ON M5S 3B3, Canada; ml.seehausen@mail.utoronto.ca (M.L.S.); stefana.dranga@mail.utoronto.ca (S.D.); vehudson8@gmail.com (V.H.); changnorman.liu@mail.utoronto.ca (N.L.); jane.michener@mail.utoronto.ca (J.M.); emma.thurston@mail.utoronto.ca (E.T.); charleneawi@gmail.com (C.W.); s.smith.a@utoronto.ca (S.M.S.); sc.thomas@utoronto.ca (S.C.T.)
* Correspondence: nigel.gale@mail.utoronto.ca;

Academic Editor: Lukas Van Zwieten

Abstract: The combination of biochar (BC) with compost has been suggested to be a promising strategy to promote plant growth and performance, but although "synergistic" effects have been stated to occur, full-factorial experiments are few, and explicit tests for synergism are lacking. We tested the hypothesis that a combination of BC and spent mushroom substrate (SMS) has a positive synergistic effect on plant growth and physiological performance in a nutrient-limited growing media. A greenhouse experiment with a full factorial design was conducted using mixed-wood BC (3.0 kg·m^{-2}) and SMS (1.5 kg·m^{-2}) (the combination was not co-composted) as organic soil amendments for the annual *Abutilon theophrasti* and the perennial *Salix purpurea*. Several measurements related to plant growth and physiological performance were taken throughout the experiment. Contrary to the hypothesis, we found that the combination of BC + SMS had neutral or antagonistic interactive effects on many plant growth traits. Antagonistic effects were found on maximum leaf area, above- and belowground biomass, reproductive allocation, maximum plant height, chlorophyll fluorescence, and stomatal conductance of *A. theophrasti*. The effect on *S. purpurea* was mostly neutral. We conclude that the generalization that BC and compost have synergistic effects on plant performance is not supported.

Keywords: biochar; compost; spent mushroom substrate; soil amendments; synergistic effects; *Abutilon theophrasti*; *Salix purpurea*

1. Introduction

Charcoal derived from wildfire plays an integral role in the re-vegetation and function of ecosystems following disturbances. Soil pyrogenic carbon provides numerous services to plants, sorbing many growth-inhibitory compounds (e.g., phenolics, salts, metals) [1,2], basifying acidified soils concomitant with improving nutrient retention and exchange [3], delivering a pulse of water-soluble elements available for plants and soil fauna [4,5], and improving soil water holding capacity [6]. When charcoal is applied as a soil amendment, the term "biochar" (BC) is used [7]. In managed systems, BC has received much attention for its potential to improve soil fertility and to mitigate climate change by carbon sequestration (e.g., [8–10]). In recent meta-analyses, Jeffrey et al. [11] and Biederman and Harpole [12] show that BC increases crop yields by an average of 10% and 30%, respectively, while Thomas and Gale [13] report a mean biomass growth response of 41% for woody plants.

Similar to natural charcoals, BCs contain high specific surface area, cation exchange capacities and microporosity, properties that have improved plant growth and function across systems. These properties greatly depend on the feedstock (e.g., waste products from forestry and agriculture such as woodchips, sawdust, corn-stover, manure, etc.), the pyrolysis process (e.g., fast, slow, and gasification), and the temperature during pyrolysis [3,14]. BCs derived from woody feedstock demonstrate particularly important ameliorative properties that include high ability to sorb soil contaminants [15], retain nutrients and water [16–18], increase soil pH, and provide pulses of nutrients that are potentially limiting in certain systems (e.g., phosphorus in Pluchon et al. [19]).

Compost is another soil amendment heralded for multiple benefits as an organic fertilizer that results in well-documented increases in plant growth [20,21]. Spent mushroom substrate (SMS), also called mushroom compost, is a by-product of the mushroom industry that is commonly used as a soil amendment, and has shown particularly strong positive effects on plant growth and soil function [22]. Like BC, mushroom compost generally limes soils and increases the availability and uptake of nutrients—especially N, P, and K [23,24]. However, negative growth responses have occasionally been reported possibly due to high salinity or metal concentration in compost [25]. Mushroom compost has improved fruit yield in several agricultural crops in the relative short term (one growing season) at high dosages (4.0–8.0 kg·m^{-2}) [24,26]; however, the effects of mushroom compost on the growth and performance of woody plants/crops have only been minimally investigated [27].

In the broader scientific literature, "synergistic" effects are essentially always defined as a deviation from an additive model, such as that described by analysis of variance; this definition is likewise well established in the plant nutrition literature ([28], p. 39). It has been repeatedly suggested that the combination of BC with compost may be a promising strategy to promote plant growth and performance, having positive synergistic effects on soil properties and plant growth responses (e.g., [29–31]). Synergistic effects of a BC-compost blend on plant growth and performance are thought to be mediated by sorption of nutrients by the porous BC matrix [29], stimulation of microbial colonization [32], degradation of possible noxious pyrogenic substances [33], improvement of the BC surface reactivity through accelerated oxidative ageing [34,35] and dissolved organic carbon sorption [36]. The hypothesized synergy is likely to be particularly effective in offsetting potentially toxic effects of high metal and salt concentrations in SMS since BCs have demonstrated the ability to mitigate salt stress [15] and metal bioavailability [1]. Indeed, recent work by Beesley et al. [37] shows promise for BC-compost applications to soils contaminated by metals and arsenic. Multiple studies have been conducted on the effect of BC on soil properties and plant growth together with, or in comparison to compost and/or other fertilizers both in the field [31,38–44] and in greenhouses [45–49]. Although many studies report positive effects of a BC-compost mix on soil properties and plant growth, there is a lack of explicit tests for synergistic effects in the literature: we are aware of only a few prior experiments that have utilized a factorial experimental design [43,45,49].

In the present study, we tested the effect of a mixed-wood BC, SMS, and a combination of both (not co-composted prior to application), on plant growth and physiological performance of two plant species: the annual *Abutilon theophrasti* Medik (Malvaceae) and the perennial *Salix purpurea* L. (Salicaceae). We tested the hypothesis that a combination of BC and SMS has a positive synergistic effect on plant growth and physiological performance in nutrient-limited media. We expected that responses would be particularly dramatic in *Abutilon theophrasti*, an annual plant known for opportunistic nutrient uptake [50]. We expected that the response of the woody perennial *S. purpurea* would be relatively less pronounced, but of broad interest due to being a common bioenergy crop species [51].

2. Materials and Methods

2.1. Growth Conditions and Experimental Design

The experiment was conducted in a University of Toronto glasshouse situated at the St. George Campus, in Ontario, Canada. Mean temperature in the glasshouse averaged 26 °C during the day and 18 °C during night, with 40% mean relative humidity and a 16:8 h light-dark period.

A completely randomized factorial design was used with two treatment factors, BC and SMS additions, and two plant species, *Abutilon theophrasti* and *Salix purpurea* (var. 9882-41). *Abutilon theophrasti* seeds originated from agricultural weed populations in southern Illinois and were purchased through V&S Seed Supply (City, Illinois, USA). *Salix purpurea* cuttings originated from the Faculty of Environmental Science and Forestry at the State University of New York, NY, USA, and were obtained from the University of Guelph's long-term agroforestry trials in Guelph, Ontario, Canada, in January 2015. The BC used in the experiment was produced through slow pyrolysis of mixed wood from shipping and construction materials produced at a pyrolysis temperature of 700 °C for 30 min. The biochar used here has been extensively characterized elsewhere (see "old Burt's" biochar in [52]), was shown not to have elevated levels of metals or organic compounds, and additionally passed germination assays and earthworm avoidance tests in accordance with the International Biochar Initiative's procedures. Additionally, the same biochar has been utilized in two prior studies showing its potential to remove soil contaminants [53,54]. Biochar was heat-treated at 100 °C for 24 h prior to use in the experiment to alleviate potentially phytotoxic compounds leachable from BC [55]. Commercially available SMS (Premier Tech, Premier®, RiviÙre-du-Loup, Québec, Canada) was used. Chemical properties of the BC and SMS used in the experiment are listed in Table 1.

Table 1. Chemical properties of biochar and compost. Compost analysis was conducted at the Agriculture and Food Laboratory, University of Guelph, Guelph, ON. Biochar analyses were conducted at the Analytic Services Unit, Queen's University, Kingston, ON and derived from Denyes et al. [52]. Test procedures and units are reported in parenthesis. BC: biochar; SMS: spent mushroom substrate.

Attribute	Value for SMS	Value for BC
Ammonium-N (KCl-NH$_4$, mg·kg^{-1})	10	2
Nitrate-N (KCl-NH$_4$, mg·kg^{-1})	388	3.0
Phosphorus (NaHCO$_3$, mg·L^{-1})	120	31
Magnesium (NH$_4$C$_2$H$_3$O$_2$, mg·L^{-1})	820	848
Potassium (NH$_4$C$_2$H$_3$O$_2$, mg·L^{-1})	2000	2150
Manganese (NH$_4$C$_2$H$_3$O$_2$, mg·L^{-1})	4.3	752
Zinc (NH$_4$C$_2$H$_3$O$_2$, mg·L^{-1})	8.2	5.6
pH (CaCl$_2$)	6.4	10.6

A non-fertilized growth medium consisting of peat moss (70%), perlite (20%) and sand (10%) was used, in which the BC (1.5 kg·m^{-2}), SMS (3.0 kg·m^{-2}), or a combined mixture of BC and SMS (1.5 kg·m^{-2} and 3.0 kg·m^{-2}, respectively) soil amendments were completely mixed, forming the treatments of the experiment. The control treatment did not contain BC or SMS. There were 20 replicates (1 plant/pot) planted per plant species and treatment, with a total of 160 pots (2 species × 4 treatments × 20 replicates). The water content of the SMS was 41.10% ± 0.17% (n = 10 samples of 10 g each, dried for 48 h at 60 °C), therefore the dosage on dry mass basis was ~1.76 kg·m^{-2}. Dosages for BC and SMS were calculated relative to pot soil surface area (200 cm^2/4-L pot; 30 g BC and/or 60 g SMS/pot) and are in the range of dosages used in previous studies that improved plant growth (e.g., [24,26,56]). *Abutilon theophrasti* seeds were germinated in a non-soil growing mix seven days prior to planting. Cuttings of *S. purpurea* were shortened to 30 cm in length and were inserted into pots to a depth of 5–7 cm. Planting took place on 21 January 2015 and plants were grown for 91 days until 22 April 2015. The placement of the pots was completely randomized on one glasshouse table.

2.2. Plant Growth and Physiological Performance

To calculate leaf area of both plant species, the length of every leaf on each individual plant was measured to the nearest 0.5 cm on a weekly basis from 2 February 2015 onwards (12 days after planting). A species-specific allometric equation was applied to estimate area/leaf, and sum these estimates across all leaves. For *A. theophrasti*, the equation described in Thomas et al. [57] was used: $A = 0.6519 \times L^{1.9523}$, where A is leaf area in cm^2 and L is leaf length in cm. An allometric equation was developed for *S. purpurea* following the methods described in Thomas et al. [57]: $A = 0.1882 \times L^2$.

To investigate physiological plant responses to the treatments, leaf-level gas-exchange parameters were measured on 18 March 2015 (44 days after planting—approximately halfway through the experiment) between 09:00 am–02:00 pm local time. Light-saturated photosynthetic rate and stomatal conductance were measured using an LI-6400xt Portable Photosynthesis System (Li-Cor, Lincoln, NE, USA). Triplicate measurements were taken on the most recently developed fully-expanded leaf of each plant using a 6-cm^2 leaf cuvette. The system flow rate was set to 400 mmol/s, and CO_2 concentration of the sample set to 400 ppm. Relative humidity in the chamber was maintained at approximately 50%. A red-blue light source (Licor 6000-02B) was used to maintain a photosynthetic photon flux density of 1000 $\mu mol \cdot m^{-2} \cdot s^{-1}$ during measurements. Gas flux rates were monitored during measurements to ensure steady-state values prior to recording data.

Dark-adapted photosynthetic yield (the ratio of variable to maximal fluorescence: Fv/Fm) was measured on 16 March 2015 (52 days after planting) between 09:00 am–02:00 pm local time with a Walz MINI-PAM Fluorometer (Heinz Walz, Effeltrich, Germany), using the saturation pulse method [58]. Plants were dark acclimated prior to measurements for several hours at a light level of < 1 $\mu mol \cdot m^{-2} \cdot s^{-1}$ of incident photosynthetically active radiation. Measurements were taken on three recently fully-expanded leaves of *A. theophrasti* and three random leaves of about the same age and size for *S. purpurea*.

Reproductive performance of *A. theophrasti* was measured by determining the proportion of flowering plants (number of flowering plants/total number of plants) at seven dates between 19 March and 15 April 2015 (days 57–84 after planting) and the proportion of plants with fruits at six dates between 23 March and 15 April 2015 (days 61–84 after planting).

At harvest, plant height was measured for *A. theophrasti* and total branch length for *S. purpurea*. For the latter measurement, all branches were measured to the closest 0.5 cm and measurements were summed for all individual plants. Stem diameter increase at soil surface of *S. purpurea* was calculated by subtracting the initial stem diameter measured at the onset of the experiment from the stem diameter at harvest. All plants were then cut at soil surface and roots were thoroughly washed with water to free them from the growth medium. Biomass was dried for 48 h at 60 °C, after which above- and belowground biomass was measured and the root mass fraction (belowground biomass/aboveground biomass) calculated. For *A. theophrasti*, reproductive allocation of each plant was also calculated as mass of reproductive parts (flowers and fruits)/aboveground biomass.

2.3. Statistical Analysis

Effects of BC, SMS, and their interaction on plant growth and physiological metrics were tested by submitting data to analysis of variance (ANOVA), using the *glm* procedure (generalized linear model) in the statistical software R [59]. A significant positive BC × SMS interaction term was considered evidence for synergism, and a significant negative BC × SMS interaction term evidence for antagonism.

To test the effect of treatment (BC, SMS, BC + SMS) on leaf area over the growing period, repeated measures ANOVAs were employed using time, a parabolic term of time ($time^2$), treatment, the interaction of time and treatment, and the interaction of the parabolic term of time and treatment as explanatory variables. The dates of leaf area sampling were used in the random statement of the model and a variance components covariance structure was specified (PROC MIXED; SAS Inc., Cary, NC, USA, 2015 [60]). Leaf area data were log transformed to meet the assumptions of normality and homoscedasticity of residuals.

Treatment effects on photosynthetic rate, stomatal conductance, chlorophyll fluorescence, above- and belowground biomass, plant height, root mass fraction, reproductive allocation, branch length, and stem diameter gain were analyzed using a one-way ANOVA with treatment as an explanatory variable respectively for each dependent variable (PROC GLM; SAS Inc. 2015). A log transformation was used for stomatal conductance of *S. purpurea*, as well as root mass fraction and belowground biomass of *A. theophrasti* to meet the assumptions of normality and homoscedasticity of residuals. Because these assumptions could not be met for residuals with a transformation of *A. theophrasti* chlorophyll fluorescence data, non-parametric Wilcoxon Scores (rank sums) followed by a Kruskal–Wallis Test were used (PROC NPAR1WAY; SAS Inc. 2015).

Repeated measures logistic regression with a binomial distribution was used to analyze the effect of soil treatments on the proportion of *A. theophrasti* with flowers and fruits over time separately (PROC GLIMMIX; SAS Inc. 2015). Time, treatment, and the interaction of time and treatment were taken as explanatory variables in both analyses and a second-order polynomial term of time and the interaction of this term with treatment were added as explanatory variables to analyze the proportion of flowering plants.

In all cases, model assumptions, such as normality and homoscedasticity of residuals, were met and Tukey's range test was used for comparisons of means, unless stated otherwise.

3. Results

3.1. Effects of Biochar and Compost on Plant Growth and Performance

Relative to the control treatment, the application of BC alone had a significant positive effect on belowground biomass, maximum plant height and Fv/Fm of *A. theophrasti*, and on the proportion of root biomass of *S. purpurea*, but a significant negative effect on stomatal conductance of *S. purpurea* (Table 2). The SMS main effect was positive and significant for belowground biomass, reproductive allocation, maximum plant height, Fv/Fm, and stomatal conductance of *A. theophrasti*. In *S. purpurea*, the SMS main effect was negative for maximum leaf area, above- and belowground biomass, total branch length, and Fv/Fm (Table 2). The BC × SMS interaction term was significant and negative for maximum leaf area, above- and belowground biomass, reproductive allocation, maximum plant height, Fv/Fm, and stomatal conductance of *A. theophrasti*. The only cases of significant positive BC × SMS interactions were for the root mass fraction of *A. theophrasti* and aboveground biomass of *S. purpurea* (Table 2). In the latter case, the main effects of BC and SMS were negative (significantly so for the latter), thus the significant positive interaction terms indicate that addition of one soil amendment mitigates the negative effects of the other.

Table 2. Two-way analysis of variance (ANOVA) table of *F*- and *p*-values for the effects (+ = positive; − = negative) of biochar (BC), spent mushroom substrate (SMS), and their interaction (BC × SMS) on various measurements of plant growth and physiological performance relative to control conditions without soil amendments for (a) *Abutilon theophrasti* and (b) *Salix purpurea*. Degrees of freedom (*df*) of the error term = df_{error}, *df* of each effect = 1.

Measurement	BC			SMS			BC × SMS			df_{error}
	F	*p*	Effect	*F*	*p*	Effect	*F*	*p*	Effect	
(a) *Abutilon theophrasti*										
Maximum leaf area	0.042	0.8380		0.113	0.7370		23.522	<0.0001	−	76
Aboveground biomass	3.869	0.0528		3.451	0.0671		31.657	<0.0001	−	76
Belowground biomass	5.758	0.0189	+	8.543	0.0046	+	12.671	0.0006	−	76
Root mass fraction	0.193	0.6614		0.095	0.7593		6.701	0.0115	+	76
Reproductive allocation	0.238	0.6273		17.682	0.0001	+	13.077	0.0005	−	76
Maximum plant height	6.037	0.0163	+	6.745	0.0113	+	35.076	<0.0001	−	76
Chlorophyll fluorescence	8.998	0.0037	+	11.351	0.0012	+	7.981	0.0060	−	76
Photosynthetic rate	1.491	0.2300		0.348	0.5590		1.996	0.1660		36
Stomatal conductance	1.743	0.1951		19.533	0.0001	+	10.455	0.0026	−	36
(b) *Salix purpurea*										
Maximum leaf area	2.074	0.1542		8.479	0.0048	−	0.080	0.7777		70
Aboveground biomass	4.661	0.0351	−	15.669	0.0002	−	6.097	0.0166	+	57
Belowground biomass	0.052	0.8203		8.060	0.0064	−	0.810	0.3723		53
Root mass fraction	10.325	0.0022	+	1.694	0.1988		3.633	0.0621		53
Total branch length	3.038	0.0869		15.764	0.0002	−	0.816	0.3704		55
Stem diameter gain	1.672	0.2010		2.169	0.1470		2.450	0.1230		55
Chlorophyll fluorescence	0.023	0.8799		9.822	0.0026	−	0.158	0.6926		66
Photosynthetic rate	0.045	0.8330		1.326	0.2570		2.565	0.1180		36
Stomatal conductance	4.208	0.0476	+	0.998	0.3245		1.353	0.2525		36

3.2. Leaf Area Growth

Mean total leaf area of *A. theophrasti* was significantly influenced by all explanatory variables, including the interaction time^2 × treatment (Table 3). Total leaf area of plants treated with BC and SMS, respectively, reached a maximum of about 200 cm^2 around 40 days into the experiment.

However, leaf area of plants treated with BC + SMS or growing under control conditions increased more slowly than plants in the two other treatments and reached a maximum of only about 130 cm^2 (Figure 1a). A comparison of means revealed that only leaf area of plants in the BC + SMS treatment was significantly reduced compared to plants with BC or SMS treatments; however, none of the soil amendment treatments were significantly different from the control.

The interaction time^2 × treatment was not significant in the case of leaf area of *S. purpurea* and was therefore excluded from the analysis. Leaf area increase of *S. purpurea* over time was significantly described by time and time^2 (Table 3). However, the significant linear interaction time × treatment showed that while initially the increasing trend in leaf area was similar in all treatments, the highest average leaf area per plant (of about 280 cm^2) was reached by plants in control treatments (Figure 1b). The significant treatment effect followed by comparisons of means showed that overall, only the leaf area of plants in the BC + SMS treatment was significantly lower than that of plants growing under control conditions.

Table 3. Results of statistical analyses testing the effect of different explanatory variables on leaf area of *Abutilon theophrasti* and *Salix purpurea* (repeated measures analysis of variance), and on the proportion of *A. theophrasti* flowers (logistic regression).

Explanatory Variable	A. theophrasti Leaf Area			S. purpurea Leaf Area			A. theophrasti Flowers		
	F	df	p	F	df	p	F	df	p
Time	1541.8	1, 76	<0.0001	612.0	1, 71	<0.0001	20.4	1, 10	0.0011
Time^2	1482.9	1, 609	<0.0001	1034.6	1, 380	<0.0001	21.2	1, 10	0.0010
Treatment	7.1	3, 609	0.0001	4.9	3, 380	0.0024	4.3	3, 10	0.354
Time × treatment	14.2	3, 609	<0.0001	4.0	3, 380	0.0080	4.1	3, 10	0.0395
Time^2 × treatment	21.1	3, 609	<0.0001		removed		3.9	3, 10	0.0444

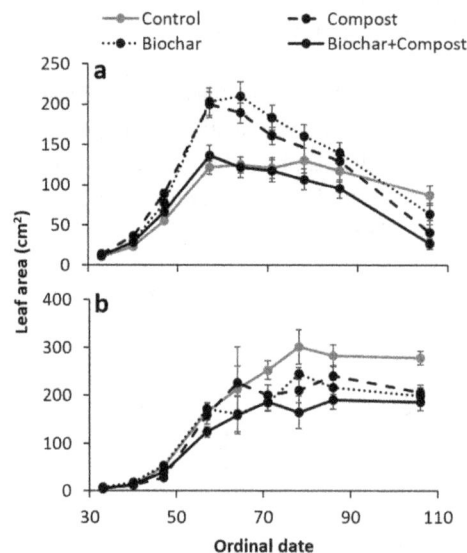

Figure 1. Mean (± standard error) leaf area of (**a**) *Abutilon theophrasti* and (**b**) *Salix purpurea* over time (ordinal date) during the greenhouse experiment.

3.3. Physiological Performance

Photosynthetic rate, stomatal conductance, and chlorophyll fluorescence of *A. theophrasti* were significantly influenced by the treatments (Table 4a). All three physiological traits showed significant increases for *A. theophrasti* in all soil amendment treatments when compared to plants in control conditions. Compared to the control, photosynthetic rate was increased ~3-fold in all treatments, stomatal conductance increased ~2.5-fold by SMS and ~1.7-fold by BC and BC + SMS, and an ~5% increase in Fv/Fm in all treatments was observed. Additionally, SMS significantly increased stomatal conductance of *A. theophrasti* by about 70% when compared to BC or BC + SMS (Figure 2a–c). For *S. purpurea*, Fv/Fm and stomatal conductance, but not photosynthetic rate were significantly influenced by the treatments (Table 4a). However, Tukey's range test only revealed a significantly reduced stomatal conductance in the BC + SMS treatment when compared to controls, but no significant treatment effects on Fv/Fm (Figure 2d–f).

Table 4. Results of the analysis of variance (ANOVA) testing the influence of treatments and control on measurements taken on *Abutilon theophrasti* and *Salix purpurea*.

Measurement	Abutilon theophrasti			Salix purpurea		
	F	df	p	F	df	p
(a) Physiological response						
Photosynthetic rate	25.4	3	<0.0001	1.3	3	0.2855
Stomatal conductance	15.1	3	<0.0001	3.6	3	0.0224
Chlorophyll fluorescence *	21.9	3	<0.0001	3.3	3	0.0246
(b) Measurements taken at harvest						
Aboveground biomass	13.0	3	<0.0001	8.8	3	<0.0001
Belowground biomass	12.5	3	<0.0001	3.0	3	0.0398
Root mass fraction	2.9	3	0.0393	5.2	3	0.0031

* values based on non-parametric Wilcoxon Scores (Rank Sums) followed by a Kruskal–Wallis Test.

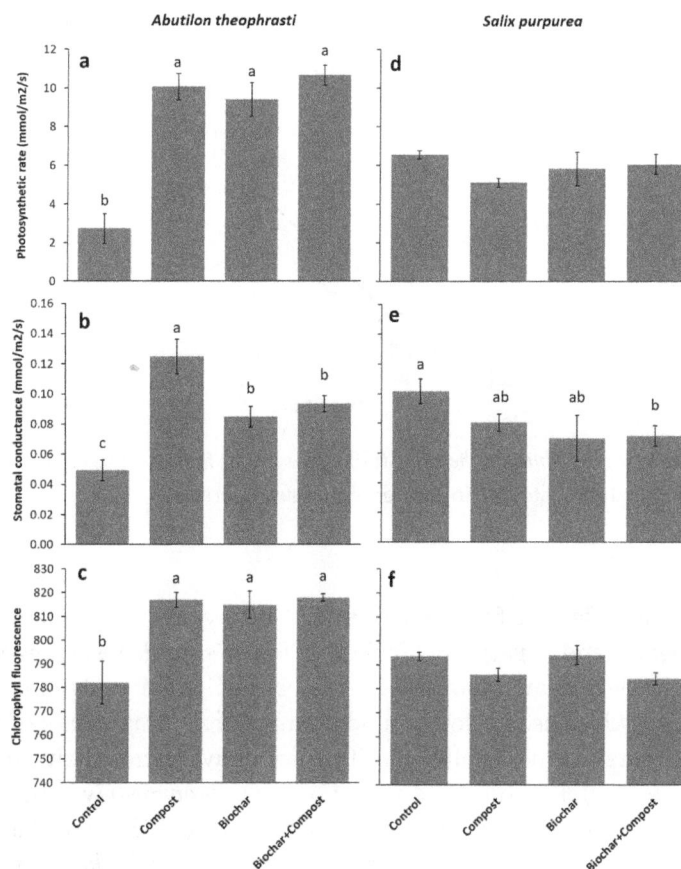

Figure 2. Mean (± standard error) light-saturated photosynthetic rate, stomatal conductance under light-saturated conditions, and ratio of variable to maximal chlorophyll fluorescence for *Abutilon theophrasti* (**a–c**) and *Salix purpurea* (**d–f**) at different soil treatments. Bars in each panel with the same lower case letters do not differ significantly at $p < 0.05$ according to Tukey's range test.

3.4. Reproductive Performance of A. theophrasti

The development of *A. theophrasti* flowers over time was best described by the interaction time^2 × treatment (Table 3; Figure 3a). Flowers were first produced by plants in BC treatments, followed by plants with BC + SMS. Flowers were produced somewhat later in SMS treatments. Plants in control conditions flowered ~7 days later than those in soil amendment treatments (Figure 3a). However, while the maximum proportion of flowers was around 40% for BC, BC + SMS, and controls, the highest proportion of flowering plants was reached in the SMS treatment (80%) (Figure 3a).

Fruit formation by *A. theophrasti* significantly increased over time ($F_{1,14} = 34.03$; $p < 0.0001$) and was significantly influenced by treatments ($F_{3,14} = 15.70$; $p < 0.0001$), but there was no time × treatment interaction. Similar to flower production, the first fruits were formed by *A. theophrasti* in BC treatments, followed by BC + SMS, and the SMS treatment, and 5–9 days later by plants in control conditions. Plants in all treatment groups produced significantly more fruits than controls (~35% increase relative to controls). Fruits were formed on 95% of plants in the SMS treatment, which was not significantly different from plants in the BC + SMS treatment, in which fruits were formed on 80% of plants. Significantly fewer fruits (65%) were formed by plants in the BC treatment group when compared to plants on SMS treatment (Figure 3b).

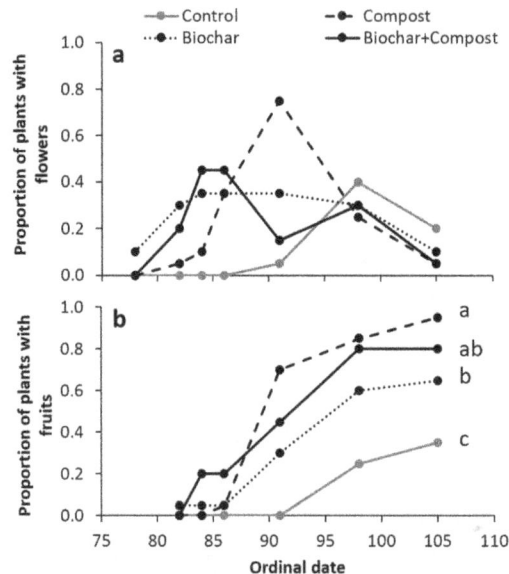

Figure 3. Mean proportion of *Abutilon theophrasti* plants with (**a**) flowers and (**b**) fruits at different soil treatments over time (ordinal date) during the greenhouse experiment.

3.5. Biomass at Harvest

Treatments had a significant effect on above- and belowground biomass, and proportion of roots for both *A. theophrasti* and *S. purpurea* (Table 4b). Tukey's range test showed that BC and SMS significantly increased aboveground biomass ~2.5-fold, but BC + SMS only ~2-fold for *A. theophrasti*. All treatments significantly increased belowground biomass of *A. theophrasti* ~2-fold when compared to controls. However, there was no significant difference between treatments in root mass fraction for *A. theophrasti* (Figure 4a–c). Reproductive allocation of *A. theophrasti* was significantly influenced by treatments ($F_{3,76} = 10.33$; $p < 0.0001$) and was ~4× higher in SMS, as well as 2.5× higher in BC when compared to the control (Table 5). Final plant height of *A. theophrasti* was also significantly influenced by treatments ($F_{3,76} = 17.41$; $p < 0.0001$). Compared to the control, plants were ~60% taller in SMS and BC treatments, but only 35% taller in the BC + SMS treatment. Comparisons of means showed that aboveground biomass of *S. purpurea* was significantly decreased by the treatments when compared to controls (reduction of ~33%), and that the root mass fraction was significantly higher in the BC and the BC + SMS treatment when compared to controls. However, using Tukey's range test, no significant differences between means were found for belowground biomass of *S. purpurea* (Figure 4d–f). Mean total branch length of plants at harvest was significantly influenced by treatments ($F_{3,55} = 6.57$; $p = 0.0007$) and was significantly lower in SMS as well as in BC + SMS treatments (by ~30%), when compared to controls (Table 5). Stem diameter increase of *S. purpurea* was not significantly influenced by the treatments ($F_{3,55} = 2.10$; $p = 0.1111$; Table 5).

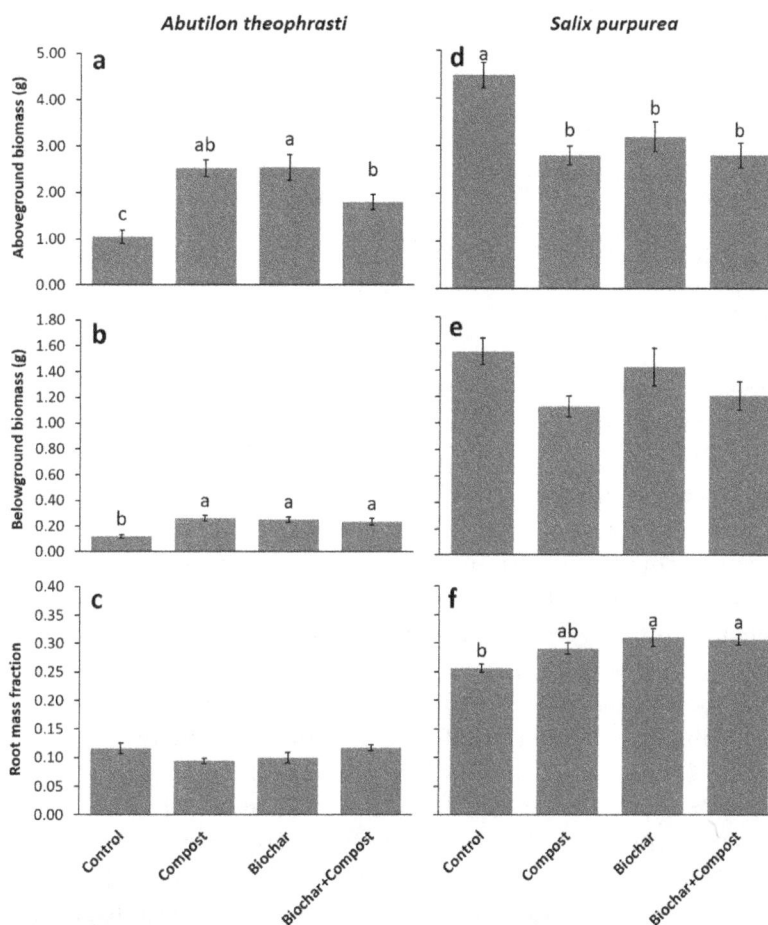

Figure 4. Mean (± standard error) above- and belowground biomass, and root mass fraction at different soil treatments for *Abutilon theophrasti* (**a–c**) and *Salix purpurea*; (**d–f**). Bars in each panel with the same lower case letters do not differ significantly at $p < 0.05$ according to Tukey's range test.

4. Discussion

We found no evidence to support the general hypothesis that BC and SMS had a positive synergistic effect on plant growth and performance in either plant species. Contrary to this hypothesis, we found that the mixture of BC and SMS generally had neutral or antagonistic interactive effects on plant growth and physiological performance. In *A. theophrasti*, most of the growth and physiological metrics taken in this study displayed antagonistic effects, with maximum leaf area and plant height being most impaired by the mixture of BC and SMS (Table 2). For *A. theophrasti* a weak synergistic effect of BC and SMS was found only in the case of root mass fraction, but this is not a performance metric. For *S. purpurea*, a synergistic effect of BC and SMS on aboveground biomass was found, but this effect was in the context of negative effects of both soil amendments when applied alone (Table 2).

Table 5. Measurements (mean ± standard error) taken at harvest of *Abutilon theophrasti* and *Salix purpurea* at different soil treatments. Numbers in each column followed by the same lower case letters do not differ significantly at $p < 0.05$ according to Tukey's range test.

Treatment	*Abutilon theophrasti*		*Salix purpurea*	
	Plant Height (cm)	Reproductive Allocation (g)	Total Branch Length (cm)	Stem Diameter Gain (Mm)
Control	45.8 ± 3.5 c	0.06 ± 0.02 c	102.6 ± 6.8 a	1.40 ± 0.17 a
Compost	72.7±3.8 a	0.23 ± 0.02 a	84.2 ± 9.6 b	0.93 ± 0.15 a
Biochar	74.3±2.4 ab	0.15 ± 0.03 b	69.3 ± 5.5 ab	1.32 ± 0.16 a
Biochar + Compost	62.0±2.6 b	0.16 ± 0.02 bc	63.3 ± 6.0 b	1.34 ± 0.15 a

Full factorial designs are necessary to test for interaction terms and thus infer synergistic or antagonistic effects. Despite the claim of "synergistic effects" of BC and compost on plant performance (and soil function) [30], we are aware of only a few prior studies that used a true full factorial design involving BC + compost mixtures [43,45,49], and analyses presented did not explicitly test for synergism. Other studies examining BC and compost as soil amendments have not presented strong evidence for synergistic effects based on a conventional definition or analysis. The majority of published studies have used a "substitutive" design for BC plus compost treatments, where the total dosage of soil amendment is kept constant, but the ratios changed (e.g., [37,39,41,46–48]; others have used experimental designs that do not involve additive combinations [38,42,46], or that do not include a BC-only treatment [31]. In the "substitutive" cases, the BC + compost treatments are modified in terms of both the BC dosage and addition of compost, and one thus cannot infer the independent effects of either change (analogous to the case of "substitutive" designs in plant competition studies [61]).

Why would BC + compost mixtures produce antagonistic effects on plant performance? As an organic fertilizer, compost (including SMS) provides plants with important nutrients and facilitates subsequent uptake. Similarly, unweathered BC also provides a short-term pulse of certain important nutrients, in particular K and P [18,62]. If growth responses to BC and SMS treatments are mainly due to nutrient provision, then a saturation of plant nutrient demands will often result in a non-additive positive effect on plant growth (e.g., [63]), and may lead to a decline in plant growth at high levels [28] (p. 46). Alternatively, the combination of BC and SMS may have resulted in an oversupply of toxic elements (e.g., Al, Pb, Mn, As), or nutrients required at only very low levels (e.g., Mn, B) [3,23,25,64,65]. For example, high amounts of Mn are present in the BC used in our experiment (Table 1); while Mn is important for plant growth as a micronutrient [66], it can be toxic to plants at amounts that are highly variable between plant species [67] and environmental conditions [68–70].

In addition, biochar has a high sorption ability [16] that can lead to a reduced availability of nutrients, in particular mineralized N or available P, for plants under certain circumstances [48,71,72]. Thus, when compared to the SMS treatment, the availability of nutrients to the plants in the BC + SMS treatment might have been reduced, leading to the observed reduction in plant growth and physiological performance. Indeed, Kammann et al. [49] report antagonism in *Chenopodium quinoa* when BC was combined with manure compost in a sandy-loam soil supplemented with nitrogen fertilizer, concomitant with strong nitrate sorption.

One potential way forward in offsetting antagonistic effects BC + compost applications is to co-compost prior to application. Several studies have demonstrated increased supply of nutrients (K, P, Ca, NH_4, NO_3) when manure-based composts were co-composted with wood BCs to maturity prior to application, ultimately increasing crop yield and biomass [43,44,49]. In Kammann et al. [49], co-composting produced a synergistic effect, rather than the antagonistic effect observed in the BC + compost treatment, due to enriched nitrate and phosphate. In the present study, co-composting prior to application may have alleviated antagonistic effects. Much additional work needs to be conducted to explore the mechanisms responsible for synergism in co-composted BC mixtures.

Positive effects of BC and SMS on plant growth have been found for both annual and perennial plants [12,13,20,21]. Fast-growing annual plants generally show higher nutritional demands than woody perennials [73]. Prior studies have specifically found very high responses of *A. theophrasti* to nutrient additions [50]. This species did show more pronounced responses to the soil amendments than *S. purpurea*. In particular, *A. theophrasti* showed a ~3-fold increase in photosynthetic rate in response to all three soil amendment treatments, while *S. purpurea* showed no significant response (Figure 2). Since *S. purpurea* was potted as cuttings, it is possible that nutrients in initial cuttings were translocated to offset nutrient deprivation. Our data are consistent with patterns of high interspecific variation in responses to biochar (e.g., [13,19]), and suggest that certain early-successional species are particularly responsive (see [74]).

Biochar effects on plant reproduction have received relatively little attention in the literature (e.g., [15,75]). The reproductive performance of *A. theophrasti* was strongly influenced by the soil treatments. All soil amendment treatments induced earlier flowering and fruit production when compared to plants in control treatments. Interestingly, the first flowers and first fruits were formed by plants in BC treatments (BC alone or in combination with SMS). BC can release volatile compounds such as ethylene into the soil [76,77]. Ethylene is known to be a gaseous plant hormone, which is responsible for inducing the formation of flowers in certain plants (e.g., [78–81]), and may therefore explain the earlier flowering of plants with BC treatments. However, most flowers and fruits were eventually produced by plants in compost treatments, which is also reflected in the higher reproductive allocation of plants receiving this treatment.

In the present study, we used a mixed-wood BC at 1.5 $kg·m^{-2}$ and a mushroom compost at ~1.8 $kg·m^{-2}$. These dosages are not excessively high, and are broadly consistent with studies that found significant effects of compost [26] and BC [54] on growth of plants in temperate environments. However, the effect of both BC and compost on the performance of plants is highly variable and depends on feedstock, dosage, and soil type, and the fertility of the plant it is applied to [11,82,83]. The SMS compost used here is slightly more acidic when compared to manure-based composts used in other similar studies [43,44,49]. The BC used here was produced at moderately high pyrolysis temperatures and has high surface area and porosity, characteristics that enable its use as a sorptive agent of soil pollutants in other studies [53,54]. Highly-sorptive BCs from high pyrolysis temperatures are desirable in restoration contexts where soil pollutants limit plant growth. Comparative studies testing a variety of BC and compost types across a spectrum of plant growth forms and soil types are necessary to untangle the mechanisms responsible for interactive effects of BC + compost.

5. Conclusions

The generalization that BC and compost mixtures have synergistic effects on plant growth and performance is not supported by this study. On the contrary, we have shown that even antagonistic effects of BC and compost mixtures can occur. Full factorial design experiments followed by an appropriate statistical analysis including the interaction term of the factors are needed to accurately test for synergism. More broadly, investigating the underlying mechanisms of synergistic or antagonistic effects of BC and other fertilizers and soil conditioners on plant growth is important in realizing the potential of biochar as a tool in ecosystem management.

Acknowledgments: We thank all students in the 2015 Stresses in the Forest Environment class (Faculty of Forestry, University of Toronto), and also W. Merrit, L. Sujeeun, S. Dehdashti, and A. Mansour for taking measurements on plants in the greenhouse and laboratory. Naresh Thevasthasan at the University of Guelph is thanked for assistance in obtaining *Salix purpurea* cuttings, and Julian Cleary and Adam Martin at the University of Toronto for reviewing an earlier version of this paper. This project was supported in part by grants from the Canadian Natural Science and Engineering Research Council to S.M.S. and S.C.T.

Author Contributions: M.L.S., N.V.G., and S.C.T. designed the experiment. M.L.S., N.V.G., S.D., V.H., N.L., J.M., E.T., and C.W. maintained the experiment and collected the data, and M.L.S., N.V.G., and S.C.T. analyzed the data. M.L.S., N.V.G., S.D., V.H., N.L., J.M., E.T., C.W., S.M.S., and S.C.T. wrote the paper.

Conflicts of Interest: The authors declare no conflict of interest.

References

1. Park, J.H.; Choppala, G.K.; Bolan, N.S.; Chung, J.W.; Chuasavathi, T. Biochar reduces the bioavailability and phytotoxicity of heavy metals. *Plant Soil* **2011**, *348*, 439–451. [CrossRef]
2. Lashari, M.S.; Liu, Y.; Li, L.; Pan, W.; Fu, J.; Pan, G.; Zheng, J.; Zheng, J.; Zhang, X.; Yu, X. Effects of amendment of biochar-manure compost in conjunction with pyroligneous solution on soil quality and wheat yield of a salt-stressed cropland from Central China Great Plain. *Field Crops Res.* **2013**, *144*, 113–118. [CrossRef]
3. Kloss, S.; Zehetner, F.; Dellantonio, A.; Hamid, R.; Ottner, F.; Liedtke, V.; Schwanninger, M.; Gerzabek, M.H.; Soja, G. Characterization of slow pyrolysis biochars: Effects of feedstocks and pyrolysis temperature on biochar properties. *J. Environ. Qual.* **2012**, *41*, 990–1000. [CrossRef] [PubMed]
4. Wardle, D.A.; Zackrisson, O.; Nilsson, M.C. The charcoal effect in Boreal forests: Mechanisms and ecological consequences. *Oecologia* **1998**, *115*, 419–426. [CrossRef]
5. DeLuca, T.H.; MacKenzie, M.D.; Gundale, M.J.; Holben, W.E. Wildfire-Produced charcoal directly influences nitrogen cycling in ponderosa pine forests. *Soil Sci. Soc. Am. J.* **2012**, *70*, 448–453. [CrossRef]
6. Bruun, E.W.; Petersen, C.T.; Hansen, E.; Holm, J.K.; Hauggaard-Nielsen, H. Biochar amendment to coarse sandy subsoil improves root growth and increases water retention. *Soil Use Manag.* **2014**, *30*, 109–118. [CrossRef]
7. Lehmann, J.; Joseph, S. Biochar for environmental management: An introduction. In *Biochar for Environmental Management: Science and Technology*; Lehmann, J., Joseph, S., Eds.; Earthscan: London, UK, 2009; pp. 1–12.
8. Lehmann, J. Bio-Energy in the black. *Front. Ecol. Environ.* **2007**, *5*, 381–387. [CrossRef]
9. Sohi, S.; Krull, E.; Lopez-Capel, E.; Bol, R. A review of biochar and its use and function in soil. *Adv. Agron.* **2010**, *105*, 47–82.
10. Woolf, D.; Amonette, J.E.; Street-Perrott, F.A.; Lehmann, J.; Joseph, S. Sustainable biochar to mitigate global climate change. *Nat. Commun.* **2010**, *1*, 1–9. [CrossRef] [PubMed]
11. Jeffery, S.; Verheijen, F.G.A.; van der Velde, M.; Bastos, A.C. A quantitative review of the effects of biochar application to soils on crop productivity using meta-analysis. *Agric. Ecosyst. Environ.* **2011**, *144*, 175–187. [CrossRef]
12. Biederman, L.A.; Harpole, W.S. Biochar and its effects on plant productivity and nutrient cycling: A meta-analysis. *GCB Bioenergy* **2013**, *5*, 202–214. [CrossRef]
13. Thomas, S.C.; Gale, N. Biochar and forest restoration: A review and meta-analysis of tree growth responses. *New For.* **2015**, *46*, 931–946. [CrossRef]
14. Zhao, L.; Cao, X.; Mašek, O.; Zimmerman, A. Heterogeneity of biochar properties as a function of feedstock sources and production temperatures. *J. Hazard. Mater.* **2013**, *256*, 1–9. [CrossRef] [PubMed]
15. Thomas, S.C.; Frye, S.; Gale, N.; Garmon, M.; Launchbury, R.; Machado, N.; Melamed, S.; Murray, J.; Petroff, A.; Winsborough, C. Biochar mitigates negative effects of salt additions on two herbaceous plant species. *J. Environ. Manag.* **2013**, *129*, 62–68. [CrossRef] [PubMed]
16. Atkinson, C.J.; Fitzgerald, J.D.; Hipps, N.A. Potential mechanisms for achieving agricultural benefits from biochar application to temperate soils: A review. *Plant Soil* **2010**, *337*, 1–18. [CrossRef]
17. Major, J.; Rondon, M.; Molina, D.; Riha, S.J.; Lehmann, J. Maize yield and nutrition during 4 years after biochar application to a Colombian savanna oxisol. *Plant Soil* **2010**, *333*, 117–128. [CrossRef]
18. Sackett, T.E.; Basiliko, N.; Noyce, G.L.; Winsborough, C.; Schurman, J.; Ikeda, C.; Thomas, S.C. Soil and greenhouse gas responses to biochar additions in a temperate hardwood forest. *GCB Bioenergy* **2015**, *7*, 1062–1074. [CrossRef]
19. Pluchon, N.; Gundale, M.J.; Nilsson, M.C.; Kardol, P.; Wardle, D.A. Stimulation of boreal tree seedling growth by wood-derived charcoal: Effects of charcoal properties, seedling species and soil fertility. *Funct. Ecol.* **2014**, *28*, 766–775. [CrossRef]
20. Sainz, M.J.; Taboada-Castro, M.T.; Vilarino, A. Growth, mineral nutrition and mycorrhizal colonization of red clover and cucumber plants grown in a soil amended with composted urban wastes. *Plant Soil* **1998**, *205*, 85–92. [CrossRef]
21. Soumare, M. Effects of a municipal solid waste compost and mineral fertilization on plant growth in two tropical agricultural soils of Mali. *Bioresour. Technol.* **2003**, *86*, 15–20. [CrossRef]

22. Maynard, A.A. Sustained vegetable production for three years using composted animal manures. *Compost Sci. Util.* **2013**, *2*, 88–96. [CrossRef]

23. Stewart, D.P.C.; Cameron, K.C.; Cornforth, I.S.; Main, B.E. Release of sulphate, potassium, calcium and magnesium from spent mushroom compost under laboratory conditions. *Biol. Fert. Soils* **1997**, *26*, 146–151. [CrossRef]

24. Stewart, D.P.C.; Cameron, K.C.; Cornforth, I.S. Effects of spent mushroom substrate on soil chemical conditions and plant growth in an intensive horticultural system: A comparison with inorganic fertiliser. *Aust. J. Soil Res.* **1998**, *36*, 185–199. [CrossRef]

25. Stamatiadis, S.; Werner, M.; Buchanan, M. Field assessment of soil quality as affected by compost and fertilizer application in a broccoli field (San Benito County, California). *Appl. Soil Ecol.* **1999**, *12*, 217–225. [CrossRef]

26. Polat, E.; Uzun, H.I.; Topçuoglu, B.; Önal, K.; Onus, A.N.; Karaca, M. Effects of spent mushroom compost on quality and productivity of cucumber (*Cucumis sativus* L.) grown in greenhouses. *Afr. J. Biotechnol.* **2009**, *8*, 176–180.

27. Chong, C.; Rinker, D.L. Use of spent mushroom substrate for growing containerized woody ornamentals: An overview. *Compost Sci. Util.* **1994**, *2*, 45–53. [CrossRef]

28. Tisdale, S.L.; Nelson, W.L.; Beaton, J.D. *Soil Fertility and Fertilizers*; Collier Macmillan Publishers: London, UK, 1985.

29. Steiner, C.; Das, K.C.; Melear, N.; Lakly, D. Reducing nitrogen loss during poultry litter composting using biochar. *J. Environ. Qual.* **2010**, *39*, 1236–1242. [CrossRef] [PubMed]

30. Fischer, D.; Glaser, B. Synergisms between compost and biochar for sustainable soil amelioration. In *Management of Organic Waste*; Kumar, S., Ed.; InTech: Rijeka, Croatia, 2012; pp. 167–198.

31. Liu, J.; Schulz, H.; Brandl, S.; Miehtke, H.; Huwe, B.; Glaser, B. Short-Term effect of biochar and compost on soil fertility and water status of a Dystric Cambisol in NE Germany under field conditions. *J. Plant Nutr. Soil Sci.* **2012**, *175*, 698–707. [CrossRef]

32. Pietikäinen, J.; Kiikkilä, O.; Fritze, H. Charcoal as a habitat for microbes and its effect on the microbial community of the underlying humus. *Oikos* **2003**, *89*, 231–242. [CrossRef]

33. Tuomela, M.; Vikman, M.; Hatakka, A.; Itävaara, M. Biodegradation of lignin in a compost environment: A review. *Bioresour. Technol.* **2000**, *72*, 169–183. [CrossRef]

34. Cheng, C.-H.; Lehmann, J. Ageing of black carbon along a temperature gradient. *Chemosphere* **2009**, *75*, 1021–1027. [CrossRef] [PubMed]

35. Zimmerman, A.R. Abiotic and microbial oxidation of laboratory-produced black carbon (biochar). *Environ. Sci. Technol.* **2010**, *44*, 1295–1337. [CrossRef] [PubMed]

36. Prost, K.; Borchard, N.; Siemens, J.; Kautz, T.; Séquaris, J. Biochar affected by composting with farmyard manure. *J. Environ. Qual.* **2012**, *42*, 164–172. [CrossRef] [PubMed]

37. Beesley, L.; Inneh, O.S.; Norton, G.J.; Moreno-Jimenez, E.; Pardo, T.; Clemente, R.; Dawson, J.J. Assessing the influence of compost and biochar amendments on the mobility and toxicity of metals and arsenic in a naturally contaminated mine soil. *Environ. Pollut.* **2014**, *186*, 195–202. [CrossRef] [PubMed]

38. Major, J.; Steiner, C.; Ditommaso, A.; Falcão, N.P.; Lehmann, J. Weed composition and cover after three years of soil fertility management in the central Brazilian Amazon: Compost, fertilizer, manure and charcoal applications. *Weed Biol. Manag.* **2005**, *5*, 69–76. [CrossRef]

39. Steiner, C.; Glaser, B.; Geraldes-Teixeira, W.; Lehmann, J.; Blum, W.E.; Zech, W. Nitrogen retention and plant uptake on a highly weathered central Amazonian Ferralsol amended with compost and charcoal. *J. Plant Nutr. Soil Sci.* **2008**, *171*, 893–899. [CrossRef]

40. Beesley, L.; Moreno-Jimenez, E.; Gomez-Eyles, J.L. Effects of biochar and greenwaste compost amendments on mobility, bioavailability, and toxicity of inorganic and organic contaminants in a multi-element polluted soil. *Environ. Pollut.* **2010**, *158*, 2282–2287. [CrossRef] [PubMed]

41. Suddick, E.C.; Six, J. An estimation of annual nitrous oxide emissions and soil quality following the amendment of high temperature walnut shell biochar and compost to a small scale vegetable crop rotation. *Sci. Total Environ.* **2013**, *465*, 298–307. [CrossRef] [PubMed]

42. Schmidt, H.P.; Kammann, C.; Niggli, C.; Evangelou, M.W.; Mackie, K.A.; Abiven, S. Biochar and biochar-compost as soil amendments to a vineyard soil: Influences on plant growth, nutrient uptake, plant health and grape quality. *Agric. Ecosyst. Environ.* **2014**, *191*, 117–123. [CrossRef]

43. Agegnehu, G.; Bass, A.M.; Nelson, P.N.; Bird, M.I. Benefits of biochar, compost and biochar-compost for soil quality, maize yield and greenhouse gas emissions in a tropical agricultural soil. *Sci. Total Environ.* **2016**, *543*, 295–306. [CrossRef] [PubMed]

44. Bass, A.M.; Bird, M.I.; Kay, G.; Muirhead, B. Soil properties, greenhouse gas emissions and crop yield under compost, biochar and co-composted biochar in two tropical agronomic systems. *Sci. Total Environ.* **2016**, *550*, 459–470. [CrossRef] [PubMed]

45. Karami, N.; Clemente, R.; Moreno-Jiménez, E.; Lepp, N.W.; Beesley, L. Efficiency of green waste compost and biochar soil amendments for reducing lead and copper mobility and uptake to ryegrass. *J. Hazard. Mater.* **2011**, *191*, 41–48. [CrossRef] [PubMed]

46. Schulz, H.; Glaser, B. Effects of biochar compared to organic and inorganic fertilizers on soil quality and plant growth in a greenhouse experiment. *J. Plant Nutr. Soil Sci.* **2012**, *175*, 410–422. [CrossRef]

47. Schulz, H.; Dunst, G.; Glaser, B. Positive effects of composted biochar on plant growth and soil fertility. *Agron. Sustain. Dev.* **2012**, *33*, 817–827. [CrossRef]

48. Vandecasteele, B.; Sinicco, T.; D'Hose, T.; Nest, T.V.; Mondini, C. Biochar amendment before or after composting affects compost quality and N losses, but not P plant uptake. *J. Environ. Manag.* **2016**, *168*, 200–209. [CrossRef] [PubMed]

49. Kammann, C.I.; Schmidt, H.P.; Messerschmidt, N.; Linsel, S.; Steffens, D.; Müller, C.; Koyro, H.W.; Conte, P.; Stephen, J. Plant growth improvement mediated by nitrate capture in co-composted biochar. *Sci. Rep.* **2015**, *5*. [CrossRef]

50. Benner, B.L.; Bazzaz, F.A. Response of the annual *Abutilon theophrasti* Medic. (Malvaceae) to timing of nutrient availability. *Am. J. Bot.* **1985**, *72*, 320–323. [CrossRef]

51. Volk, T.A.; Abrahamson, L.P.; Nowak, C.A.; Smart, L.B.; Tharakan, P.J.; White, E.H. The development of short-rotation willow in the northeastern United States for bioenergy and bioproducts, agroforestry and phytoremediation. *Biomass Bioenergy* **2006**, *30*, 715–727. [CrossRef]

52. Denyes, M.J.; Matovic, D.; Zeeb, B.A.; Rutter, A. Report on the production and characterization of biochar produced at Burt's Greenhouse. 2006. Available online: http://burtsgh.com/wpr/wp-content/uploads/2013/12/Report_on_the_Production_and_Characterization_Biochar_Produced_at_BurtsGreenhouses_Final_O1.pdf (accessed on 24 January 2017).

53. Denyes, M.J.; Langlois, V.S.; Rutter, A.; Zeeb, B.A. The use of biochar to reduce soil PCB bioavailability to *Cucurbita pepo* and *Eisenia fetida*. *Sci. Total Environ.* **2012**, *437*, 76–82. [CrossRef] [PubMed]

54. Denyes, M.J.; Rutter, A.; Zeeb, B.A. In situ application of activated carbon and biochar to PCB-contaminated soil and the effects of mixing regime. *Environ. Pollut.* **2013**, *182*, 201–208. [CrossRef] [PubMed]

55. Gale, N.V.; Sackett, T.; Thomas, S.C. Thermal treatment and leaching of biochar alleviates plant growth inhibition from mobile organic compounds. *PeerJ* **2016**, *4*, e2385. [CrossRef] [PubMed]

56. Rajkovich, S.; Enders, A.; Hanley, K.; Hyland, C.; Zimmerman, A.R.; Lehmann, J. Corn growth and nitrogen nutrition after additions of biochars with varying properties to a temperate soil. *Biol. Fert. Soils* **2002**, *48*, 271–284. [CrossRef]

57. Thomas, S.C.; Jasienski, M.; Bazzaz, F.A. Early vs. asymptotic growth responses of herbaceous plants to elevated CO_2. *Ecology* **1999**, *80*, 1552–1567. [CrossRef]

58. Maxwell, K.; Johnson, G.N. Chlorophyll fluorescence—A practical guide. *J. Exp. Bot.* **2000**, *51*, 659–668. [CrossRef] [PubMed]

59. R Core Team. *R: A Language and Environment for Statistical Computing*; R Foundation for Statistical Computing: Vienna, Austria, 2014.

60. SAS Institute Inc. *SAS Studio 3.4: User's Guide*; SAS Institute Inc.: Cary, NC, USA, 2015.

61. Firbank, L.G.; Watkinson, A.R. On the analysis of competition within two-species mixtures of plants. *J. Appl. Ecol.* **1985**, *22*, 503–517. [CrossRef]

62. Novak, J.M.; Busscher, W.J.; Laird, D.L.; Ahmedna, M.; Watts, D.W.; Niandou, M.A. Impact of biochar amendment on fertility of a southeastern coastal plain soil. *Soil Sci.* **2009**, *174*, 105–112. [CrossRef]

63. Tessier, J.T.; Raynal, D.J. Use of nitrogen to phosphorus ratios in plant tissue as an indicator of nutrient limitation and nitrogen saturation. *J. Appl. Ecol.* **2003**, *40*, 523–534. [CrossRef]

64. Kloss, S.; Zehetner, F.; Oburger, E.; Buecker, J.; Kitzler, B.; Wenzel, W.W.; Wimmer, B.; Soja, G. Trace element concentrations in leachates and mustard plant tissue (*Sinapis alba* L.) after biochar application to temperate soils. *Sci. Total Environ.* **2014**, *481*, 498–508. [CrossRef] [PubMed]

65. Domene, X.; Hanley, K.; Enders, A.; Lehmann, J. Short-Term mesofauna responses to soil additions of corn stover biochar and the role of microbial biomass. *Appl. Soil Ecol.* **2015**, *89*, 10–17. [CrossRef]

66. Hänsch, R.; Mendel, R.R. Physiological functions of mineral micronutrients (Cu, Zn, Mn, Fe, Ni, Mo, B, Cl). *Curr. Opin. Plant Biol.* **2009**, *12*, 259–266. [CrossRef] [PubMed]

67. Edwards, D.G.; Asher, C.J. Tolerance of crop and pasture species to manganese toxicity. In *Proceedings of the Ninth Plant Nutrition Colloquium*; Scaife, A., Ed.; Commonwealth Agricultural Bureaux: Warwick, England, 1982; pp. 145–150.

68. Horst, W.J. The physiology of manganese toxicity. In *Manganese in Soils and Plants*; Graham, R.D., Hannam, R.J., Uren, N.C., Eds.; Kluwer Academic Publishers: Dordrecht, The Netherland, 1988; pp. 175–188.

69. Le Bot, J.; Goss, M.J.; Carvalho, G.P.R.; van Beusichem, M.L.; Kirby, E.A. The significance of the magnesium to manganese ratio in plant tissues for growth and alleviation of manganese toxicity in tomato (*Lycopersicon esculentum*) and wheat (*Triticum sativum*) plants. *Plant Soil* **1990**, *124*, 205–210. [CrossRef]

70. Wang, J.; Evangelou, B.P.; Nielsen, M.T. Surface chemical properties of purified root cell walls from two tobacco genotypes exhibiting different tolerance to manganese toxicity. *Plant Physiol.* **1992**, *100*, 496–501. [CrossRef] [PubMed]

71. DeLuca, T.H.; Gundale, M.J.; MacKenzie, M.D.; Jones, D.L. Biochar effects on soil nutrient transformations. In *Biochar for Environmental Management: Science, Technology and Implementation*; Lehmanm, J., Joseph, S., Eds.; Taylor and Francis: New York, NY, USA, 2015; pp. 421–454.

72. Vandecasteele, B.; Reubens, B.; Willekens, K.; de Neve, S. Composting for increasing the fertilizer value of chicken manure: Effects of feedstock on P availability. *Waste Biomass Valoriz.* **2015**, *5*, 491–503. [CrossRef]

73. Chapin, F.S. The mineral nutrition of wild plants. *Annu. Rev. Ecol. Syst.* **1980**, *11*, 233–260. [CrossRef]

74. Gale, N.V.; Halim, M.A.; Horsburgh, M.; Thomas, S.C. Comparative responses of early-successional plants to charcoal soil amendments. University of Toronto: Toronto, Canada, Unpublished work. 2017.

75. Conversa, G.; Bonasia, A.; Lazzizera, C.; Elia, A. Influence of biochar, mycorrhizal inoculation, and fertilizer rate on growth and flowering of Pelargonium (*Pelargonium zonale* L.) plants. *Front. Plant Sci.* **2015**, *6*, 429. [CrossRef] [PubMed]

76. Spokas, K.A.; Baker, J.M.; Reicosky, D.C. Ethylene: Potential key for biochar amendment impacts. *Plant Soil* **2010**, *333*, 443–452. [CrossRef]

77. Fulton, W.; Gray, M.; Prahl, F.; Kleber, M. A simple technique to eliminate ethylene emissions from biochar amendment in agriculture. *Agron. Sustain. Dev.* **2013**, *33*, 469–474. [CrossRef]

78. Burg, S.P.; Burg, E.A. Auxin-induced ethylene formation: Its relation to flowering in the pineapple. *Science* **1966**, *152*, 1269. [CrossRef] [PubMed]

79. Ogawara, T.; Higashi, K.; Kamada, H.; Ezura, H. Ethylene advances the transition from vegetative growth to flowering in *Arabidopsis thaliana*. *J. Plant Physiol.* **2013**, *160*, 1335–1340. [CrossRef] [PubMed]

80. Dukovski, D.; Bernatzky, R.; Han, S. Flowering induction of *Guzmania* by ethylene. *Sci. Hortic.* **2006**, *110*, 104–108. [CrossRef]

81. Abeles, F.B.; Morgan, P.W.; Saltveit, M.E., Jr. *Ethylene in Plant Biology*, 2nd ed.; Academic Press: San Diego, CA, USA, 2012.

82. Wróblewska, H. Studies on the effect of compost made of post-use wood waste on the growth of willow plants. *Mol. Cryst. Liq. Crys.* **2008**, *483*, 352–366. [CrossRef]

83. Spokas, K.A.; Cantrell, K.B.; Novak, J.M.; Archer, D.W.; Ippolito, J.A.; Collins, H.P.; Boateng, A.A.; Lima, I.M.; Lamb, M.C.; McAloon, A.J.; et al. Biochar: A synthesis of its agronomic impact beyond carbon sequestration. *J. Environ. Qual.* **2012**, *41*, 973–989. [CrossRef] [PubMed]

5

Metabolite Profiling for Leaf Senescence in Barley Reveals Decreases in Amino Acids and Glycolysis Intermediates

Liliana Avila-Ospina, Gilles Clément and Céline Masclaux-Daubresse *

INRA-AgroParisTech, Institut Jean-Pierre Bourgin, UMR1318, ERL CNRS 3559, Saclay Plant Sciences, Versailles 78000, France; liliana.avila-ospina@ips2.universite-paris-saclay.fr (L.A.-O.); gilles.clement@inra.fr (G.C.)
* Correspondence: celine.masclaux-daubresse@inra.fr

Academic Editor: Karin Krupinska

Abstract: Leaf senescence is a long developmental phase important for plant performance and nutrient management. Cell constituents are recycled in old leaves to provide nutrients that are redistributed to the sink organs. Up to now, metabolomic changes during leaf senescence have been mainly studied in Arabidopsis (*Arabidopsis thaliana* L.). The metabolite profiling conducted in barley (*Hordeum vulgare* L.) during primary leaf senescence under two nitrate regimes and in flag leaf shows that amino acids, hexose, sucrose and glycolysis intermediates decrease during senescence, while minor carbohydrates accumulate. Tricarboxylic acid (TCA) compounds changed with senescence only in primary leaves. The senescence-related metabolite changes in the flag leaf were globally similar to those observed in primary leaves. The effect of senescence on the metabolite changes of barley leaves was similar to that previously described in Arabidopsis except for sugars and glycolysis compounds. This suggests a different role of sugars in the control of leaf senescence in Arabidopsis and in barley.

Keywords: metabolome; leaf senescence; amino acids; glycolysis; TCA

---

1. Introduction

Barley (*Hordeum vulgare* L.) is a major crop cultivated worldwide. Like in other crops, nitrogen remobilization efficiency is important as a main determinant of grain yield and seed quality [1,2]. In wheat and barley, grain protein content (GPC) and leaf senescence are controlled by the same gene (*Gpc-1, NAMB-1*), a transcription factor whose homologous gene in Arabidopsis is also involved in the control of leaf senescence [3,4]. The fact that the same gene controls both leaf senescence and GPC suggests that the metabolic changes occurring during leaf senescence are essential for grain composition.

Leaf senescence is a long developmental process essential for plant physiology and metabolism [5]. The numerous genes up- or down-regulated during leaf senescence are indeed strongly related to the primary and secondary metabolisms and especially involved in the transition from anabolism to catabolism [6–9]. The transcriptome analyses show that the cellular mechanisms involved in the protection against free radicals and reactive oxygen species are overexpressed during leaf senescence. This certainly contributes to the better survival of the leaf tissues all along the senescence process and ensures efficient nutrient remobilization to the sink organs. From that point of view, leaf senescence that ends with cell death, is essential to efficient nutrient recycling and mobilization at the whole plant level. This facilitates survival of environmental changes and seed production. The tightly regulated sequence of events occurring during senescence aims at recycling and remobilizing mineral nutrients and nitrogen-containing molecules as efficiently as possible and in good accordance with the needs of the plant [10,11]. The methodical degradation of cell constituents starts at the chloroplast level and has a strong impact on the photosynthesis capacity of the leaf. Mitochondria is maintained and collapses at the last stage of senescence. However, the nature of the nutrients used to support mitochondria respiration is certainly different at the different stages of the leaf senescence. As a result of chloroplast dismantling, protein degradation and recycling processes, major modifications of the leaf metabolite contents certainly modify the source sink relationships in the plant [12,13].

The molecular mechanisms occurring during the leaf senescence have been studied extensively for a long time because they are key elements controlling plant productivity and plant adaptation to the environment [9,14]. Most of the studies have focused on the signals and the transcriptional factors involved in the regulation of this process [15]. Actors of the degradation process like proteases, nucleases, lipases and other hydrolases have also been identified and their exact roles in the cell component degradations are still under investigation [6,16–20]. Recently, the role of autophagy machinery in the degradation processes occurring during leaf senescence and for nutrient remobilization has been reported [9,21–27]. Besides cell component degradation, which could be orchestrated by autophagy, and other proteolytic processes occurring in the organelles, the cytosol and the vacuole, a panel of enzymes and transporters in charge of nutrient inter-conversions and phloem loading and unloading remain to be identified [28].

While the changes in gene expressions during leaf senescence have been extensively studied in many plant species [6,29,30], there are still few reports describing how metabolite contents are affected during leaf senescence [12,13] and data available is mainly focused on Arabidopsis leaf senescence. Because we expected that N remobilization metabolism was more active in crops like barley than in Arabidopsis due to the higher harvest index of cereals, the metabolic changes during leaf senescence were investigated in barley.

The present study aims at comparing the metabolic changes occurring in barley leaves during senescence under nitrate-sufficient and nitrate-limited conditions. It is indeed known that under nitrate limitation, N remobilization metabolism is increased. Our results highlight the different behaviors in metabolite managements during senescence depending on nitrogen resources availability.

2. Results

2.1. Characterization of Leaf Senescence in Barley

Leaf senescence in barley was first studied at vegetative stage on primary leaves. Metabolite contents were compared in three leaves (L1, L2 and L3 from the older to the younger), individually harvested on seedlings grown under low or high nitrate conditions for 20 days. Metabolite changes at reproductive stage were also studied on flag leaves harvested at different time points (T1, T3 and T5) chosen according to the expression of leaf-senescence molecular markers [27]. Figure 1 shows chlorophyll decreases in both flag leaves and primary leaves. In seedlings, the younger leaf L3 can be considered as a mature leaf as it displayed the highest chlorophyll content under both low and high nitrate conditions and because no difference in chlorophyll content could be observed between the tip, middle and base of the leaf. Under low nitrate conditions, chlorophyll levels were lower than under high nitrate, reflecting the lower nitrogen concentrations in plants grown under low nitrate. L1 and L2 showed lower chlorophyll concentrations at the tip of the leaves compare to their bases, thus showing the senescence progresses. The mature status of L3 and the senescence of L1 and L2 have been previously confirmed in the study of Avila-Ospina et al. [27]. The cytosolic glutamine synthetase and *NAC13* genes were indeed found less expressed in L3 than in L2 and L1. Similarly, chlorophyll contents (this work) and senescence markers [27] showed the progress of flag leaf senescence from T1 to T5.

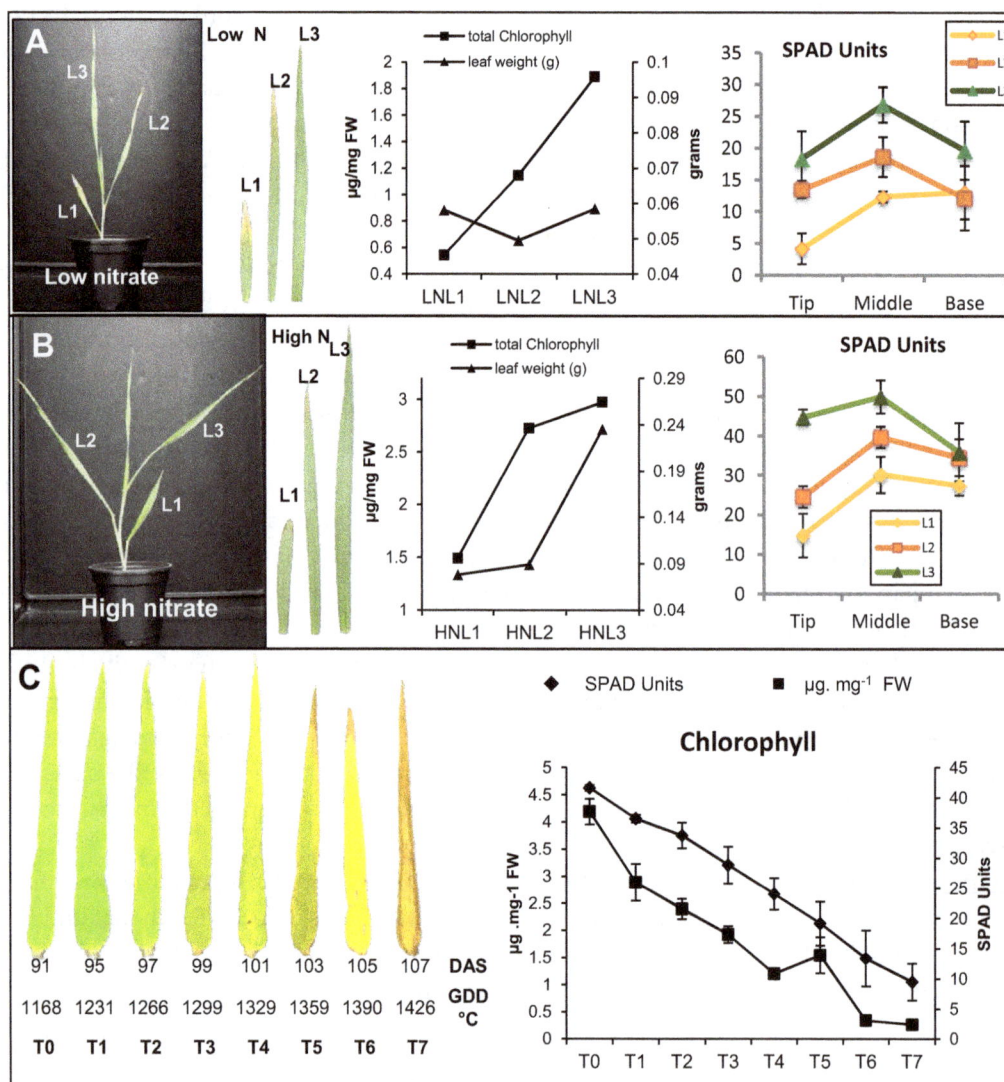

Figure 1. Natural leaf senescence in barley. Leaf senescence was monitored in primary leaves from seedlings grown under low (**A**, LN) and high (**B**, HN) nitrate conditions in growth chambers, and in flag leaves of plants grown in the field. Leaf ranks harvested from seedlings at 20 DAS were numbered from the oldest (L1) to the youngest (L3). Flag leaves were harvested at different time points after heading (from T0 to T7). Changes in chlorophyll contents in primary leaves were measured through spectrophotometry after extraction (black square, **A**,**B**) (see [27]). Chlorophyll contents at the tip, middle and base of primary leaves were estimated using SPAD measurements. Chlorophyll contents in flag leaves were measured by SPAD (black circle, **C**) and spectrophotometry (black square, **C**). The data presented are the mean ± SD of four biological replicates. DAS (days after sowing); GDD °C (growing degree days in °C); T (harvesting time). Seedling experiment was performed twice showing similar results.

2.2. Senescence-Associated Changes Occurring at the Vegetative Stage in Barley Leaves

Metabolite-relative contents were investigated using GC-MS analyser. Raw data were obtained on the basis of peak areas relative to the ribitol standard per mg fresh weight. These values can be considered as metabolite-relative concentrations although they do not represent any mole of gram units per fresh weight. In order to analyse more easily the senescence-related metabolic changes, we then normalized all these metabolite concentrations to their respective concentrations in the L3 leaf (Supplementary Data Set 1). As such normalisation was performed separately for the LN and the HN leaves, it only permits the investigation of the senescence effect but it cannot reveal any nitrate effect. The Log_2 ratios of the metabolite fold changes were calculated (Figure 2) and they are represented on a heat map to give an easier view of the senescence related changes (Figure 2). Only the metabolites with significantly different concentrations in leaves L1 and L2 relative to L3 are presented in the Figures 2 and 3.

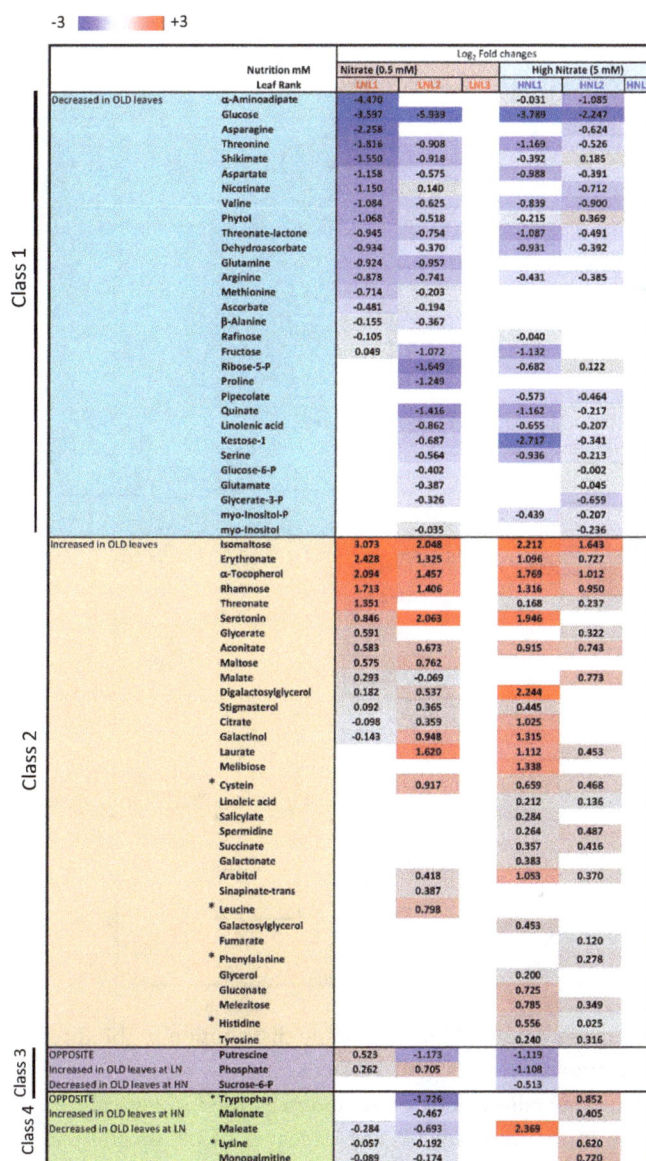

Nutrition mM		Nitrate (0.5 mM)			High Nitrate (5 mM)		
Leaf Rank		LNL1	LNL2	LNL3	HNL1	HNL2	HNL3
Class 1 — Decreased in OLD leaves	α-Aminoadipate	-4.470			-0.031	-1.085	
	Glucose	-3.597	-5.939		-3.789	-2.247	
	Asparagine	-2.258				-0.624	
	Threonine	-1.816	-0.908		-1.169	-0.526	
	Shikimate	-1.550	-0.918		-0.392	0.185	
	Aspartate	-1.158	-0.575		-0.988	-0.391	
	Nicotinate	-1.150	0.140			-0.712	
	Valine	-1.084	-0.625		-0.839	-0.900	
	Phytol	-1.068	-0.518		-0.215	0.369	
	Threonate-lactone	-0.945	-0.754		-1.087	-0.491	
	Dehydroascorbate	-0.934	-0.370		-0.931	-0.392	
	Glutamine	-0.924	-0.957				
	Arginine	-0.878	-0.741		-0.431	-0.385	
	Methionine	-0.714	-0.203				
	Ascorbate	-0.481	-0.194				
	β-Alanine	-0.155	-0.367				
	Rafinose	-0.105			-0.040		
	Fructose	0.049	-1.072		-1.132		
	Ribose-5-P		-1.649		-0.682	0.122	
	Proline		-1.249				
	Pipecolate				-0.573	-0.464	
	Quinate		-1.416		-1.162	-0.217	
	Linolenic acid		-0.862		-0.655	-0.207	
	Kestose-1		-0.687		-2.717	-0.341	
	Serine		-0.564		-0.936	-0.213	
	Glucose-6-P		-0.402			-0.002	
	Glutamate		-0.387			-0.045	
	Glycerate-3-P		-0.326			-0.659	
	myo-Inositol-P				-0.439	-0.207	
	myo-Inositol		-0.035			-0.236	
Class 2 — Increased in OLD leaves	Isomaltose	1.073	2.048		2.212	1.643	
	Erythronate	2.428	1.325		1.096	0.727	
	α-Tocopherol	2.094	1.457		1.769	1.012	
	Rhamnose	1.713	1.406		1.316	0.950	
	Threonate	1.351			0.168	0.237	
	Serotonin	0.846	2.063		1.946		
	Glycerate	0.591				0.322	
	Aconitate	0.583	0.673		0.915	0.743	
	Maltose	0.575	0.762				
	Malate	0.293	-0.069			0.773	
	Digalactosylglycerol	0.182	0.537		2.244		
	Stigmasterol	0.092	0.365		0.445		
	Citrate	-0.098	0.359		1.025		
	Galactinol	-0.143	0.948		1.315		
	Laurate		1.620		1.112	0.453	
	Melibiose		1.338				
	* Cystein		0.917		0.659	0.468	
	Linoleic acid				0.212	0.136	
	Salicylate				0.284		
	Spermidine				0.264	0.487	
	Succinate				0.357	0.416	
	Galactonate				0.383		
	Arabitol		0.418		1.053	0.370	
	Sinapinate-trans		0.387				
	* Leucine		0.798				
	Galactosylglycerol				0.453		
	Fumarate					0.120	
	* Phenylalanine					0.278	
	Glycerol				0.200		
	Gluconate				0.725		
	Melezitose				0.785	0.349	
	* Histidine				0.556	0.025	
	Tyrosine				0.240	0.316	
Class 3 — OPPOSITE Increased in OLD leaves at LN Decreased in OLD leaves at HN	Putrescine	0.525	-1.173		-1.119		
	Phosphate	0.262	0.705		-1.108		
	Sucrose-6-P				-0.513		
Class 4 — OPPOSITE Increased in OLD leaves at HN Decreased in OLD leaves at LN	* Tryptophan		-1.726			0.852	
	Malonate		-0.467			0.405	
	Maleate	-0.284	-0.693		2.369		
	* Lysine	-0.057	-0.192			0.620	
	Monopalmitine	-0.089	-0.174			0.720	

Figure 2. Metabolite changes during primary leaf senescence of plants grown under low (LN) or high nitrate (HN) conditions. Values for each metabolite were first measured as peak area in GC-MS analyses normalized according to the ribitol internal standard and to the mg FW (see Supplementary Data Set 1). ANOVA and Newman-Keuls (SNK) comparisons were performed in order to select the metabolites whose relative contents significantly changed with leaf senescence. Only these significantly different metabolites are presented in the figure. The Log_2 Fold changes from L3 were then calculated independently for LN and HN leaves on the mean of eight biological repeats. The values range from low (blue) to high (red) metabolite-relative content compared to the youngest leaf (L3). The scale bar of red and blue shades ranges from −3 to +3. Class 1 represents metabolites decreased during leaf senescence, Class 2 metabolites accumulated during senescence, Class 3 metabolites accumulated in LN leaves and decreased in HN leaves during leaf senescence and Class 4 metabolites decreased in LN leaves and increased in HN leaves during leaf senescence. Asterisks identify amino acids.

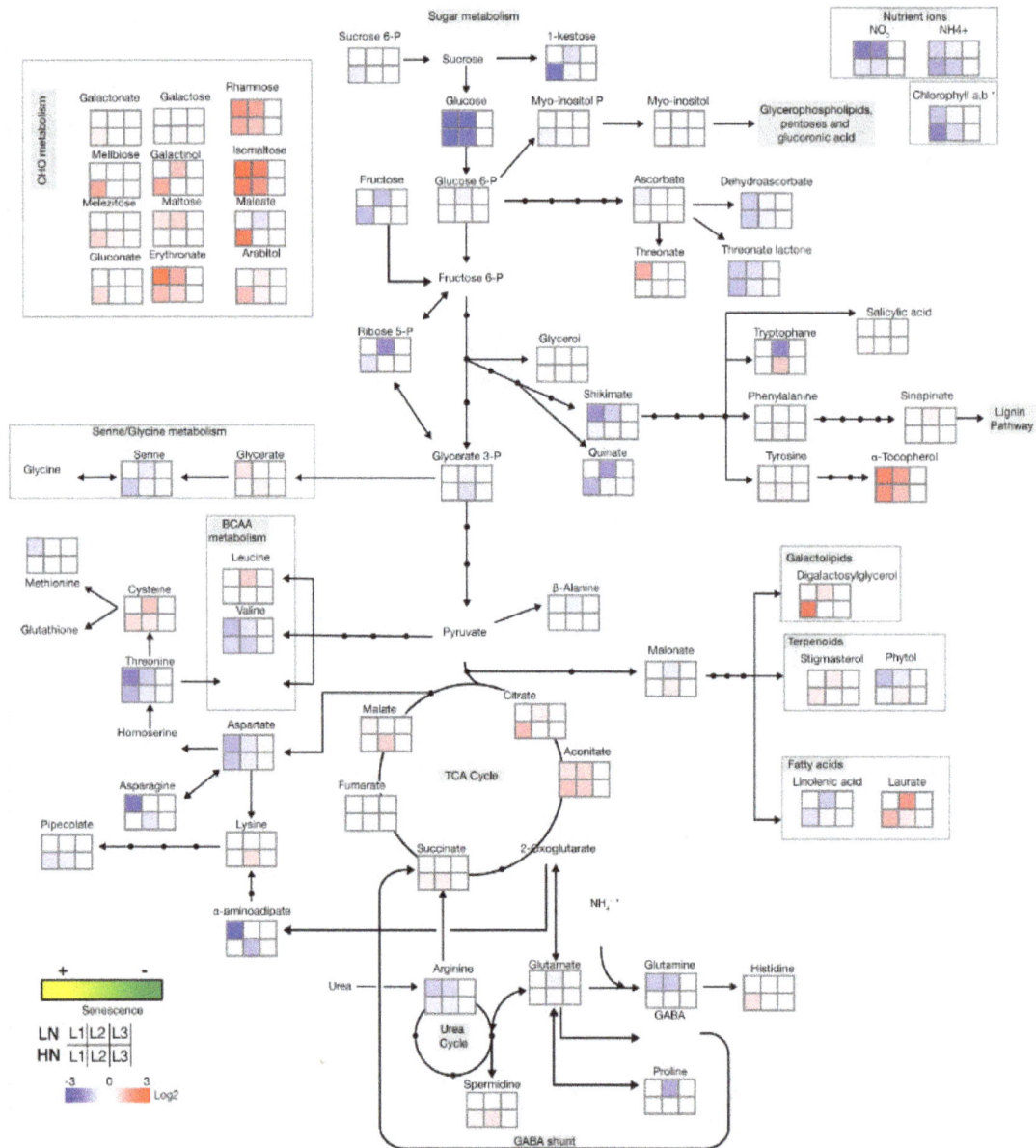

Figure 3. Heat map representation of the metabolite distribution in the three L1, L2 and L3 leaves of barley seedlings grown under LN and HN conditions. Figure 3 uses the same Log₂ Fold change values and the same false colour presentation as used in Figure 2 to represent the relative metabolite distribution in the three L1, L2 and L3 leaves of LN or HN plants. Only metabolites significantly different with senescence are shown. As in Figure 3, only senescence effect is presented, as normalization to L3 was performed separately on HN plants and on LN plants. The values range from low (blue) to high (red) metabolite-relative content compared to the youngest leaf (L3). The scale bar of red and blue shades ranges from −3 to +3. The stage of leaf senescence is schematically indicated by shades of colours from yellow (more senescent old leaf) to green (less senescent young leaf).

Leaf senescence effects were then illustrated by different metabolite classes (Figure 2). Class-1 and Class-2 include the metabolites down- or up-accumulated in old leaves in at least one of the two nitrate conditions respectively. Class-3 and Class-4 include metabolites that change in a different way depending on the nitrate nutrition.

The metabolic map of the senescence-related changes in Figure 3 uses the same false colours as in Figure 2. In agreement with the results reported previously [27], the chlorophyll contents were also represented. From Figures 2 and 3, we can then observe that most of the metabolite changes that occurred with leaf senescence were similar in the plants grown under low or high nitrate, although the magnitudes of the metabolite changes were different under LN compared to HN. In Figure 2, we can observe that the Class 1, that groups the metabolites repressed during leaf senescence, mainly contains 10 amino acids among which glutamate, glutamine, asparagine and serine as known as abundant amino acids in plants. Class 1 also contains phosphate sugars and glycolysis sugars, while the Class 2 that represents the metabolites accumulated during senescence mainly contains minor carbohydrates and organic acids. Class 2 also contains minor (less abundant) amino acids as lysine, histidine, phenylalanine and leucine, tyrosine and cysteine. Figure 3 presents these changes on a metabolic map. Globally, we can see the accumulation of the carbohydrates (CHOs), lipids, TCA (tricarboxylic acid) compounds and minor amino acids in old leaves while glycolysis compounds (glycerate3-P, glucose, fructose, glucose 6-P) and major amino acids decreased in old leaves. Interestingly, tryptophan, lysine, phenylalanine and histidine only increased in old leaves under high nitrate conditions, suggesting they are less remobilized than the others when nitrate is available. In addition it is interesting to notice that decreases in glutamine, asparagine, arginine and threonine were higher under low nitrate than under high nitrate conditions. This suggests that these amino acids are major nitrogen sources for remobilization. Oxidative-stress related compounds are known to be important for the redox homeostasis and especially during leaf senescence. We can observe in Figure 2 a large increase in α-tocopherols in old leaves. However, the decreases in both ascorbate and dehydroascorbate in old leaves suggest that the antioxidants do not all vary in a similar way and do not play the same roles depending on the tissues and stress conditions ([31]).

2.3. Senescence-Associated Changes Occurring at the Reproductive Stage in Barley Leaves

In flag leaves, the senescence-related metabolic changes were normalized to their concentrations in the youngest one (harvested at T1). The Log_2 ratios of their fold change relative to T1 leaves were then calculated (Supplementary Data Set 2) and the metabolites were grouped in several classes depending on their senescence-related behaviours. Class-1 includes 25 metabolites with a significantly lower concentration in the old T5 leaves (Down during senescence). Class-2 includes 17 metabolites with significantly higher concentrations in the old T5 leaves (Up during senescence). Class-3 includes 11 metabolites with significantly lower concentrations in mature T3 leaves compared to young (T1) and old (T5) leaves. In contrast, Class-4 included two metabolites with significantly higher concentrations in the mature T3 leaves compared to young (T1) and old (T3) leaves. The metabolite changes occurring with flag leaf senescence are represented on the metabolic map in Figure 4 using the same false colours as in Supplementary Data Set 2.

The senescence-related changes observed in the flag leaves are mostly similar to those observed in seedling leaves. As such, in flag leaves like in primary leaves, most of the CHOs (except galactinol) increased with ageing, while glycolysis compounds (sucrose, hexose and hexose6-P) and most of the amino acids decreased. Like in primary leaves, aspartate, glutamine, alanine, serine and threonine sharply decreased in flag leaf with ageing. Only methionine, lysine and β-alanine slightly increased during flag leaf senescence.

Despite all these similarities, a major difference between the flag and the primary leaves concerned the TCA cycle compounds. While almost all the TCA compounds increased with senescence in primary leaves, none of them show any significant change with senescence in the flag leaves. It seems thus that the mitochondrial respiration is not affected in a similar manner by ageing at reproductive and vegetative stages. The maintenance of mitochondrial respiration during flag leaf senescence could explain that TCA compounds are poorly affected by ageing [32]. However, pipecolate and α-aminoadipate, which are both markers of the lysine anapleurotic respiratory pathway, increased during senescence in the flag leaf, suggesting a specific role of this pathway in flag leaf compared to seedling leaves [33,34].

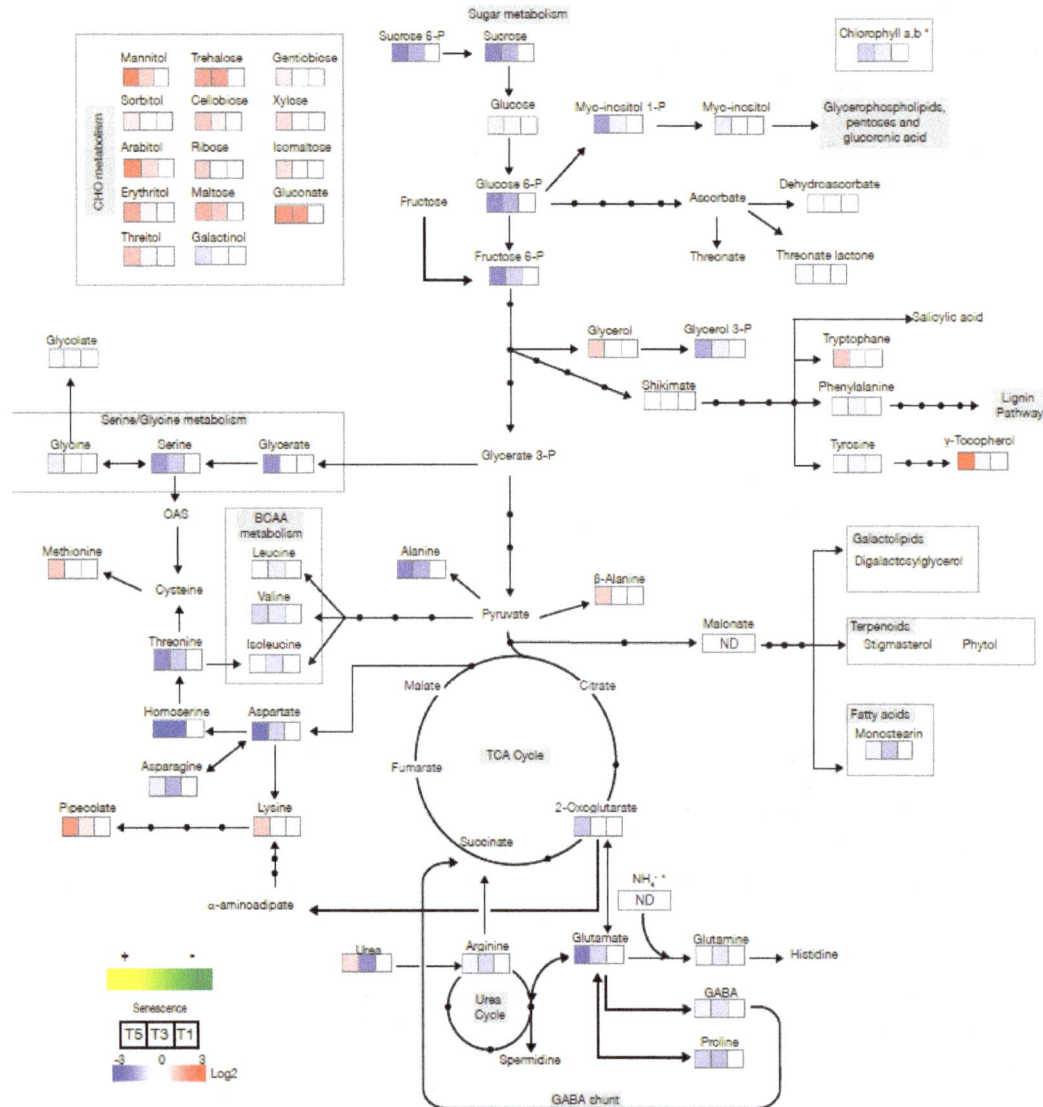

Figure 4. Heat map representation of the metabolite distribution in the three T1, T3 and T5 leaves of barley plants grown in the field. Metabolite concentrations were determined as the peak area in GC-MS analyses normalized to ribitol and to sample mg fresh weight (see Supplementary Data Set 2). The Log_2 of the ratios of the mean of the metabolite contents in each leaf relative to the youngest (T1) leaf were used to represent the metabolite distribution as a heat map. The shades of red or blue colours represent values ranging from −3 to 3 according to the scale bar. The stage of leaf senescence is schematically indicated by shades of colours from yellow (more senescent old leaf) to green (less senescent young leaf). The Log_2 Fold changes from T1 were calculated on the mean of four biological repeats. Only significant metabolites are presented. Significance was tested using ANOVA and Newman-Keuls (SNK) comparisons.

2.4. Nitrate Regime Affects Metabolite Contents in Primary Leaves but Poorly Modify Senescence Effect

The normalization used in Figure 3 only allowed the analyses of the senescence effect in primary leaves, and stated above, most of the senescence-related changes were similar in plants grown under low or high nitrate conditions.

In order to better compare the metabolite contents under low and high nitrate conditions, we then calculated the Log_2 of the ratio of the metabolite levels in LN leaves relative to HN leaves (Figure 5). The number of metabolites that decreased under low nitrate (31) was higher than the number of metabolites over-accumulated under LN (22).

Nutrition mM	Log₂ Fold change		
Leaf Rank	Ratio LN/HN		
	L1	L2	L3
Class-1			
Homoserine	-16.447	-15.912	-2.514
Glutamine	-2.947	-3.118	-2.950
Aspartate	-2.334	-2.202	-1.998
Serine	-2.238	-1.317	-1.164
Threonine	-1.839	-1.545	-1.039
Glutamate	-1.576	-1.529	-1.130
Alanine	-1.574	-1.623	-1.348
beta-Alanine	-2.604	-2.100	
Leucine	-1.984	-19.101	
Arginine		-14.431	-1.760
Glycine	-1.816		
Valine	-1.312		
Phenylalanine	-1.093		
Asparagine		-13.855	
Cysteine			-1.159
Citrate	-2.037	-1.286	-0.822
Malate	-1.979	-2.070	-1.234
Fumarate	-1.707	-2.283	-2.425
Maleate	-3.327	-2.876	
Aconitate	-3.218	-1.543	
Glycerate		-1.119	-1.200
Erythronate		-1.117	-1.775
Pyruvate	-1.340		-1.646
Phytol	-1.647	-1.424	
Maltitol	-18.458		
Ribose-5-P	-13.608	-16.017	-1.521
Glycerate-3-P	-3.450		
Glycerol-2-P	-3.150		
Fructose-6-P	-2.832		
Sedoheptulose	-1.006	-1.345	-3.576
Melezitose			-17.571
Class-2			
2-4-dihydroxybutanoate	1.099	1.489	2.235
Ribonate	19.936	18.070	
Ascorbate	2.114	2.314	
Oxalate	1.114		
Gluconate	15.210		
Malonate			1.162
Proline			1.975
Lysine			2.123
gamma-Tocopherol	16.541	13.302	
Erythritol	16.373	11.674	
Galactinol		2.045	2.415
Threitol	17.073		
Xylitol	17.309		
Arabitol			15.757
Melibiose	16.093	3.356	
Isomaltose	19.937	18.866	
Digalactosylglycerol		1.492	3.372
Glucose	1.291		2.243
Gentiobiose	2.042		1.494
Fructose	1.928		
Lactose	3.813		
alpha-Aminoadipate		2.265	2.675
myo-Inositol-1-P	1.377	1.738	-1.019

Figure 5. Primary leaf metabolite changes in low nitrate conditions (LN) compared to high nitrate (HN) conditions. Values for each metabolite were first measured as peak area normalized to the ribitol internal standard and to the mg FW (see Supplementary Data Set 1). The Log₂ of the fold changes in LN leaves from HN leaves were calculated for each leaf L1, L2 and L3. Data used to calculate the Log₂ Fold changes are the mean of eight biological repeats. The values go from low (blue) to high (red) metabolite-relative contents in LN leaves compared to HN leaves. The scale bar of red and blue shades range from −5 to +5. Only metabolites significantly different in LN leaf compared to HN leaf are shown. Significance was determined on the normalized peak areas using T-test ($n = 8$; $p < 0.05$).

It was not surprising that almost all the amino acids (15 over 19) were less abundant in the LN leaves than in the HN ones, especially in older leaves. Only proline and lysine were more abundant in the LN leaves relative to HN, but this was only observed in the L3. The organic acids appeared also globally less abundant under LN as well as the glycolysis intermediates. The over-accumulated metabolites under LN were mostly sugar acids, sugar alcohol and hexoses. The global picture of nitrate limitation effects is the depletion of amino acids, organic acids and glycolysis intermediates, and the accumulation of minor CHO. Such effects of nitrate limitation are in good accordance with results obtained in Arabidopsis [35,36]. Differences between LN and HN were stronger and more numerous in the older leaf L1 than in the younger L3. This suggests that N limitation effect is increased with ageing. This could be due to the fact that nutrient recycling metabolism is more active in old leaves under LN than under HN.

3. Discussion

In the present study, we have analysed how metabolite contents change during leaf senescence in barley leaves. We used both primary leaves and flag leaves to consider leaf senescence at vegetative and reproductive stages. The effect of nitrate availability was also investigated by growing plants under two different nitrate regimes corresponding to ample N condition and N limitation. The nitrogen status of the plants has been described previously [27] and was confirmed by the differences of nitrate and amino acid contents between the HN and the LN plants.

The main result is that the senescence effects observed under the two nitrate regimes and in both primary and flag leaves are globally similar. It shows the decrease of the major amino acids and of the glycolysis sugars and the increase of the minor carbohydrates. Comparing metabolite changes in flag leaves and primary leaves, we can see that mainly only TCA compounds and few minor amino acids do not behave the same way. While TCA compounds increase with leaf senescence in primary leaves, they are unchanged in flag leaves. The reason of such difference could be that growth conditions were quite different or that the position of the leaves on the plant modifies their respiration activities. Another difference resides in the pipecolate and α-aminoadipate changes. Indeed these two compounds are linked to the degradation of lysine and to its anapleurotic role in the mitochondrial respiratory pathway [33,34]. While pipecolate and α-aminoadipate decreased in primary leaves with senescence, they sharply accumulated in flag leaves with ageing, suggesting that lysine was used for mitochondria respiration in seedlings but not so much in the flag leaves. The anapleurotic respiration of amino acids has been mainly observed in Arabidopsis in response to dark stress [32,34,37] and Chrobok et al. [32] showed that the whole metabolic pathway of lysine degradation was up-regulated in Arabidopsis individual darken leaves. It could then be possible that the differences of light environment between primary and flag leaves are the reason of these discrepancies and that lysine level could be related to both glutamate content and TCA replenishment.

Metabolic changes during leaf senescence have been mainly investigated in Arabidopsis [11,13,32]. They showed many similar features to what we found here in barley, such as the accumulation of minor CHO and the decrease in most of the amino acids like glutamine, glutamate, aspartate, serine, glycine and arginine. However, a considerable difference between the two plant species is related to the increase of glycolysis sugars in Arabidopsis while glycolysis sugars decrease with senescence in both primary and flag leaves of barley. The role of sucrose and glucose in the control of leaf senescence had been debated for long [38] and this question has mainly focused on Arabidopsis. The absence of hexose and sucrose accumulation with leaf senescence in barley then suggests that on this point the regulation of leaf senescence by sugars in barley is not the same as in Arabidopsis. In addition, the fact that glycolysis sugars fluctuate differently in barley and in Arabidopsis leaves with senescence suggests that their use for TCA replenishment or their neosynthesis through neoglucogenesis is different in the two plant species.

4. Material and Methods

4.1. Plant Material and Growth Conditions

Hordeum vulgare L. cultivar Golden Promise—a two-rowed spring barley cultivar—was grown in a growth chamber (16 h/8 h photoperiod—25/17 °C). Seeds were sown on sand and five-day old plants were transferred into polyvinyl chloride (PVC) tubes (6ø—45 cm units) containing sand as a substrate. Plants were watered eight times per day with a nutritive solution containing 0.5 mM NO_3^- (250 µM KH_2PO_4; 250 µM $MgSO_4$; 250 µM K_2SO_4; 250 µM KNO_3; 125 µM CaN_2O_6; 250 µM $CaCl_2$; 0.04 µM $(NH_4)_6Mo_7O_{24}$; 24.3 µM H_3BO_3; 11.8 µM $MnSO_4$; 3.48 µM $ZnSO_4$; 1 µM $CuSO_4$; 0.001% Sequestrene 138 FE 100 Syngenta) named high nitrate treatment (HN) or a 5 mM NO_3^- (250 µM KH_2PO_4; 250 µM $MgSO_4$; 4 mM KNO_3; 500 µM CaN_2O_6; 200 µM NaCl; 0.04 µM $(NH_4)_6Mo_7O_{24}$; 24.3 µM H_3BO_3; 11.8 µM $MnSO_4$; 3.48 µM $ZnSO_4$; 1 µM $CuSO_4$; 0.001% Sequestrene 138 FE 100 Syngenta) named low nitrate treatment (LN) [27]. Individual leaf ranks, L1 to L4 of high nitrate grown plants (HN) and L1 to L3 of low nitrate grown plants (LN) (from bottom to top leaves) were harvested, immediately frozen in liquid nitrogen and stored at -80 °C for further experiments. Chlorophyll contents measured spectrometrically or using SPAD before harvest are presented in . Two plantings were performed and following analyses were carried out on two plant cultures.

In field experiments, spring barley (Hordeum vulgare L.) Cultivar Carina was used. The experiments were performed at Hohenschulen research farm at 15.5 km west of Kiel (https://www.hohenschulen.uni-kiel.de/en) during the 2013 growing season, June being nearly wet and July warm and relatively dry. Spring barley was sown using a drill on 2 April 2013. The barley was managed organically and organic manure equal to 70 kg·N·ha^{-1} was added. There were four replicate plots 150 m^2 each. Plants were grown in a concentration of 300 plants/m^2 with 12.5 cm of row distance. Crop was spreaded with 1.5 L/ha of Ariane C (Dow agrosciences) and 20 g/ha of Trimmer SX (FCS) (herbicides) on 14 May 2013. Subsequently, it was added 0.3 L/ha of Moddus (Syngenta; Maintal; Germany) and Ethephon (Bayer CropSc.; Frankfurt; Germany) (growth regulators), 0.5 L/ha of Gladio (Syngenta) (fungicide), 5 kg/ha of $MgSO_4$ and 10 L/ha of Mn-EDTA on 5 June 2013. At last, 150 kg/ha Kierserit (KALI) (25% MgO, 20% S) and 30 kg/ha KAS (76% NH_4NO_3, 24% $CaCO_3$) were added on 7 June 2013. Flag leaves from the main shoots were harvested from each plot between 10:00 and 12:00 and immediately frozen in liquid nitrogen and stored at -80 °C for further experiments. Several harvests were performed 95 days after sowing (DAS) (T0), 99 DAS (T1) and from T1 every 2 days for 2 weeks until flag leaves showed very low SPAD values. Metabolite profiling was performed on leaves harvested 95, 99 and 103 days after sowing; T1, T3 and T5 by reference to Avila-Ospina et al. [27].

The plant material studied here is the same as previously used to monitor leaf senescence markers [27]. It is important to note that none of the results presented here have ever been presented previously.

4.2. Chlorophyll and Anion Determinations

Chlorophyll content was determined spectrophotometrically in crude leaf extracts according to [27] or using the SPAD-502 chlorophyll meter (KONICA-MINOLTA, Carrière sur Seine, France). Anion concentrations were determined using Dionex HPLC (HPLC Dionex DX 120; Thermo Fischer Scientific, Courtaboeuf, France) on the same extract as used for metabolite profiling.

4.3. Metabolite Profiling Using GC-MS

Extraction, derivatization, analysis, and data processing were performed according to Fiehn [39]. The ground frozen samples (20 mg FW) were resuspended in 1 mL of frozen (-20 °C) Water:Acetonitrile:Isopropanol (2:3:3) containing ribitol at 4 µg·mL^{-1} and extracted for 10 min at 4 °C with shaking at 1400 rpm in an Eppendorf thermomixer. Insoluble material was removed by centrifugation at 20,000× g for 5 min. 100 µL of supernatant were collected and dried for 4 h in a Speed-Vac and stored at -80 °C. Just before derivatization, samples were dried again for

1.5 h in a Speed-Vac. Three blank tubes underwent the same steps as the samples. After drying, 10 μL of 20 mg·mL^{-1} methoxyamine in pyridine were added to the samples. The reaction was performed for 90 min at 28 °C under continuous shaking in an Eppendorf thermomixer. Then 90 μL of N-methyl-N-trimethylsilyl-trifluoroacetamide (MSTFA) (Aldrich Saint Quentin Fallavier; France; 394866-10x1ml) was added and the reaction continued for 30 min at 37 °C. After cooling, 45 μL of the derivatized sample was transferred to an Agilent (Les Ulis, France) vial for injection.

Metabolites were analysed by GC-MS 4 h after derivatization. One microliter of the derivatized samples was injected in splitless mode on an Agilent 7890A gas chromatograph coupled to an Agilent 5975C mass spectrometer. The column was an Rtx-5SilMS from Restek (30 m with 10 m Integraguard column). The liner (Restek 20994) was changed before each series of analyses and 10 cm of column was cut. The oven temperature ramp was 70 °C for 7 min, then 10 °C/min to 325 °C for 4 min (run length 36.5 min). Helium constant flow was 1.5231 mL/min. Temperatures were as follows: injector, 250 °C; transfer line, 290 °C; source: 250 °C; and quadripole, 150 °C. Samples and blanks were randomized. Amino acid standards were injected at the beginning and end of the analysis to monitor the derivatization stability. An alkane mix (C10, C12, C15, C19, C22, C28, C32 and C36) was injected in the middle of the queue for external calibration. Five scans per second were acquired.

Metabolites, analysed using a gas chromatography mass spectrometry technique were annotated and their levels on a fresh weight basis were normalized with respect to the ribitol internal standard.

4.4. Metabolomic Data Processing

Raw Agilent datafiles were converted to NetCDF format and analysed with AMDIS [40]. A home retention index/mass spectra library, built from the NIST, Golm, and Fiehn databases and standard compounds, was used for metabolite identification. Peak areas were then determined using the QuanLynx software (Waters; Guyancourt, France) after conversion of the NetCDF file to MassLynx format. Statistical analysis was done with TMEV [41]; univariate analysis by permutation (1 way-ANOVA and 2 way-ANOVA) was first used to select the significant metabolites. Multivariate analysis (hierarchical clustering and principal component analysis; PCA) was then done in order to establish the metabolite clusters. Only metabolites showing repeatable and significant differences (based on a t-test) according to leaf age or number reflecting different senescence stages and nitrate growth conditions were included.

5. Conclusions

The senescence effects of leaf metabolism in both primary and flag leaves are globally similar. The decrease of the major amino acids and the increase of the minor carbohydrates are similar to what found in Arabidopsis. However, glycolysis sugars show decrease in barley leaves with ageing by contrast with Arabidopsis. This suggests that the regulation of leaf senescence by sugars in barley is not the same as in Arabidopsis.

Acknowledgments: The authors thank Joël Talbotec for help with plant material growth. The authors also thank the whole team at Observatoire du Végétal (http://www-ijpb.versailles.inra.fr/fr/plateformes/Observatoire-du-vegetal.html) for providing all the facilities to grow plants and perform metabolite profiling. The authors thank Laurence Cantrill for advices and correcting the English. The PhD fellowship for L. Avila-Ospina was supported by the FP7 Marie Curie Actions—Networks for Initial Training Project FP7-MC-ITN No. 264394 acronym CropLife.

Author Contributions: Experiments have been carried out by L.A.-O. and G.C.; Data mining was performed by L.A.-O. with contribution of G.C. and C.M.-D. Writing was performed by L.A.-O. and C.M.-D.

Conflicts of Interest: The authors declare no conflict of interest.

References

1. Chardon, F.; Noël, V.; Masclaux-Daubresse, C. Manipulating NUE in Arabidopsis and crop plants to improve yield and seed quality. *J. Exp. Bot.* **2012**, *63*, 3401–3412. [CrossRef] [PubMed]

2. Kichey, T.; Hirel, B.; Heumez, E.; Dubois, F.; Le Gouis, J. In winter wheat (*Triticum aestivum* L.), post-anthesis nitrogen uptake and remobilisation to the grain correlates with agronomic traits and nitrogen physiological markers. *Field Crop Res.* **2007**, *102*, 22–32. [CrossRef]

3. Uauy, C.; Distelfeld, A.; Fahima, T.; Blechl, A.; Dubcovsky, J. A *NAC* gene regulating senescence improves grain protein, zinc, and iron content in wheat. *Science* **2006**, *314*, 1298–1301. [CrossRef] [PubMed]

4. Distelfeld, A.; Avni, R.; Fischer, A. Senescence, nutrient remobilization, and yield in wheat and barley. *J. Exp. Bot.* **2014**, *65*, 3783–3798. [CrossRef] [PubMed]

5. Masclaux-Daubresse, C.; Reisdorf-Cren, M.; Orsel, M. Leaf nitrogen remobilisation for plant development and grain filling. *Plant Biol.* **2008**, *10*, 23–36. [CrossRef] [PubMed]

6. Guo, Y.; Cai, Z.; Gan, S. Transcriptome of Arabidopsis leaf senescence. *Plant Cell Environ.* **2004**, *27*, 521–549. [CrossRef]

7. Hollmann, J.; Gregersen, P.L.; Krupinska, K. Identification of predominant genes involved in regulation and execution of senescence-associated nitrogen remobilization in flag leaves of field grown barley. *J. Exp. Bot.* **2014**, *65*, 2963–3973. [CrossRef] [PubMed]

8. Buchanan-Wollaston, V. The molecular biology of leaf senescence. *J. Exp. Bot.* **1997**, *48*, 181–199. [CrossRef]

9. Avila-Ospina, L.; Moison, M.; Yoshimoto, K.; Masclaux-Daubresse, C. Autophagy, plant senescence, and nutrient recycling. *J. Exp. Bot.* **2014**, *65*, 3799–3811. [CrossRef] [PubMed]

10. Himelblau, E.; Amasino, R.M. Nutrients mobilized from leaves of Arabidopsis thaliana during leaf senescence. *J. Plant Physiol.* **2001**, *158*, 1317–1323. [CrossRef]

11. Diaz, C.; Lemaitre, T.; Christ, A.; Azzopardi, M.; Kato, Y.; Sato, F.; Morot-Gaudry, J.F.; Le Dily, F.; Masclaux-Daubresse, C. Nitrogen recycling and remobilization are differentially controlled by leaf senescence and development stage in Arabidopsis under low nitrogen nutrition. *Plant Physiol.* **2008**, *147*, 1437–1449. [CrossRef] [PubMed]

12. Diaz, C.; Purdy, S.; Christ, A.; Morot-Gaudry, J.-F.; Wingler, A.; Masclaux-Daubresse, C. Characterization of markers to determine the extent and variability of leaf senescence in Arabidopsis. A metabolic profiling approach. *Plant Physiol.* **2005**, *138*, 898–908. [CrossRef] [PubMed]

13. Watanabe, M.; Balazadeh, S.; Tohge, T.; Erban, A.; Giavalisco, P.; Kopka, J.; Mueller-Roeber, B.; Fernie, A.; Hoefgen, R. Comprehensive dissection of spatio-temporal metabolic shifts in primary, secondary and lipid metabolism during developmental senescence in *Arabidopsis thaliana. Plant Physiol.* **2013**, *62*, 1290–1310. [CrossRef] [PubMed]

14. Masclaux-Daubresse, C.; Daniel-Vedele, F.; Dechorgnat, J.; Chardon, F.; Gaufichon, L.; Suzuki, A. Nitrogen uptake, assimilation and remobilization in plants: Challenges for sustainable and productive agriculture. *Ann. Bot.* **2010**, *105*, 1141–1157. [CrossRef] [PubMed]

15. Garapati, P.; Xue, G.P.; Munne-Bosch, S.; Balazadeh, S. Transcription factor ATAF1 in Arabidopsis promotes senescence by direct regulation of key chloroplast maintenance and senescence transcriptional cascades. *Plant Physiol.* **2015**, *168*, 1122–1139. [CrossRef] [PubMed]

16. Roberts, I.N.; Caputo, C.; Criado, M.V.; Funk, C. Senescence-associated proteases in plants. *Physiol. Plantarum* **2012**, *145*, 130–139. [CrossRef] [PubMed]

17. Donnison, I.S.; Gay, A.P.; Thomas, H.; Edwards, K.J.; Edwards, D.; James, C.L.; Thomas, A.M.; Ougham, H.J. Modification of nitrogen remobilization, grain fill and leaf senescence in maize (*Zea mays*) by transposon insertional mutagenesis in a protease gene. *New Phytol.* **2007**, *173*, 481–494. [CrossRef] [PubMed]

18. Buchanan-Wollaston, V.; Earl, S.; Harrison, E.; Mathas, E.; Navabpour, S.; Page, T.; Pink, D. The molecular analysis of leaf senescence—a genomics approach. *Plant Biotechnol. J.* **2003**, *1*, 3–22. [CrossRef] [PubMed]

19. Sakamoto, W.; Takami, T. Nucleases in higher plants and their possible involvement in DNA degradation during leaf senescence. *J. Exp. Bot.* **2014**, *65*, 3835–3843. [CrossRef] [PubMed]

20. Diaz-Mendoza, M.; Velasco-Arroyo, B.; Gonzalez-Melendi, P.; Martínez, M.; Isabel, D. C1A cysteine protease-cystatin interactions in leaf senescence. *J. Exp. Bot.* **2014**, *65*, 3825–3833. [CrossRef] [PubMed]

21. Doelling, J.H.; Walker, J.M.; Friedman, E.M.; Thompson, A.R.; Vierstra, R.D. The APG8/12-activating enzyme APG7 is required for proper nutrient recycling and senescence in Arabidopsis thaliana. *J. Biol. Chem.* **2002**, *277*, 33105–33114. [CrossRef] [PubMed]

22. Hanaoka, H.; Noda, T.; Shirano, Y.; Kato, T.; Hayashi, H.; Shibata, D.; Tabata, S.; Ohsumi, Y. Leaf senescence and starvation-induced chlorosis are accelerated by the disruption of an Arabidopsis autophagy gene. *Plant Physiol.* **2002**, *129*, 1181–1193. [CrossRef] [PubMed]

23. Guiboileau, A.; Avila-Ospina, L.; Yoshimoto, K.; Soulay, F.; Azzopardi, M.; Marmagne, A.; Lothier, J.; Masclaux-Daubresse, C. Physiological and metabolic consequences of autophagy deficiency for the management of nitrogen and protein resources in Arabidopsis leaves depending on nitrate availability. *New Phytol.* **2013**, *199*, 683–694. [CrossRef] [PubMed]

24. Guiboileau, A.; Yoshimoto, K.; Soulay, F.; Bataillé, M.; Avice, J.; Masclaux-Daubresse, C. Autophagy machinery controls nitrogen remobilization at the whole-plant level under both limiting and ample nitrate conditions in Arabidopsis. *New Phytol.* **2012**, *194*, 732–740. [CrossRef] [PubMed]

25. Avice, J.-C.; Etienne, P. Leaf senescence and nitrogen remobilization efficiency in oilseed rape (*Brassica napus* L.). *J. Exp. Bot.* **2014**, *65*, 3813–3824. [CrossRef] [PubMed]

26. Prins, A.; van Heerden, P.D.R.; Olmos, E.; Kunert, K.J.; Foyer, C.H. Cysteine proteinases regulate chloroplast protein content and composition in tobacco leaves: A model for dynamic interactions with ribulose-1,5-biphosphate carboxylase/oxygenase (rubisco) vesicular bodies. *J. Exp. Bot.* **2008**, *59*, 1935–1950. [CrossRef] [PubMed]

27. Avila-Ospina, L.; Marmagne, A.; Talbotec, J.; Krupinska, K.; Masclaux-Daubresse, C. The identification of new cytosolic glutamine synthetase and asparagine synthetase genes in barley (*Hordeum vulgare* L.), and their expression during leaf senescence. *J. Exp. Bot.* **2015**, *66*, 2013–2026. [CrossRef] [PubMed]

28. Havé, M.; Marmagne, A.; Chardon, F.; Masclaux-Daubresse, C. Nitrogen remobilisation during leaf senescence: Lessons from Arabidopsis to crops. *J. Exp. Bot.* **2016**. [CrossRef] [PubMed]

29. Sekhon, R.S.; Childs, K.L.; Santoro, N.; Foster, C.E.; Buell, C.R.; de Leon, N.; Kaeppler, S.M. Transcriptional and metabolic analysis of senescence induced by preventing pollination in maize. *Plant Physiol.* **2012**, *159*, 1730–1744. [CrossRef] [PubMed]

30. Palmer, N.A.; Donze-Reiner, T.; Horvath, D.; Heng-Moss, T.; Waters, B.; Tobias, C.; Sarath, G. Switchgrass (*Panicum virgatum* L.) flag leaf transcriptomes reveal molecular signatures of leaf development, senescence, and mineral dynamics. *Funct. Integr. Genom.* **2015**, *15*, 1–16. [CrossRef] [PubMed]

31. Foyer, C.H.; Noctor, G. Ascorbate and glutathione: The heart of the redox hub. *Plant Physiol.* **2011**, *155*, 2–18. [CrossRef] [PubMed]

32. Chrobok, D.; Law, S.R.; Brouwer, B.; Lindèn, P.; Ziolkowska, A.; Liebsch, D.; Narsai, R.; Szal, B.; Moritz, T.; Rouhier, N.; et al. Dissecting the metabolic role of mitochondria during developmental leaf senescence. *Plant Physiol.* **2016**. [CrossRef] [PubMed]

33. Boex-Fontvieille, E.; Gauthier, P.; Gilard, F.; Hodges, M.; Tcherkez, G. A new anaplerotic respiratory pathway involving lysine biosynthesis in isocitrate dehydrogenase-deficient Arabidopsis mutants. *New Phytol.* **2013**, *99*, 673–682. [CrossRef] [PubMed]

34. Araujo, W.L.; Tohge, T.; Ishizaki, K.; Leaver, C.J.; Fernie, A.R. Protein degradation—An alternative respiratory substrate for stressed plants. *Trends Plant Sci.* **2011**, *16*, 489–498. [CrossRef] [PubMed]

35. Lemaitre, T.; Gaufichon, L.; Boutet-Mercey, S.; Christ, A.; Masclaux-Daubresse, C. Enzymatic and metabolic diagnostic of nitrogen deficiency in Arabidopsis thaliana Wassileskija accession. *Plant Cell Physiol.* **2008**, *49*, 1056–1065. [CrossRef] [PubMed]

36. Balazadeh, S.; Schildhauer, J.; Araujo, W.L.; Munne-Bosch, S.; Fernie, A.R.; Proost, S.; Humbeck, K.; Mueller-Roeber, B. Reversal of senescence by N resupply to N-starved Arabidopsis thaliana: Transcriptomic and metabolomic consequences. *J. Exp. Bot.* **2014**, *65*, 3975–3992. [CrossRef] [PubMed]

37. Araujo, W.L.; Ishizaki, K.; Nunes-Nesi, A.; Larson, T.R.; Tohge, T.; Krahnert, I.; Witt, S.; Obata, T.; Schauer, N.; Graham, I.A.; et al. Identification of the 2-hydroxyglutarate and isovaleryl-CoA dehydrogenases as alternative electron donors linking lysine catabolism to the electron transport chain of Arabidopsis mitochondria. *Plant Cell* **2010**, *22*, 1549–1563. [CrossRef] [PubMed]

38. Wingler, A.; Masclaux-Daubresse, C.; Fischer, A.M. Sugars, senescence, and ageing in plants and heterotrophic organisms. *J. Exp. Bot.* **2009**, *60*, 1063–1066. [CrossRef] [PubMed]

39. Fiehn, O. Metabolite profiling in Arabidopsis. *Methods Mol. Boil.* **2006**, *323*, 439–447.

40. AMDIS. Available online: http://chemdata.nist.gov/mass-spc/amdis/ (accessed on 31 May 2016).

41. TMEV. Available online: http://mev.tm4.org/#/welcome (accessed on 1 January 2017).

6

Using FACE Systems to Screen Wheat Cultivars for Yield Increases at Elevated CO_2

James Bunce

Crop Systems and Global Change Laboratory, USDA-ARS, Beltsville Agricultural Research Center, 10300 Baltimore Avenue, Beltsville, MD 20705-2350, USA; James.Bunce@ars.usda.gov

Academic Editor: Peter Langridge

Abstract: Because of continuing increases in atmospheric CO_2, identifying cultivars of crops with larger yield increases at elevated CO_2 may provide an avenue to increase crop yield potential in future climates. Free-air CO_2 enrichment (FACE) systems have most often been used with multiple replications of each CO_2 treatment in order to increase confidence in the effect of elevated CO_2. For screening of cultivars for yield increases at elevated CO_2, less precision about the CO_2 effect, but more precision about cultivar ranking within CO_2 treatments is appropriate. As a small-scale test of this approach, three plots, each of four cultivars of wheat, were grown in single FACE and control plots over two years, and the cultivar rankings of yield at elevated and ambient CO_2 were compared. Each replicate plot was the size used in traditional cultivar comparisons. An additional test using four smaller replicate plots per cultivar within one FACE and one ambient plot was used to compare nine cultivars in another year. In all cases, elevated CO_2 altered the ranking of cultivars for yield. This approach may provide a more efficient way to utilize FACE systems for the screening of CO_2 responsiveness.

Keywords: wheat; elevated CO_2; free-air carbon dioxide enrichment; screening; yield

---

1. Introduction

The CO_2 concentration in the atmosphere continues to increase rapidly, and may increase yields of C_3 crops, helping to meet the expected increased demand for agricultural products. Intraspecific variation in the responsiveness of yield has been reported in many C_3 crops species: barley [1], common bean [2], cow pea [3], oat [4], rape [5], rice [6,7], soybean [8,9], and wheat [10,11]. Such variation could allow the achievement of larger field yield increases as CO_2 rises, if new cultivars could be developed to better exploit the rising CO_2 [12]. However, with the possible exceptions of elevated CO_2 differentially prolonging vegetative growth in soybean [9,13], and increasing leaf mass per area in wheat [14], traits accounting for intraspecific differences in the CO_2 response of yield have mostly not yet been identified. Additionally, many of the studies of intraspecific variation in response were conducted in controlled environment chambers or glasshouses, and tests of whether the ranking of CO_2 responses of yield extrapolates to field conditions have seldom been conducted.

In the field, open top chambers are too small to compare more than just a few cultivars at a time, while free-air carbon dioxide enrichment (FACE) systems can be considerably larger. The two FACE experiments which have compared the largest number of cultivars have been for up to 18 cultivars of soybeans in Illinois [8] and eight cultivars of rice in Japan [6]. The FACE rings were 20 m in diameter for soybean (but only one-half of the plot was used for the 18 cultivars), and 17 m diameter for rice. The sub-plot size per cultivar was 2.3 m^2 for soybeans, and ≥ 3 m^2 in rice, with yield assessed for 1.15 m^2 area in soybeans, and 0.36 to 0.95 m^2 in rice. In these experiments, there were four enriched and four ambient FACE rings, with one subplot of each cultivar in each ring. While this experimental design has replication for the CO_2 effect, variation among rings within each CO_2 treatment would tend to obscure cultivar differences in CO_2 response. More precision about cultivar differences within a CO_2 treatment could be provided by replication of cultivar subplots within a CO_2 treatment plot. This raises the question of the size of the subplots required to accurately rank cultivars for yield within a CO_2 plot. We explored this approach with a comparison of wheat cultivars, first using a subplot size typical of current wheat variety test trials in our region, and in a second experiment using smaller subplots. Yield trials are generally not replicated in the same location and year, but over years and locations, and that approach is also probably best for elevated CO_2 yield trials as well. However, the intent here was not to conduct multi-year yield trials at elevated CO_2, but to test the feasibility of the approach in a FACE system. Once a substantial number of lines with strong and weak yield responses to elevated CO_2 have been identified, identification of traits responsible for such differences in response would be a next step toward selecting crops for future CO_2 environments.

2. Results

In the first experiment, which compared responses of four cultivars in two different years, significant differences among cultivars occurred for seed yield both at ambient and elevated CO_2, at $p = 0.05$ in both years (Table 1). The range of mean yields within CO_2 treatments was 25% to 30% of the overall mean for both years and CO_2 treatments. In the first year, the highest and lowest yielding cultivars, Pioneer 25 R40 and Choptank, respectively, were the same at both CO_2 levels. The second and third ranked cultivars, Pioneer 25 R32 and Jamestown switched rank between CO_2 treatments (Table 1). In the second year, the ranking of the yields Pioneer 25 R32 and Jamestown responded to the CO_2 treatment in the same way as in the first year, that is, the ranking of Jamestown increased at elevated CO_2, and the ranking of Pioneer 25 R32 deceased at elevated CO_2 (Table 1). In the second year, the ranking of Pioneer 25 R32 decreased even below that of Choptank.

Table 1. The mean seed yield (g dry mass m^{-2}) and the ranking of mean yield of four wheat cultivars grown at elevated and ambient CO_2 in 2013 and 2014; the probability of significant differences in yield within a year and CO_2 treatment; and the mean coefficient of variation (CV). Within columns, values followed by different letters were significantly different at $p = 0.05$, by ANOVA.

Year	2013				2014			
CO_2 Treatment	Ambient		Elevated		Ambient		Elevated	
Cultivar:	Rank	Yield	Rank	Yield	Rank	Yield	Rank	Yield
Pioneer 25 R40	1	482 a	1	530 a	1	452 a	1	445 a
Pioneer 25 R32	2	458 ab	3	425 c	2	428 a	4	350 c
Jamestown	3	442 b	2	475 b	3	356 b	2	410 b
Choptank	4	350 c	4	410 c	4	348 b	3	391 b
Probability	0.021		0.028		0.030		0.041	
Mean CV (%)	12.2		9.4		10.1		11.7	

In the second experiment, which compared responses of nine lines in one year, significant differences among lines in yield occurred at both CO_2 levels (Table 2). The range of mean yields among lines was larger relative to the overall mean yield at ambient CO_2 (60%) than at elevated CO_2 (29%). The mean yield of each line at elevated CO_2 was not significantly correlated with its mean yield at ambient CO_2, nor was the ranking of yields significantly correlated for the two CO_2 treatments (Figure 1). The ranking of line 1 increased by three places at elevated CO_2 compared with ambient CO_2, and the ranking of two other lines, 4 and 7, increased by two places at elevated CO_2 (Table 2). The ranking of line 5 decreased by five places at elevated CO_2, line 11 decreased by three places, and line 2 decreased by two places.

Table 2. The mean seed yield (g dry mass m^{-2}) and the ranking of mean yield of nine wheat lines grown at elevated and ambient CO_2 in 2015; the probability of significant differences in yield within a year and CO_2 treatment; and the mean coefficient of variation (CV). Within columns, values followed by different letters were significantly different at $p = 0.05$, by ANOVA. The identity of the nine lines is presented in Table 3.

CO_2 Treatment:	Ambient		Elevated	
Line:	Rank	Yield	Rank	Yield
1	8	353 c	5	513 ab
2	1	567 a	3	538 a
3	2	465 b	1	570 a
4	4	418 b	2	560 a
5	3	430 b	8	460 c
6	7	357 c	6	503 b
7	9	320 c	7	495 bc
8	5	410 b	4	535 a
11	6	365 c	9	423 c
Probability		0.031		0.048
Mean CV (%)		12.4		10.3

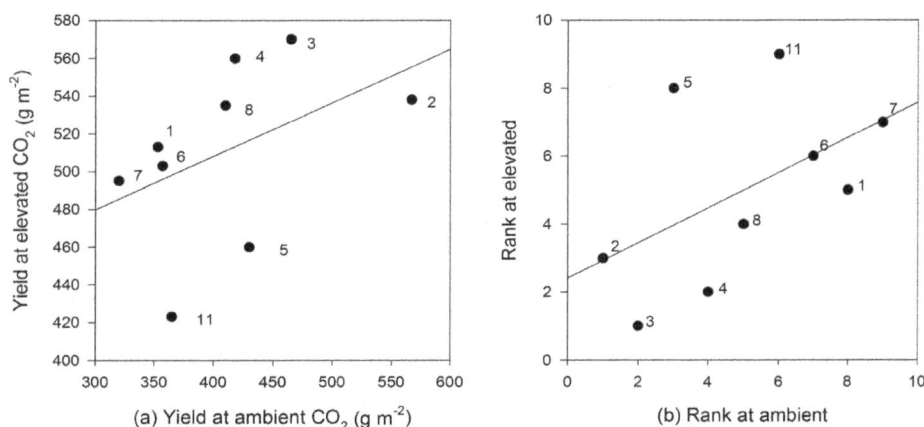

(a) Yield at ambient CO_2 (g m^{-2})

(b) Rank at ambient

Figure 1. The mean grain yield of nine lines of wheat grown at elevated and ambient CO_2 (**a**), and the ranking of grain yields at elevated and ambient CO_2 (**b**). Linear regressions had an R^2 value of 0.200 for yield, and 0.267 for the ranking of yields, which are not significant at $p = 0.05$. The numbers identify the lines, which are listed in Table 3.

Table 3. The nine lines of wheat tested in the second experiment. All lines are from the wheat association mapping initiative breeding population developed by Centro Internacional de Mejoramiento de Maíz y Trigo [15].

Line Number	Identification
1	FCT/3/AZ//MUS/4/DOVE/BUC
2	LEA/TAN/4/TSH/3/KAL/BB//TQFN/5/PAVON/6/SW89.3064
3	MILAN/3/JUP/BJY//URES
4	PRL/SARA/TSI/VEE#5
5	CLC89//ESDA/KAUZ/3/BJY/COC//PRL/BOW
6	KAMBARA2
7	CN079//PF70354/MUS/3/PASTOR/4/BAV92
8	CN079//PF70354/MUS/3/PASTOR/4/CROC_1/AE.SQUARROSA(224)//OPATA
11	CHIH95.1.10

The three lines which had the largest increases in yield ranking at elevated CO_2 in the second experiment (lines 1, 4, and 7) averaged 34% increases in the number of seed heads and 15% increases in seeds per head in the elevated CO_2 plot. The three lines with the largest decreases in yield ranking at elevated CO_2 in this experiment (lines 5, 11, and 2) averaged 7% increases in seed head number, and 1% increases in seeds per head in the elevated CO_2 plot.

The coefficient of variation for averaged 11% across lines within each plot, for both experiments (Tables 1 and 2), that is both for a subplot size of 3.0 m^2 in the first experiment, and 0.20 m^2 in the second experiment. The coefficient of variation averaged the same at ambient and elevated CO_2 in both experiments (Tables 1 and 2). Homogeneity of variance tests also indicated no significant differences in variance among lines in either CO_2 treatment in either experiment.

3. Discussion

The first experiment used a plot size typical of local variety test plots for wheat [16] in Maryland, and indicated that there were significant differences in yield among the four cultivars, both at ambient and at elevated CO_2. However, the comparison of cultivar rankings indicated that Jamestown had a better yield response to elevated CO_2 than did Pioneer 25 R32 in both years, and that Choptank also had a better yield response than Pioneer 25 R32 in the second year. Thus, this experimental design revealed cultivar differences in yield response to elevated CO_2 while using only a single elevated CO_2 plot and a single ambient CO_2 plot. If a few years of local yield trials at ambient CO_2 were available, one might not need a concurrent ambient CO_2 trial in order to determine which cultivars had superior yield responses to elevated CO_2. However, the relatively large size (4 m^2) of the cultivar subplots used in this experiment would greatly limit the number of cultivars that could be compared even using larger FACE systems.

The second experiment tested whether much smaller subplot sizes could also be used to indicate differences among wheat lines in yield responses to elevated CO_2. The lack of increase in variability among subplots for the smaller size, as measured by the coefficient of variation, indicated no loss in precision in comparing yield among lines when using the smaller subplot size. The mean coefficient of variation of 11% for both the larger and smaller sub-plots is similar to typical coefficients of variation in standard wheat yield trials in Maryland [17]. For example, among 57 lines compared at Beltsville in the Maryland State Wheat Trials in the same year as our second experiment, the mean coefficient of variation was 12.8%.

In our second experiment, nine lines were compared in an elevated CO_2 plot of only 36 m^2, with $n = 4$. A FACE ring 20 m diameter could theoretically be used to compare responses of about 90 wheat lines, with the subplot size and number of replicate subplots used here in the second experiment. At this ring size, horizontal uniformity of CO_2 concentration may become an issue [18], and an area distributed FACE system [19], as used here, may be more effective. Identification of lines with larger yield increases at elevated CO_2 is only a first step to improving the response of a crop species to future CO_2 conditions. Identification of traits responsible for differences in yield is an important next step. The preliminary data presented here in the second experiment suggest that both increases in the number of seed heads per m^2 and in the number of seeds per head may be important parameters in yield increases at elevated CO_2 in wheat. The approach used here of having replicated subplots within one elevated CO_2 treatment plot may provide a more efficient route to identifying lines with improved yield responses to elevated CO_2 under open-air field conditions.

4. Materials and Methods

All experiments were conducted at the South Farm of the Beltsville Agricultural Research Center, Beltsville MD (39°02' N, 76° 94' W, elevation 30 m). In one experiment, four locally adapted cultivars of wheat (*Triticum aestivum* L.)—Jamestown, Choptank, Pioneer 25 R40, and Pioneer 25 R32—were grown in one elevated CO_2 plot and one ambient CO_2 plot for two years. In this experiment, each plot was 60 m^2 in area, and there were three subplots of 4 m^2 area of each cultivar, with cultivars randomly assigned to subplots. Row width was 15 cm, and plant density was about 120 plants m^{-2}. The two plots were in the same field, within 50 m of each other, at locations randomly assigned each year. In a second experiment, nine lines from the WAMI breeding population [15] listed in Table 1, selected for differences in tillering, were grown in one elevated and one ambient plot, each 36 m^2 in area. For each CO_2 treatment, there were four subplots, each 0.25 m^2 in area per genotype, arranged randomly. Each subplot had four 12.5 cm wide rows, 50 cm in length, with about 120 plants m^{-2}. The borders of each subplot, and the perimeter 0.75 m of the whole plots were planted with Choptank wheat, which was not sampled. In both experiments, the plots were tilled and planted in mid-October, following soybean crops, and harvested in mid-June. A 10-10-10 nitrogen, potassium, phosphorus fertilizer was applied when regrowth began in spring, at a rate providing 25 g N m^{-2}. No significant pest problems occurred. As is usual for this climate, frequent precipitation prevented any significant soil water deficits.

CO_2 enrichment began at planting, using an area-distributed FACE system, which reduces the horizontal variation in mean CO_2 compared with perimeter ring FACE systems [19]. CO_2 enrichment was applied continuously except when either air or soil temperatures were below 0 °C. The daytime target enrichment was 190 μmol mol^{-1} above the ambient concentration, and 220 μmol mol^{-1} above the ambient concentration at night. These treatments acknowledge that the CO_2 concentrations which crops will experience in the future will probably be more increased at night than during the daytime because of enhanced photosynthesis and respiration per ground area. Mean CO_2 concentrations during periods when temperatures were above 0 °C over the three years of these experiments averaged 422 and 628 μmol mol^{-1} for the ambient and elevated plots, respectively. The midday ambient CO_2 concentration averaged 392 μmol mol^{-1}.

The crops were harvested at grain maturity. In the experiments with the 4 m^2 subplots, total above ground dry mass, head number, tiller number, mean mass per grain, and total grain dry mass were obtained from a bordered 3 m^2 area within each subplot. In the experiment with 0.25 m^2 subplots, bordered 0.20 m^2 areas were harvested from each subplot, with the same plant parameters measured.

Because there was only one replicate plot of each CO_2 treatment each year, no assessment of the overall CO_2 effect was made in either experiment. Using the replication of the genotype subplots within each CO_2 treatment, analysis of variance was used to test whether genotypic differences in yield existed within each plot, separately for each CO_2 treatment and year. The coefficient of variation (the standard deviation divided by the mean $\times$ 100%) averaged across genotypes was compared for the experiments with the large and the small subplot sizes. The coefficient of variation should increase when plots become too small relative to the spatial variation in environmental properties. Homogeneity of variance tests were also conducted among lines within each CO_2 treatment in both experiments. An altered ranking of genotype yields at ambient versus elevated CO_2 within a year was taken as preliminary evidence of a differential response to CO_2 enrichment among genotypes. In the experiment with two years of observations, we examined the consistency of changes in rank with CO^2 across years. A high correlation among lines for yield or the ranking of yield at ambient vs. elevated CO_2 could indicate little variation among lines in their CO_2 response, so these correlations were examined in the second experiment, which had only one year of yield data.

Acknowledgments: I thank Matthew Reynolds, CIMMYT for providing the nine lines used in experiment 2, and Lewis Ziska, USDA for advice.

Author Contributions: The author planned, executed, and analyzed the experiments, and wrote the article.

Conflicts of Interest: The author declares no conflict of interest.

Abbreviations

The following abbreviations are used in this manuscript:

FACE   Free-air carbon dioxide enrichment

References

1.   Clausen, S.K.; Frenck, G.; Linden, L.G.; Mikkelsen, T.N.; Lunde, C.; Jorgensen, R.B. Effects of single and multifactor treatments with elevated temperature, CO_2 and ozone on oilseed rape and barley. *J. Agron. Crop Sci.* **2011**, *197*, 442–453. [CrossRef]

2.   Bunce, J.A. Contrasting responses of seed yield to elevated carbon dioxide under field conditions within Phaseolus vulgaris. *Agric. Ecosyst. Environ.* **2008**, *128*, 219–234. [CrossRef]

3.   Ahmed, F.E.; Hall, A.E.; Madore, M.A. Interactive effects of high temperature and elevated carbon dioxide concentration on cowpea [(*Vigna unguiculata* (L.) Walp.]. *Plant Cell Environ.* **1993**, *16*, 835–842. [CrossRef]

4.   Johannessen, M.M.; Mikkelsen, T.N.; Nersting, L.G.; Gullord, M.; von Bothmer, R.; Jorgenses, R.B. Effect of increased atmospheric CO_2 on varieties of oat. *Plant Breed.* **2005**, *124*, 253–256. [CrossRef]

5.   Johannessen, M.M.; Mikkelsen, T.N.; Jorgensen, R.B. CO_2 exploitation and genetic diversity in winter varieties of oilseed rape (*Brassica napus*); varieties of tomorrow. *Euphytica* **2002**, *128*, 75–86. [CrossRef]

6.   Hasegawa, T.; Tokida, T.; Nakamura, H.; Zhu, C.; Usui, Y.; Yoshimoto, M.; Fukuoka, M.; Fukuoka, M.; Wakatsuki, H.; Katayanagi, N.; et al. Rice cultivar responses to elevated CO_2 at two free–air CO_2 enrichment (FACE) site in Japan. *Funct. Plant Biol.* **2013**, *40*, 148–159. [CrossRef]

7.   Yang, L.; Liu, H.; Wand, Y.; Zhu, J.; Huang, J.; Liu, G.; Dong, F.; Wang, Y. Impact of elevated CO_2 concentration on inter–subspecific hybrid rice cultivar Liangyoupeijiu under fully open–air field conditions. *Field Crop Res.* **2009**, *112*, 7–15. [CrossRef]

8.   Bishop, K.A.; Betzelberger, A.M.; Long, S.P.; Ainsworth, E.A. Is there potential to adapt soybean (*Glycine max* Merr.) to future CO_2? An analysis of the yield response of 18 genotypes in free–air CO_2 enrichment. *Plant Cell Environ.* **2015**, *38*, 1765–1774. [CrossRef] [PubMed]

9.   Bunce, J.A. Variable responses to CO_2 of the duration of vegetative growth and yield within a maturity group in soybeans. *Am. J. Plant Sci.* **2016**, *7*, 1759–1764. [CrossRef]

10.   Batts, G.R.; Ellis, R.H.; Morison, J.I.L.; Nkemka, P.N.; Gregory, P.J.; Hadley, P. Yield and partitioning in crops of contrasting cultivars of winter wheat in response to CO_2 and temperature in field studies using temperature gradient tunnels. *J. Agric. Sci.* **1998**, *130*, 17–27. [CrossRef]

11.   Tausz-Posch, S.; Seneweera, S.; Norton, R.M.; Fitzgerald, G.J.; Tausz, M. Can a wheat cultivar with high transpiration efficiency maintain its yield advantage over a near–isogenic cultivar under elevated CO_2? *Field Crop Res.* **2012**, *133*, 160–165. [CrossRef]

12.   Ziska, L.H.; Bunce, J.A.; Shimono, H.; Gealy, D.R.; Baker, J.T.; Newton, P.C.D.; Reynolds, M.P.; Jagadish, K.S.V.; Zhu, C.; Howden, M.; et al. Security and climate change: On the potential to adapt global crop production by active selection to rising atmospheric carbon dioxide. *Proc. R. Soc. B: Biol. Sci.* **2012**, *279*, 4097–4105. [CrossRef] [PubMed]

13.   Bunce, J.A. Elevated carbon dioxide effects on reproductive phenology and seed yield among soybean cultivars. *Crop Sci.* **2015**, *55*, 339–343. [CrossRef]

14.   Thilakarathne, C.L.; Tausz-Posch, S.; Cane, K.; Norton, M.; Tausz, M.; Seneweera, S. Intraspecific variation in growth and yield response to elevated CO_2 in wheat depends on the differences in leaf mass per unit area. *Funct. Plant Biol.* **2013**, *40*, 185–194. [CrossRef]

15.   Lopes, M.S.; Dreisigacker, S.; Pena, R.J.; Sukumaran, S.; Reynolds, M.P. Genetic characterization of the wheat association mapping initiative (WAMI) panel for dissection of complex traits in spring wheat. *Theor. Appl. Genet.* **2015**, *128*, 453–464. [CrossRef] [PubMed]

16.   2016–17 Maryland Wheat and Barley Variety Test Entry Form. Available online: http://www.psla.umd.edu/sites/default/files/_docs/MD_CROPS/2016_17Wheat&BarleyVariety_Test_Application_Form (accessed on 23 February 2017).

17.   Small Grains in Maryland. Available online: http://www.psla.umd.edu/extension/extension-project-pages/small-grains-maryland (accessed on 23 February 2017).

18.   Nakamura, H.; Tokida, T.; Yoshimoto, M.; Sakai, H.; Fukuoka, M.; Hasegawa, T. Performance of the enlarged rice–FACE system using pure CO_2 installed in Tsukuba, Japan. *J. Agric. Meteorol.* **2013**, *68*, 15–23. [CrossRef]

19.   Bunce, J.A. Performance characteristics of an area distributed free air carbon dioxide enrichment (FACE) system. *Agric. For. Meteorol.* **2011**, *151*, 1152–1157. [CrossRef]

7

Effect of Sowing Method and N Application on Seed Yield and N Use Efficiency of Winter Oilseed Rape

Klaus Sieling *, Ulf Böttcher and Henning Kage

Agronomy and Crop Science, Institute of Crop Science and Plant Breeding, Christian-Albrechts-University of Kiel, Hermann-Rodewald-Str. 9, D-24118 Kiel, Germany; boettcher@pflanzenbau.uni-kiel.de (U.B.); kage@pflanzenbau.uni-kiel.de (H.K.)
* Correspondence: sieling@pflanzenbau.uni-kiel.de

Academic Editor: Bertrand Hirel

Abstract: In northern Europe, replacing winter barley with winter wheat as the preceding crop for winter oilseed rape (*Brassica napus* L.; WOSR) often results in a delayed WOSR sowing and poor autumn growth. Based on data from a field experiment running in 2009/2010, 2010/2011, and 2012/2013, this study aims (i) to investigate how a delayed sowing method affects seed yield, N offtake with the seeds, and apparent N use efficiency (NUE) of WOSR; (ii) to test the ability of autumn and spring N fertilization to compensate for the negative effects of a delayed sowing method; and (iii) to estimate the minimum autumnal growth for optimal seed yield. In order to create sufficiently differentiated canopies, a combination of four sowing methods (first week of August until the third week of September) and four autumn N treatments (0, 30, 60, and 90 kg·N·ha^{-1}) was established. Each of these 16 different canopies was fertilized with 5 N amounts (0/0, 40/40, 80/80, 120/120, 140/140 kg·N·ha^{-1}) in spring in order to estimate separate N response curves. Above-ground N accumulation in autumn and seed yield and N offtake by the seeds were determined. Plant establishment after mid-September significantly decreased seed yield. Autumn N fertilization of at least 30 kg·N·ha^{-1} increased seed yield and N offtake by the seeds without any significant interaction with sowing method and spring N supply. However, the pathway(s) remain(s) unclear. Spring N fertilization up to 130 kg·N·ha^{-1} (estimated by a Linear-Plateau N response curve) increased seed yield. NUE decreased with increasing N supply, where WOSR used autumn N to a lesser extent than spring N. An above-ground N uptake of at least 10–15 kg·N·ha^{-1} at the end of autumn growth was required to achieve high seed yields. From an environmental point of view, optimal autumn growth should be attained by choosing an adequate sowing method, not by applying additional N in autumn.

Keywords: oilseed rape; seed yield; sowing method; N fertilization; autumn N uptake; N use efficiency

1. Introduction

Farmers in northern Europe often replace winter barley with winter wheat as the preceding crop for winter oilseed rape (*Brassica napus* L.; WOSR) causing a delayed WOSR sowing due to the later wheat harvest. Luteman and Dixon [1] and Sieling et al. [2] observed a reduced autumn and winter growth by delayed plant establishment, especially if it was later than 10 September, increasing the risk of winter kill [3–5]. While Luteman and Dixon [1] found only small yield penalties, Uzan et al. [6] reported a decline in seed yield of canola cultivars with a delay in sowing date under Mediterranean environment conditions due to shortening the length of the reproductive period and consequently the potential grain-filling period. Also, Scott et al. [7] showed large yield penalties if sowing occurred after mid-September.

In order to compensate for the poor autumn growth, an additional nitrogen (N) supply of about 30–50 kg·N·ha^{-1} in autumn is intended to ensure crop N supply, adequate crop growth before winter and, consequently, a good overwinter survival. Autumn N, especially in early-sown oilseed rape, leads to a better above-ground growth [2] and enhances N accumulation before winter [8]. It is, however, debatable whether this increases seed yield and overwintering. Results of Ogilvy and Bastiman [9] revealed that, although plots receiving nitrogen in the seedbed or at the two-leaf stage were more vigorous before winter compared with unfertilized plants, neither the number of plants established, the survival over winter, nor the seed yield were affected by this treatment. According to Engström et al. [10] no effects on seed yield occurred after applying 30 or 60 kg·N·ha^{-1} at sowing to WOSR in Sweden. In addition, Sieling and Kage [11] observed that WOSR yield increased by 0.2 t·ha^{-1} after an autumn N supply of 40 kg·N·ha^{-1}. In experiments in France, severe N deficiencies in autumn, described in terms of the nitrogen nutrition index, reduced shoot biomass, tap root biomass, leaf area index, and radiation-use efficiency compared with N sufficient treatments [12]. However, no difference in seed yield occurred. The authors assumed that sufficient N release due to natural mineralization provided enough growth in autumn to ensure sufficient regrowth in spring. In other experiments, autumn N gave a small yield response where the preceding cereal straw was baled or incorporated instead of burning [13,14]. In addition, N application to boost crop growth rate may lead to plants being more susceptible to freezing in winter due to vigorous vegetative growth prior to the first killing frost or to increased leaching risk of N not taken up by the crop.

Spring N fertilization clearly increases WOSR seed yield. WOSR demands high amounts of N fertilizer often exceeding 200 kg N·ha^{-1} to achieve maximum yields in high yielding environments with yield levels >4 t·ha^{-1}. Scott et al. [15] and Zhao et al. [16] suggested a maximum yield response to a N rate around 200 kg·N·ha^{-1}, which was in good agreement with the results of Bilsborrow et al. [17] who obtained $>85\%$ of the maximum recorded yield with an application of 150 kg·N·ha^{-1}. Several approaches from, for example, France and Germany take the amount of N in the WOSR canopy at the end of autumn growth and/or at the beginning of spring growth into account [18–20]. In one study of insignificant leaf losses over winter, a well-developed rapeseed canopy showed a higher leaf area index at the beginning of spring growth than a poor canopy, which, in consequence, allows for reduced N fertilization rates [21]. Henke et al. [19] showed that even if canopy N is lost over winter due to frost, optimum N requirements in spring decreased with increasing canopy N in late autumn. These results assumed a negative interaction between autumn N supply, promoting above-ground dry matter accumulation and N uptake, and spring N application [2].

WOSR has the ability to take up more N than winter cereals between sowing and spring [22]; however, relatively little of this N ends up in the seed [23]. Compared to the N requirement, N offtake by the WOSR seeds is comparatively low, leading to low nitrogen harvest indices and high N balance surpluses (fertilizer N minus N offtake by the seeds), which can lead to high N leaching rates during periods of heavy rainfall [23–27]. Thus, along with an increasing rate of N-fertilization, the rate of fertilizer-N recovery (apparent N use efficiency (NUE)) declines substantially [26]. Due to only small yield effects, N fertilization in autumn is less N efficient in spring, thus most of the autumn N remains in the system. In field experiments reported by Sieling and Kage [11], 40 kg·N·ha^{-1} in autumn increased N offtake by the seeds by about 4 kg·N·ha^{-1}, thus 36 kg·N·ha^{-1} were left in the soil changing the N balance.

In general, it is assumed that WOSR requires at least six to eight leaves, a root collar diameter of 1 cm, a tap root length of 20 cm, and about 1 g of dry matter (DM) per plant to secure optimal winter survival [28]. Under the climate conditions of northern Europe, sowing date and N availability in autumn are the main factors influencing WOSR growth before winter, which can be managed by the farmers [2]. However, no data are available on which to base thresholds for optimal autumn growth.

Based on data from a three-year field experiment, this study aims (i) to investigate how delayed sowing affects seed yield, N offtake with the seeds, and N use efficiency of WOSR; (ii) to test the ability of autumn and spring N fertilization to compensate for the negative effects of a delayed sowing method; and (iii) to estimate the minimum autumnal growth for optimal seed yield. We hypothesize that delaying sowing reduces seed yield, N offtake, and N use efficiency, while autumn N supply may increase these factors, especially in late sown canopies. In addition, spring N fertilization may be reduced if additional N was applied in autumn.

2. Results

The experimental design provided an increased seed density at later sowing in order to compensate for a decreased emergence rate, thus confounding both effects. However, in considering the average of the three years, plant densities were quite similar for all sowing methods ranging between 37 and 47 plants·m^{-2} at the end of autumn growth and between 36 and 40 plants·m^{-2} at the beginning of spring growth. Consequently, winter survival also remained unaffected. Therefore, no significant interaction between sowing date and seed density occurred.

2.1. Seed Yield

In considering the average of all three years and all other treatments, sowing in September significantly decreased WOSR seed yield ($p < 0.05$) (Table 1). While both sowing methods in August resulted in similar seed yields of about 5 t·ha^{-1}, drilling in the first and third September week caused a yield loss of 0.3 t·ha^{-1} and 1.2 t·ha^{-1}, respectively. Autumn N application significantly increased seed yield by 0.6 t·ha^{-1}, in comparison to the unfertilized and the 90 kg·N·ha^{-1} treatments (Table 1). The sowing method by autumn N interaction was not significant at $p = 0.05$.

Table 1. Sowing method by autumn N application interaction effects on winter oilseed rape (*Brassica napus* L.; WOSR) seed yield (t·ha^{-1}) (based on the average of 2009/2010, 2010/2011, 2012/2013 and all spring treatments).

Sowing Method [‡]	Autumn N Application (kg·N·ha^{-1})				
	0	30	60	90	Mean
SD 1	4.72	4.87	5.02	5.12	4.93 [a,†]
SD 2	4.74	4.94	5.16	5.18	5.01 [a]
SD 3	4.05	4.74	4.86	4.98	4.66 [b]
SD 4	3.42	3.71	3.84	4.06	3.76 [c]
Mean	4.23 [c]	4.56 [b]	4.72 [a,b]	4.83 [a]	

[‡]—SD 1—first week of August; SD 2—third week of August; SD 3—first week of September; SD 4—third week of September. [†]—Different letters indicate significant differences within a factor at $p = 0.05$. The sowing method by autumn N interaction was not significant ($p > 0.05$). SD—sowing method.

Spring N supply boosted seed yield (Figure 1a; for estimated model parameters, see Table 2). However, N amount that exceeded 116–136 kg·N·ha^{-1} depending on the sowing method failed to show any further increase. The comparison of the function parameter revealed that the sowing method only influenced the level of the yield plateau ranging between 4.23 t·ha^{-1} if sown in the third week of September (SD 4) and 5.56 t·ha^{-1} at the third week of August (SD 2). In contrast, the N application rate at the intersection of the linear model and the plateau yield (Nopt), as well as the linear slope, remained unaffected (Table 2) indicating that the sowing method (SD) by spring N interaction was not significant.

A similar pattern occurred when comparing spring N effects at different autumn N levels (Figure 1b; for estimated model parameters, see Table 2). Applying N in autumn increased the plateau yield from 4.48 t·ha^{-1} in the unfertilized control up to 5.12, 5.20, and 5.32 t·ha^{-1} in the 30, 60, and 90 kg·N·ha^{-1} treatment, respectively, but did not affect Nopt and slope (Table 2). No significant interactions between both N application dates occurred.

2.2. N Offtake by the Seeds

Considering the N offtake by the seeds revealed similar results as observed for seed yield. Delaying sowing reduced N offtake ranging between 112 kg·N·ha^{-1} sown at the latest date and 141 kg·N·ha^{-1}, if crop establishment was made in the third week of August. The additional 30 kg·N·ha^{-1} applied in autumn increased N offtake by 9 kg·N·ha^{-1}, thus leaving 21 kg·N·ha^{-1} (= 70% of the applied N amount) in the system (Table 3). A further increase in the autumn N supply up to 60 and 90 kg·N·ha^{-1} increased N offtake by 14 and 18 kg·N·ha^{-1}, thus boosting the N surplus by 46 (= 77%) and 72 kg·N·ha^{-1} (= 80%), respectively. Again, no significant interactions could be identified.

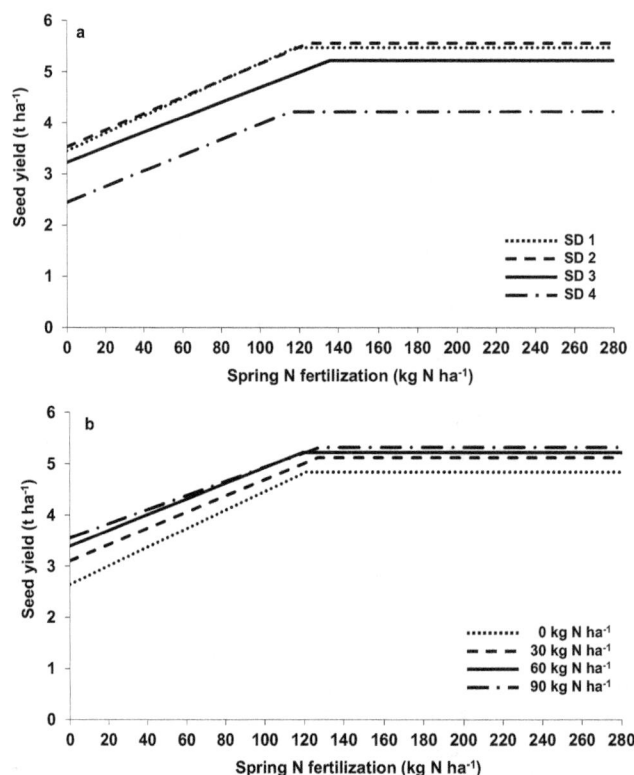

Figure 1. Effect of sowing method (SD) by spring N fertilization interaction (**a**) and autumn N by spring N fertilization (**b**) on seed yield of oilseed rape (2010, 2011, 2013; for function parameters, see Table 2). (SD 1—first week of August; S—third week of August; S—first week of September; S—third week of September).

Table 2. Estimated model parameters of the linear-plateau functions quantifying the relationship between spring N fertilization (kg·N·ha^{-1}) and seed yield (t·ha^{-1}) or N offtake with the seeds (kg·N·ha^{-1}) and between N uptake at the end of autumn growth and seed yield (t·ha^{-1}) (2009/2010, 2010/2011, 2012/2013).

Sowing Method [§]	Autumn N (kg·N·ha^{-1})	Spring N (kg·N·ha^{-1})	n	R^2	p	Plateau [†]	Nopt [‡]	Slope
Seed yield (Figure 1a)								
SD 1			240	0.65	<0.0001	5.47 [a,b $]	117.7 [ns]	0.0172 [ns]
SD 2			239	0.66	<0.0001	5.56 [a]	125.2 [ns]	0.0162 [ns]
SD 3			240	0.49	<0.0001	5.22 [b]	136.2 [ns]	0.0147 [ns]
SD 4			240	0.27	<0.0001	4.23 [c]	116.4 [ns]	0.0153 [ns]

Table 2. *Cont.*

Sowing Method [§]	Autumn N (kg·N·ha^{-1})	Spring N (kg·N·ha^{-1})	n	R^2	p	Plateau [†]	Nopt [‡]	Slope
	Seed yield (Figure 1b)							
	0		240	0.41	<0.0001	4.48 [b]	121.4 [ns]	0.0185 [ns]
	30		240	0.44	<0.0001	5.12 [a]	127.3 [ns]	0.0159 [ns]
	60		239	0.39	<0.0001	5.20 [a]	119.1 [ns]	0.0152 [ns]
	90		240	0.41	<0.0001	5.32 [a]	127.7 [ns]	0.0138 [ns]
N offtake by the seeds (Figure 2a)								
SD 1			240	0.73	<0.0001	162.8 [b]	156.5 [ns]	0.5344 [a]
SD 2			239	0.75	<0.0001	171.1 [a]	195.3 [ns]	0.4360 [b]
SD 3			240	0.65	<0.0001	159.6 [b]	182.2 [ns]	0.4257 [b]
SD 4			240	0.39	<0.0001	131.3 [c]	176.1 [ns]	0.3428 [c]
N offtake by the seeds (Figure 2b)								
	0		240	0.56	<0.0001	148.5 [b]	181.4 [ns]	0.4568 [ns]
	30		240	0.62	<0.0001	157.7 [a]	190.4 [ns]	0.4189 [ns]
	60		239	0.56	<0.0001	158.3 [a]	177.2 [ns]	0.4056 [ns]
	90		240	0.63	<0.0001	164.6 [a]	190.8 [ns]	0.3909 [ns]
	Seed yield (Figure 4)							
		0	190	0.42	<0.0001	3.99 [c]	64.9 [a]	0.0245 [c]
		80	190	0.41	<0.0001	4.80 [b]	13.1 [b]	0.1329 [b]
		160	192	0.44	<0.0001	5.40 [a]	6.8 [c]	0.3487 [a]
		240	191	0.43	<0.0001	5.47 [a]	6.7 [c]	0.3697 [a]
		280	191	0.42	<0.0001	5.52 [a]	6.4 [c]	0.4319 [a]

[†]—Plateau—plateau yield and N offtake, respectively. [‡] Nopt—N application rate (kg·N·ha^{-1}) and N uptake at the end of autumn growth, respectively, at the intersection of the linear model and the plateau (i.e., minimum N amount required to achieve maximum yield). [§]—SD 1—first week of August; SD 2—third week of August; SD 3—first week of September; SD 4—third week of September. [§]—Different letters indicate significant differences within one parameter and one factor at $p = 0.05$; ns-not significant at $p = 0.05$.

Table 3. Sowing method by autumn N application interaction effects on N offtake by WOSR seed yield (kg·N·ha^{-1}) (based on average of 2009/2010, 2010/2011, 2012/2013 and all spring treatments).

Sowing Method [§]	Autumn N Application (kg·N·ha^{-1})				
	0	30	60	90	Mean
SD 1	133	156	140	143	138 [a,b,†]
SD 2	134	138	145	146	141 [a]
SD 3	116	136	140	143	133 [b]
SD 4	100	110	114	123	112 [c]
Mean	121 [c]	130 [b]	135 [a,b]	139 [a]	

[§]—SD 1—first week of August; SD 2—third week of August; SD 3—first week of September; SD 4—third week of September. [†]—Different letters indicate significant differences within a factor at $p = 0.05$. The sowing method by autumn N interaction was significant at $p = 0.05$.

Since spring N fertilization increased not only seed yield, but also seed N concentration (data not shown), the N rate required to achieve maximum yield (Nopt) was higher for N offtake (160–195 kg·N·ha^{-1}) than for seed yield (about 116–136 kg·N·ha^{-1}) (Figure 2a; for estimated model parameters, see Table 2) at all sowing methods. In addition, the linear slope decreased with delayed sowing.

In combination with spring N, autumn N application was positively correlated with only the level of the plateau ranging between 149 and 166 kg·N·ha^{-1} (Figure 2b; for estimated model parameters, see Table 2), while Nopt and the slope were similar. Autumn N did not affect seed N concentration; in consequence, increased N offtake was mainly due to higher seed yields.

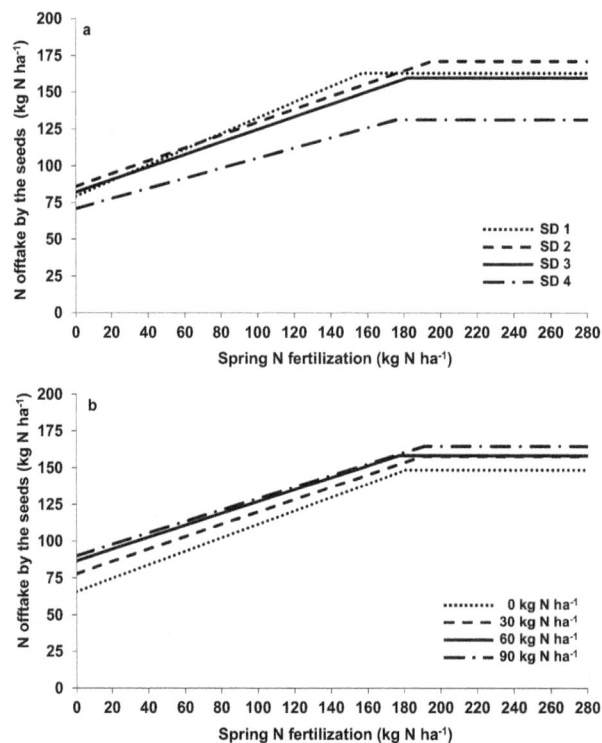

Figure 2. Effect of sowing method (SD) by spring N fertilization interaction (**a**) and autumn N by spring N fertilization (**b**) on N offtake by the seeds of oilseed rape (2010, 2011, 2013; for function parameters, see Table 2). (SD 1—first week of August; SD 2—third week of August; SD 3—first week of September; SD 4—third week of September).

2.3. Seed Oil Concentration and Oil Yield

Seed oil concentration was highest at early sowing and fell if sowing was delayed (Figure 3). Likewise, spring N significantly decreased oil concentration while autumn N had no effects. The slope was steepest if plant establishment occurred in August. In consequence, oil yield showed a similar pattern as seed yield with significant effects of sowing method (2.17, 2.19, 2.03, and 1.61 t·ha^{-1} in SD 1, SD 2, SD 3, and SD 4, respectively), autumn N (1.84, 1.99, 2.06, and 2.10 t·ha^{-1} in the 0, 30, 60, and 90 kg·N·ha^{-1} treatment, respectively) and spring N application (1.44, 1.99, 2.20, 2.18, and 2.18 t·ha^{-1} in the 0, 80, 160, 240, and 280 kg·N·ha^{-1} treatment, respectively). No significant interactions were observed.

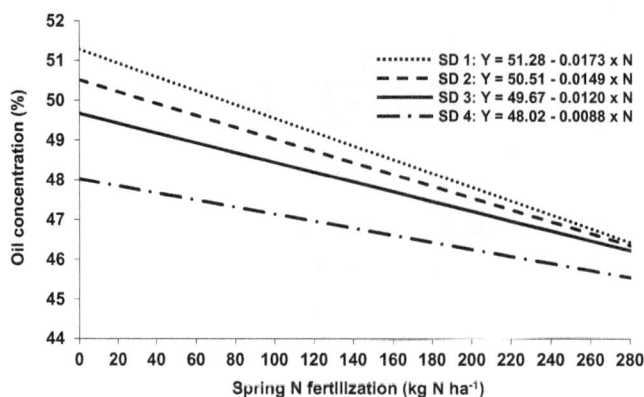

Figure 3. Effect of sowing method (SD) by spring N fertilization interaction on seed oil concentration of the seeds of oilseed rape (2010, 2011, 2013). (SD 1—first week of August; SD 2—third week of August; SD 3—first week of September; SD 4—third week of September).

2.4. Apparent N Use Efficiency

The apparent N use efficiency (NUE) describes the additional N offtake by the seeds due to the fertilization in relation to the respective N supply. Since no significant interaction between autumn and spring N application occurred, NUE of autumn N was presented based on the average of the spring N treatments and vice versa (Table 4). NUE of autumn N supply decreased with increasing N amount from 25% in the 30 kg·N·ha^{-1} treatment down to 20% in the plots which had received 90 kg·N·ha^{-1}; however, this reduction was not significant at $p = 0.05$. The date of plant establishment significantly affected NUE, being lowest at SD 1 (11%) and highest at SD 3 (38%). In contrast, the early sown plots utilized spring applied N to a larger extent than the later sown ones (Table 4). Increasing spring N application significantly reduced the respective NUE from 46% (80 kg·N·ha^{-1}) to 28% (280 kg·N·ha^{-1}). The sowing method by autumn N interaction and the sowing method by spring N interaction were not significant ($p > 0.05$).

Table 4. Sowing method by N application interaction effects on apparent N use efficiency (%) (2009/2010, 2010/2011, 2012/2013).

	Sowing Method [§]				
	SD 1	SD 2	SD 3	SD 4	Mean
Autumn N application (kg·N·ha^{-1})					
0	-	-	-	-	-
30	8.4	13.0	44.4	32.6	24.6 [ns,†]
60	14.0	18.5	40.0	22.5	23.7 [ns]
90	11.1	13.2	30.3	25.0	19.9 [ns]
Mean	11.1 [c]	14.9 [c]	38.2 [a]	26.7 [b]	
Spring N application (kg·N·ha^{-1})					
0	-	-	-	-	-
80	53.1	47.8	41.3	41.9	46.0 [a]
160	47.0	43.6	42.6	34.3	41.9 [a]
240	35.9	35.1	31.6	26.2	32.2 [b,c]
280	31.9	31.5	28.1	22.1	28.4 [c]
Mean	42.0 [a]	39.5 [a,b]	35.9 [b]	31.1 [c]	

[§]—SD 1—first week of August; SD 2—third week of August; SD 3—first week of September; SD 4—third week of September. [†]—ns—not significant ($p > 0.05$). Different letters indicate significant differences within one parameter and one factor at $p = 0.05$. The sowing method by autumn N interaction and the sowing method by spring N interaction were not significant ($p > 0.05$).

2.5. Relation between Above-Ground N Uptake in Autumn and Seed Yield

The combination of four sowing methods and four autumn N treatments resulted in 16 different WOSR canopies before winter. The estimation of the N uptake at the end of autumn growth allowed to determine the least required N application needed for optimal seed yield. Presuming a Linear-Plateau approach for each spring N treatment separately, the minimum autumn N uptake depended on the amount of spring N fertilization (Figure 4). A comparison of the estimated model parameters (Table 4) revealed similar curves if spring N supply was at least 160 kg·N·ha^{-1}. In this case, WOSR canopies should have taken up at least about 7 kg·N·ha^{-1} to achieve maximum seed yield. In contrast, canopies without spring N supply required 65 kg·N·ha^{-1} accumulated in the above-ground biomass. Larger canopies did not result in higher seed yield.

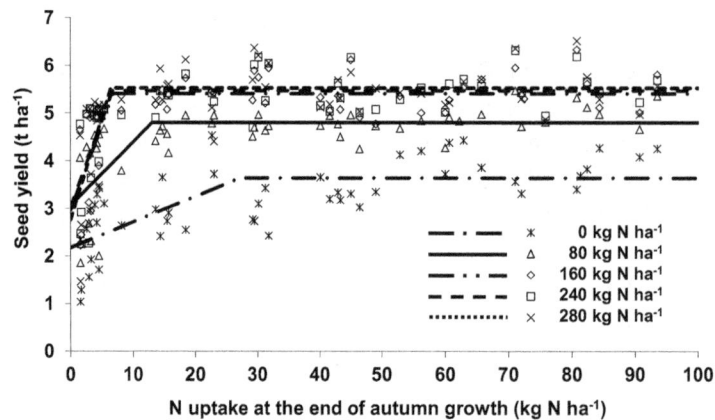

Figure 4. Relationship between N uptake of oilseed rape at the end of autumn growth and seed yield at different spring N fertilization levels (2010, 2011, 2013; for function parameters see Table 2).

3. Discussion

The experimental design allowed for an investigation of the effects of sowing method, autumn N, and spring N supply, as well as their interactions, on yield and N offtake by the seeds of WSOR. Due to varying sowing methods and autumn N supplies, 16 different WOSR canopies were established in autumn, each combined with increasing spring N fertilization in order to estimate N response curves separately for each canopy. The results confirmed the current recommendation that under the climatic conditions of northern Europe WOSR should ideally be sown between the middle and end of August, whereas plant establishment after mid-September clearly decreased seed yield, which was in agreement with [7]. It should be noted that, if the weather conditions in the following spring are favorable, even an WOSR crop sown in the third week of September with two to four leaves at the end of autumn growth was able to achieve more than 5 t·ha^{-1} [29]; however, since long-term weather forecasts are not yet available, the risk of yield penalties increases with shifting WOSR sowing to the end of September.

In combination with late sowing, farmers often apply additional N in autumn to ensure adequate WOSR growth before winter. No significant interaction between the sowing method and autumn as well as spring N fertilization on seed yield could be observed in our experiments (Table 1). Applying at least 30 kg·N·ha^{-1} in autumn increased WOSR yield at all sowing methods averaged over three years, although unfertilized WOSR canopies established in August already accumulated about 100 g·DM·m^{-2} and 34 kg·N·ha^{-1} at the end of autumn growth compared to 3 g·DM·m^{-2} and <2 kg·N·ha^{-1} in the late sown ones [29]. In addition, the N amount in early sown canopies clearly exceeded the threshold for optimal autumn N uptake, as shown in Figure 4. However, the pathway(s) which allowed autumn N to significantly increase seed yield for the early sowing method treatment require(s) further study. On the one hand, it can be argued that despite the often relatively small N uptake before winter, the additional N supply in autumn enables the plants to better utilize (N) resources in spring, thus indicating a physiological pathway. On the other hand, autumn N might act as advanced spring N fertilization if N is not leached out of the potential rooting zone during winter. In this case, an interaction between autumn and spring N treatments should be expected; however, it was not observed in our experiments (Figure 1b, Figure 2b, and Table 2). In contrast, Henke et al. [19] provided an approach which took autumn N uptake into account when estimating spring N fertilization, thus assuming an autumn N by spring N interaction. The authors suggested reducing the spring N amount if N in the canopy exceeds 50 kg·N·ha^{-1} at the end of autumn growth. With only small increases in N offtake by the seeds (Table 3) and the resulting small NUE (Table 4) in mind, autumn N should not exceed 30 kg·N·ha^{-1}.

In order to describe the N effects, a "Linear Response and Plateau" approach (LRP) was chosen, since using quadratic N response curves resulted in quite high N amounts required to achieve yield maximum due to a less steep slope compared to wheat [30,31]. Although the quadratic approach seems to be safer in terms of preventing financial losses under conditions of uncertainty, the risk of high positive N balances rises and the introduction of penalty functions may therefore favor the LPR model, especially against the background of high N surpluses in WOSR growing systems [31].

Based on the LRP approach, the minimum amount of spring N required for maximum seed yield ranged, based on average of the three experimental years, between 120 and 140 kg·N·ha^{-1}, regardless of the sowing method and the autumn N supply, which was lower than local recommendation according to the "Düngeverordnung" [32], the national implementation of the EU Nitrate Directive, as well as results from other authors [26,33]. Since N fertilization also increased seed N concentration, optimal N supply for the maximum N offtake by the seeds was higher (about 180–190 kg·N·ha^{-1}, Figure 2a,b and Table 2). In addition, delayed sowing methods also led to a higher seed N concentration, whereas autumn N had no effects [29].

N-use efficiency (NUE) for autumn (spring) N was calculated by comparing the seed yield according to the average of the spring (autumn) N treatments in relation to the amount of applied N in autumn (spring). In general, NUE of WOSR is lower compared to wheat or barley [23,30,34], which often has been associated with a low foliar N remobilization, thus with a high N concentration in WOSR leaves dropping off during the growth period, especially after flowering [35–37]. NUE decreased with increasing N amounts (Table 4). WOSR utilized autumn N to a lower extent than spring applied N [9,11]. Autumn NUE was highest when WOSR was sown in the first week of September, which was mainly due to the lower N offtake of the unfertilized plots compared to earlier sowing, while the autumn N fertilized ones showed similar values (Table 3). Therefore, NUE alone is not suitable for evaluating the sustainability of a cropping system as the N offtake and, in consequence, the N balance (N surplus) have to be taken into account. From an environmental point of view, N application in autumn, if any, should not exceed 30 kg·N·ha^{-1}, since most of the N remained in the system, increasing the risk of losses in the subsequent crops [27]. According to the presented results using the LPR approach, applying 120–130 kg·N·ha^{-1} in spring were sufficient to achieve the highest seed yields. However, using quadratic N response curves, 180–200 kg·N·ha^{-1} were required to get the economic optimal seed yield [29], representing an application rate that was similar to that used by the farmers.

The estimation of the minimum amount of N taken up by the canopy at the end of autumn in order to achieve maximum seed yield was based on the LPR approach, where only the spring N fertilization was additionally considered. Including autumn N supply did not improve the relationship (except if the canopy remained unfertilized in spring), indicating that the above-ground N accumulation at the end of autumn growth adequately covered the effects of autumn N application. Optimal autumn N uptake depended on the amount of spring N. If there are plans for 160 kg·N·ha^{-1} or more to be applied in spring, an N uptake of 10 kg·N·ha^{-1} before winter should enable the canopies to produce optimal seed yield. Even under the less favorable conditions occurring in autumn 2010, 13 kg·N·ha^{-1} seemed to be sufficient [29]. Assuming a quadratic relationship, about 40 kg·N·ha^{-1} were required to achieve 95% of the yield maximum [29] being similar to the 50 kg·N·ha^{-1} suggested by Henke et al. [19] to represent an average rapeseed canopy which will be fertilized in spring according to the official recommendation. However, a threshold of 1 g single plant DM for minimum autumn growth which is often used by the farmers results in about 20 kg·N·ha^{-1} assuming 40 plants·m^{-2} and 5% N concentration in the above-ground biomass.

4. Materials and Methods

4.1. Site and Soil

The field experiment was carried out on a pseudogleyic sandy loam (Luvisol: 100 g·kg^{-1} clay, pH 6.6, 86 mg·kg^{-1} P, 79 mg·kg^{-1} K, 150 g·kg^{-1} Mg, 13.8 g·kg^{-1} C$_{org}$, 1.1 g·kg^{-1} N$_{org}$) at the Hohenschulen Experimental Farm (10.0° E, 54.3° N, 30 m above see level) of the Kiel University, located in northern Germany 15 km west of Kiel (Schleswig-Holstein).

The climate of NW Germany can be described as humid. Total annual rainfall averages 750 mm at the experimental site, with ca. 400 mm received during April–September, the main growing season, and ca. 350 mm during October–March. In all three years, the mean air temperature in October was below the long-term average. Especially in autumn 2010, the low temperature in August until November, in combination with very high rainfall in August and September, resulted in poor autumn growth in all treatments (Table 5).

Table 5. Monthly rainfall (mm) and mean air temperature (°C) at Hohenschulen, Germany.

	Total Rainfall (mm)				Mean Air Temperature (°C)			
	2009/2010	2010/2011	2012/2013	Long-Term Mean	2009/2010	2010/2011	2012/2013	Long-Term Mean
August	69	105	56	59	17.8	16.1	16.7	17.9
September	28	123	59	61	14.3	12.6	12.7	14.2
October	63	55	71	75	7.7	8.7	8.6	9.8
November	109	113	39	58	7.3	3.8	5.1	4.9
December	49	29	72	62	0.2	−4.6	0.0	1.6
January	21	37	68	48	−4.1	0.4	0.5	1.2
February	43	39	24	50	−1.7	−0.1	−0.7	2.0
March	56	19	22	44	3.6	3.2	−1.1	3.4
April	17	3	18	44	7.5	10.4	6.1	8.0
May	50	36	141	62	8.8	12.2	11.4	11.9
June	50	93	100	69	14.3	15.2	13.9	14.8
July	24	150	40	100	19.5	15.7	17.6	16.9

4.2. Treatments

In 2009/2010, 2010/2011, and 2012/2013, a field experiment with winter oilseed rape (WOSR) cv. Visby was established to test the effects of sowing method (SD) and N application in autumn and spring on crop growth in autumn, seed yield, and plant N recovery. Heavy rainfall in August 2011 prevented a proper crop establishment, therefore, no oilseed rape was grown in that year. In order to create sufficiently differentiated canopies, the four sowing dates ranged between the first week of August and the third week of September (Table 6). The seed density was increased with delaying the sowing dates resulting in 35, 40, 50, and 70 seeds·m^{-2}, respectively, according to common farmers' practice. Since sowing date was confounded by seed rate, the term 'sowing method' is used throughout the paper. In addition, each sowing method received four autumn N treatments (0, 30, 60, and 90 kg·N·ha^{-1} as calcium ammonium nitrate with 27% N; for application dates, see Table 6), resulting in 16 sowing method by N supply combinations. Each of these 16 different canopies was fertilized with five N amounts (0/0, 40/40, 80/80, 120/120, and 140/140 kg·N·ha^{-1} as calcium ammonium nitrate with 27% N) in spring in order to estimate separate N response curves. N was applied in split-dressings at the beginning of spring growth and at stem elongation (for application dates, see Table 6).

Table 6. Dates of sowing, autumn, and spring N application of winter oilseed rape.

		Year		
		2009/2010	2010/2011	2012/2013
Sowing	SD 1	5 August	12 August	11 August
	SD 2	18 August	28 August	23 August
	SD 3	1 September	7 September	5 September
	SD 4	21 September	21 September	17 September
N application in autumn	SD 1	27 August	24 August	3 September
	SD 2	27 August	8 September	3 September
	SD 3	2 September	8 September	10 September
	SD 4	22 September	30 September	19 September
First N application in spring		23 March	3 March	4 March
Second N application in spring		12 April	8 April	15 April

The sowing methods as main plots were randomized with four replicates, while the N treatments were randomized within the main plots. This experimental design allowed for analyzing the sowing method effects separately for each year. The single plot size was 3 m × 12 m. Winter barley was the preceding crop; its straw remained on the plots. In general, the plots were ploughed within one day before sowing. Crop management not involving the treatments (e.g., P and K supply, soil tillage, pesticide application) were applied according to local recommendations to achieve optimal yield.

4.3. Measurements

In all sowing methods plant samples were taken at the end of autumn growth. Plants were dug up from an area of 1 m^2. After washing, above-ground biomass was analyzed for dry matter (DM) and N concentration using the near infrared spectroscopy (NIRS) analysis. Autumn N uptake was calculated by multiplying above-ground DM and the respective N concentration.

In order to minimize border effects, only a core of 1.75 m × 6 m (= 10.5 m^2) was combine harvested at maturity to determine seed yield (standardized to t·ha^{-1} with 91% DM based on the moisture content of a seed subsample). Oil and N concentration of the seeds (at 100% DM) were determined by NIRS. N offtake by the seeds results from the product of seed yield DM and N concentration.

Apparent N use efficiency (NUE) was defined as the difference between the N offtake by the seeds in the fertilized and unfertilized treatment in relation to the N amount. Since no interaction between autumn and spring N fertilization occurred, NUE of the autumn (spring) N application was calculated based on the average of the spring (autumn) N treatments.

4.4. Statistical Analysis

First, the effects of sowing methods (SD), autumn and spring N fertilization were analyzed using a mixed model approach (SAS 9.4, Proc MIXED, SAS Institute Inc., Cary, NC, US) with "year" and "replication" as random terms. Since the SD by autumn N by spring N interaction was not significant, only the interactions between two factors as average on the remaining factor are presented.

Second, the effects of spring N application on seed yield and N offtake by the seeds were estimated by using a "Linear-Plateau" approach (Proc NLIN) separately for the sowing methods and the autumn N treatments, respectively.

The 'Linear-Plateau' model is specified by three parameters a, b, and P as follows:

$$Y = a + bX + \varepsilon \rightarrow \text{if } X < \text{Nopt} \tag{1}$$

$$Y = P + \varepsilon \rightarrow \text{if } X \geq \text{Nopt} \tag{2}$$

where Y denotes the DM yield (t·ha^{-1}) or the N offtake by the seeds (kg·N·ha^{-1}), X the application rate of N (kg·N·ha^{-1}) or the N uptake in autumn, a the intercept, b the linear coefficient, and ε the error term. Nopt is the nitrogen application rate at the intersection of the linear model and the plateau yield P (Table 2). Function parameters were compared by a modified t-test based on Zar [38].

Seed oil concentration was analyzed by performing an analysis of covariance (Proc GLM).

A p-value of 0.05 was considered to be significant in the statistical analyses.

5. Conclusions

The presented results confirmed that the sowing of WOSR in the third week of August until the first week of September as being the best sowing method in northern Germany. Even in canopies which were well developed before winter, autumn N application increased seed yield. However, although economically suitable, autumn N should be critically discussed from an environmental point of view, since NUE is low and the pathway(s) of autumn N on yield is/are still unidentified.

Autumn N uptake of at least 10 kg·N·ha^{-1} was required to achieve high seed yields, if no severe stress (severe frost during winter, drought in spring) occurred.

Acknowledgments: We thank Kirsten Schulz and Cordula Weise for performing the plant sampling and the laboratory analyses.

Author Contributions: Klaus Sieling analyzed the data and wrote the paper. All authors discussed the results and commented on the manuscript.

Conflicts of Interest: The authors declare no conflict of interest.

References

1. Luteman, P.J.W.; Dixon, F.L. The effect of drilling date on the growth and yield of oil-seed rape (*Brassica napus* L.). *J. Agric. Sci.* **1987**, *108*, 195–200. [CrossRef]

2. Sieling, K.; Böttcher, U.; Kage, H. Sowing date and N application effects on tape root and above-ground dry matter of winter oilseed rape in autumn. *Eur. J. Agric.* **2017**, *83*, 40–46. [CrossRef]

3. Lääniste, P.; Jõudu, J.; Eremeev, V.; Mäeorg, E. Sowing date influence on winter oilseed rape overwintering in Estonia. *Acta Agric. Scand.* **2007**, *B 57*, 342–348. [CrossRef]

4. Balodis, O.; Gaile, Z. Winter oilseed rape (*Brassica napus* L.) autumn growth. In Proceedings of the Annual 17th International Scientific Conference Research for Rural Development, Jelgava, Latvia, 18–20 May 2011; Latvia University of Agriculture: Jelgava, Latvia; Volume 1, pp. 6–12.

5. Waalen, W.; Øvergaard, S.I.; Åssveen, M.; Gusta, L.V. Winter survival of winter rapeseed and winter turnip rapeseed in field trials as explained by PPLS regression. *Eur. J. Agron.* **2013**, *51*, 81–90. [CrossRef]

6. Uzun, B.; Zengin, Ü.; Furat, S.; Akdesir, Ö. Sowing date effects on growth, flowering, oil content and seed yield of canola cultivars. *Asian J. Chem.* **2009**, *21*, 1957–1965.

7. Scott, R.K.; Ogunremi, E.A.; Ivins, J.U.D.; Mendham, N.J. The effect of sowing date and season on growth abd yield of oilseed rape (*Brassica napus*). *J. Agric. Sci.* **1973**, *81*, 277–285. [CrossRef]

8. Dejoux, J.-F.; Meynard, J.-M.; Reau, R.; Roche, R.; Saulas, P. Evaluation of environmentally-friendly crop management systems based on very early sowing dates for winter oilseed rape in France. *Agronomie* **2003**, *23*, 725–736. [CrossRef]

9. Ogilvy, S.E.; Bastiman, B. The effect of rate and timing of autumn nitrogen on the pre-flowering dry matter production and seed yield of winter oilseed rape. *Asp. Appl. Biol.* **1992**, *30*, 413–416.

10. Engström, L.; Stenberg, M.; Aronsson, H.; Lindén, B. Reducing nitrate leaching after winter oilseed rape and peas in mild and cold winters. *Agron. Sustain. Dev.* **2011**, *31*, 337–347. [CrossRef]

11. Sieling, K.; Kage, H. Autumnal N fertilization of late sown oilseed rape after minimum tillage. In Proceedings of the 12th International Rapeseed Congress, Wuhan, China, 26–30 March 2007; pp. 375–378.

12. Colnenne, C.; Meynard, J.-M.; Roche, R.; Reau, R. Effects of nitrogen deficiencies on autumnal growth of oilseed rape. *Eur. J. Agron.* **2002**, *17*, 11–28. [CrossRef]

13. Chalmers, A.G. Autumn and spring fertiliser nitrogen requirements for winter oilseed rape. *Asp. Appl. Biol.* **1989**, *23*, 125–133.

14. Chalmers, A.G.; Darby, R.J. Nitrogen application to oilseed rape and implications for potential leaching loss. *Asp. Appl. Biol.* **1992**, *30*, 425–430.

15. Scott, R.K.; Ogunremi, E.A.; Ivins, J.U.D.; Mendham, N.J. The effect of fertilizers and harvest date on growth and yield of oilseed rape sown in autumn and spring. *J. Agric. Sci.* **1973**, *81*, 287–293. [CrossRef]

16. Zhao, F.J.; Evans, E.J.; Bilsborrow, P.E.; Syers, J.K. Influence of sulphur and nitrogen on seed yield and quality of low glucosinolate oilseed rape (*Brassica napus* L.). *J. Sci. Food Agric.* **1993**, *63*, 29–37. [CrossRef]

17. Bilsborrow, P.E.; Evans, E.J.; Zhao, F.J. The influence of spring nitrogen on yield, yield components and glucosinolate content of autumn-sown oilseed rape (*Brassica napus*). *J. Agric. Sci.* **1993**, *120*, 219–224. [CrossRef]

18. Reau, R.; Wagner, D.; Palleau, J.P. End of winter diagnosis: Winter rapeseed and nitrogen fertilization. In Proceedings of the Third Congress of the European Society for Agronomy, Podova, Italy, 18–22 September 1994; European Society for Agrnonomy: Podova, Italy; pp. 220–221.

19. Henke, J.; Sieling, K.; Sauermann, W.; Kage, H. Analysing soil and canopy factors affecting optimum nitrogen fertilization rates of oilseed rape (*Brassica napus*). *J. Agric. Sci.* **2009**, *147*, 1–8. [CrossRef]

20. Makowski, D.; Maltas, A.; Morison, M.; Reau, R. Calculating N fertilizer doses for oil-seed rape using plant and soil data. *Agron. Sustain. Dev.* **2005**, *25*, 159–161. [CrossRef]

21.	Mendham, N.J.; Shipway, P.A.; Scott, R.K. The effects of delayed sowing and weather on growth, development and yield of winter oil-seed rape (*Brassica napus* L.). *J. Agric. Sci.* **1981**, *96*, 389–416. [CrossRef]

22.	Barraclough, P.B. Root growth, macro-nutrient uptake dynamics and soil fertility requirements of a high-yielding winter oilseed rape crop. *Plant Soil* **1989**, *119*, 59–70. [CrossRef]

23.	Sieling, K.; Schröder, H.; Hanus, H. Mineral and slurry nitrogen effects on yield, N uptake, and apparent N use efficiency of oilseed rape (*Brassica napus*). *J. Agric. Sci.* **1998**, *130*, 165–172. [CrossRef]

24.	Shepherd, M.A.; Sylvester-Bradley, R. Effect of nitrogen fertilizer applied to winter oilseed rape (*Brassica napus*) on soil mineral nitrogen after harvest and on the response of a succeeding crop of winter wheat to nitrogen fertilizer. *J. Agric. Sci.* **1996**, *126*, 63–74. [CrossRef]

25.	Beaudoin, N.; Saad, J.K.; van Laethem, C.; Machet, J.M.; Maucorps, J.; Mary, B. Nitrate leaching in intensive agriculture in Northern France: Effect of farming practices, soils and crop rotations. *Agric. Ecosyst. Environ.* **2005**, *111*, 292–310. [CrossRef]

26.	Rathke, G.W.; Behrens, T.; Diepenbrock, W. Integrated nitrogen management strategies to improve seed yield, oil content and nitrogen efficiency of winter oilseed rape (*Brassica napus* L.): A review. *Agric. Ecosyst. Environ.* **2006**, *117*, 80–108. [CrossRef]

27.	Sieling, K.; Kage, H. N balance as an indicator of N leaching in an oilseed rape—Winter wheat—Winter barley rotation. *Agric. Ecosyst. Environ.* **2006**, *115*, 261–269. [CrossRef]

28.	Cramer, N. *Raps—Züchtung, Anbau und Vermarktung von Körnerraps*; Verlag Eugen Ulmer: Stuttgart, Germany, 1990.

29.	Sieling, K. Unpublished work. 2017.

30.	Sieling, K. Growth stage-specific application of slurry and mineral N to oilseed rape, wheat and barley. *J. Agric. Sci.* **2004**, *142*, 495–502. [CrossRef]

31.	Henke, J.; Breustedt, G.; Sieling, K.; Kage, H. Impact of uncertainty on the optimum nitrogen fertilization rate and agronomic, ecological and economic factors in an oilseed rape based crop rotation. *J. Agric. Sci.* **2007**, *145*, 455–468. [CrossRef]

32.	Bundesgesetzblatt. Düngeverordnung. Bekanntmachung der Neufassung der Düngeverordnung, *Bundesgesetzblatt*. Available online: https://www.bgbl.de/xaver/bgbl/start.xav?start=//*%5B@attr_id=%27bgbl107s0221.pdf%27%5D#__bgbl__%2F%2F*%5B%40attr_id%3D%27bgbl107s0221.pdf%27%5D__1487932661132 (assessed on 11 November 2007).

33.	Sieling, K.; Kage, H. Efficient N management using winter oilseed rape: A review. *Agron. Sustain. Dev.* **2010**, *30*, 271–279. [CrossRef]

34.	Sieling, K.; Schröder, H.; Finck, M.; Hanus, H. Yield, N uptake, and apparent N-use efficiency of winter wheat and winter barley grown in different cropping systems. *J. Agric. Sci.* **1998**, *131*, 375–387. [CrossRef]

35.	Rossato, L.; Lainé, P.; Ourry, A. Nitrogen storage and remobilisation in *Brassica napus* L. during the growth cycle: Nitrogen fluxes within the plant and changes in soluble protein patterns. *J. Exp. Bot.* **2001**, *52*, 1655–1663. [CrossRef] [PubMed]

36.	Ulas, A.; Behrens, T.; Wiesler, F.; Horst, W.J.; Schulte auf'm Erley, G. Does genotypic variation in nitrogen remobilisation efficiency contribute to nitrogen efficiency of winter oilseed-rape cultivars (*Brassica napus* L.)? *Plant Soil* **2013**, *371*, 463–471. [CrossRef]

37.	Girondé, A.; Poret, M.; Etienne, P.; Trouverie, J.; Bouchereau, A.; Le Cahérec, F.; Leport, L.; Orsel, M.; Niogret, M.-F.; Deleu, C.; et al. A profiling approach of the natural variability of foliar N remobilization at the rosette stage gives clues to understand the limiting processes involved in the low N use efficiency of winter oilseed rape. *J. Exp. Bot.* **2015**, *66*, 2461–2473. [CrossRef] [PubMed]

38.	Zar, J.H. *Biostatistical Analysis*, 5th ed.; Prentice Hall: Upper Saddle River, NJ, USA, 2009.

8

Residues Management Practices and Nitrogen-Potassium Fertilization Influence on the Quality of Pineapple (*Ananas comosus* (L.) Merrill) Sugarloaf Fruit for Exportation and Local Consumption

Elvire Line SOSSA [1,*], Codjo Emile AGBANGBA [2],
Sènan Gbèmawonmèdé Gwladys Stéfania ACCALOGOUN [3], Guillaume Lucien AMADJI [1],
Kossi Euloge AGBOSSOU [4] and Djidjoho Joseph HOUNHOUIGAN [3]

[1]   Faculty of Agronomic Sciences, Research Unit Eco-Pedology, University of Abomey-Calavi,
      Laboratory of Soil Sciences, 01 P.O. Box 526, Cotonou, Benin; gamadji@yahoo.fr
[2]   Laboratory of Biomathematics and Forests Estimations, Faculty of Agronomic Sciences,
      University of Abomey-Calavi, 03 P.O. Box 2819, Cotonou, Benin; agbaemile@yahoo.fr
[3]   Laboratory of Nutrition and Alimentary Sciences, Faculty of Agronomic Sciences,
      University of Abomey-Calavi, 01 P.O. Box 526, Cotonou, Benin; stefaccalogoun@yahoo.fr (S.G.G.S.A.);
      joseph.hounhouigan@gmail.com (D.J.H.)
[4]   Laboratory of Hydraulic and Water Control, Faculty of Agronomic Sciences, University of Abomey-Calavi,
      01 BP 526, Cotonou, Benin; euloge.agbossou@gmail.com
*   Correspondence: elvas2@yahoo.fr

Academic Editor: Francesco Montemurro

Abstract: Heterogeneity in pineapple fruit quality explains the low export volume of fruits from Benin to international markets. This work aims to investigate the influences of residues mulching or burying and N-K fertilization on (1) fresh fruit juice quality and the proportion of fruit meeting European standards and (2) fruit acceptability for fresh local consumption, as well as to identify morphological characteristics most related to fruit chemical quality attributes. The experimental design was a split-plot with three replications, where the main factor was N-K fertilization (T1 = 1.6 N and 1.6 K, T2 = 5.8 N and 6.6 K, T3 = 10 N and 11.6 K, T4 = 1.6 N and 11.6 K, T5 = 10 N and 1.6 K in g·plant^{-1}) and the sub-plot factor was mulching with pineapple residues (no mulching = 0, surface mulching = 10, buried = 10 in t·ha^{-1}). The results suggested that residues mulching and N-K fertilization has improved the percentage of fruit meeting European standards and local acceptability. The treatments T2B (T2 + burying) and T4B (T4 + burying) gave a higher proportion of fruits meeting European standards and were also promising for producing highly acceptable fruits by local consumers. Finally, the results revealed that the ratios of crown length: fruit length, crown length: infructescence length and crown length: median diameter were significantly associated with fruit quality, which has not yet been reported.

Keywords: pineapple; mulching; fertilization; acceptability; European standards

---

1. Introduction

Pineapple (*Ananas comosus* (L.) Merr.) is a well-appreciated fruit all over the world [1] and is cultivated all around the tropical and subtropical regions for local consumption and international export. It plays an important role in the human diet and is a good source of fiber and micronutrients, especially vitamins and minerals [2].

Benin is a country in West Africa at the coast of the Atlantic Ocean, with a suitable climate for growing tropical fruits including pineapple [3]. Pineapple production is of great importance for the Beninese economy and has contributed for 13 billion CFA francs to the Gross Domestic Product (GDP) in 2006. This contribution represented in this year about 1.2% of the total GDP and 4.3% of the agricultural GDP [4]. In 2009, it was estimated that less than 2% of the production was exported to Europe, which is far below the export potential, given that Benin has favorable production systems, coastal access, and well appreciated cultivars. This low export volume is due to the quality of the Beninese pineapple, which is heterogeneous [5]. A recent study on pineapple supply chains in Benin revealed that heterogeneity in quality attributes such as fruit weight, taste, firmness and flesh translucency was a constraint to the success of the chain [6]. Homogenous, good taste and heavier fruits are selected and bought by merchants in the producers' field who sold in domestic, border or regional markets (Kraké, Nigeria, Burkina Faso, Niger and Mali) [7]. Hence, cultural practices that guarantee good and homogeneous quality of fruit need to receive considerable attention. Heavy planting material and flowering induction at the optimum time can increase homogeneity in pineapple quality attributes [6]. The heterogeneity in pineapple quality might also be explained by the use of various fertilization practices, which influence the plant growth and consequently its yield and its quality [8–10]. Nitrogen and potassium fertilizers have a high impact on pineapple fruit yield, as well as organoleptic and sanitary quality [11]. Chemical fertilization represents a large part of total production costs [12]. Moreover, the nearly exclusive use of mineral fertilizers causes soil acidification [13]. Alternative cultivation practices that could reduce the use of N and K chemical fertilizers should be investigated. The positive effect of organic fertilizers (such as *Mucuna puriens*, *Panicum maximum* and compost of pineapple residues) and black polyethylene on pineapple fruit quality attributes were reported by some authors [12,14–17]. Pineapple residues are usually removed and burned in situ prior to being returned to the soil [18], leading to the loss of nutrients and environment pollution [19]. In Benin, pineapple is dominated by conventional monocropping with agrochemical inputs and little or no use of residues [20]. In this study, we hypothesize that the use of crop residues could reduce the need for N and K fertilizers for pineapple crops. The integration of crop residues to mineral fertilizer would improve fruit quality, meeting local acceptability and European standards as well. Stringent international regulations on quality norms and standards pose significant challenges to many small-scale pineapple producers and exporters in Benin [7]. These norms include the crown length: fruit length ratio, which should be between 50% and 150% [21]. So far, no literature has reported the relation between the crown length: Fruit length ratio and quality attributes. This work investigates (1) the influences of residues mulching or burying and N-K fertilization on fresh fruit juice quality and on the proportion of fruit meeting European standards, (2) the influences of residues mulching or burying and N-K fertilization on fruit acceptability for fresh local consumption, and (3) morphological characteristics most related to fruit chemical quality attributes.

2. Materials and Methods

2.1. Study Area

The study was carried out in the Atlantic department in southern Benin (1°59′ N and 2°15′ E). In this area, average annual rainfall is 1200 mm. The dominant soil type is a low-desaturated lateritic soil, commonly called "terre de barre" [22]. The soil physico-chemical analysis indicates that this is a soil of silty-clay-sandy texture, well drained, with an average pH of 5.6 and a C/N ratio of 11.2. These characteristics meet the requirements of pineapple cultivation soil described by [23].

2.2. Experimental Design and Management

The field experiment was conducted between 10 November 2013 and 15 April 2015. The experimental design consisted of a split plot with 3 replications, the main factor was the nitrogen-potassium (N-K) fertilization at 5 levels in $g \cdot plant^{-1}$ (T1:1.6 N and 1.6 K, T2:5.8 N and 6.6 K, T3:10 N and 11.6 K, T4:1.6 N and 11.6 K, T5:10 N and 1.6 K) and the sub-plot factor was the use of fresh pineapple residues at 3 levels (surface mulching (M) at 10 $t \cdot ha^{-1}$, 10 $t \cdot ha^{-1}$ buried residues at 10 cm deep (B), and no mulching (NM)). Each plot had an area of 12 square meters and included 48 plants.

Following soil tillage and delimitation of experimental units, the fresh pineapple residues coming from the same field were cut into pieces 10–15 cm long and were applied using a hoe. Planting material of about 300 g was sorted and planted at a density of 41,500 plants per hectare; the distance between lines or between ridges was 60 cm and the distance between plants was 40 cm.

Urea (46% N), trisulfate of phosphorus (TSP: 46% P_2O_5) and potassium sulphate (K_2SO_4: 50% K_2O, 45% SO_3) were used as mineral fertilizers. Phosphorus was applied fourteen days after planting (DAP) at a dose of 100 $kg \cdot ha^{-1}$. Five treatments resulting from the combination of different doses of nitrogen (N) and potassium (K) were done in six phases. Thus, the rates of nitrogen and potassium to be applied were split respectively into five (5) and six (6) equal portions. The first application was done at 45 DAP (1/5N + 1/6K), the second at 90 DAP (1/5N + 1/6K), the third at 135 DAP (1/5N + 1/6K), the fourth at 180 DAP (1/5N + 1/6K), the fifth at 225 DAP (1/5N + 1/6K) and the sixth at 270 DAP (1/6K). Fertilizers were applied at the base of each plant. The maintenance of the plots was done by monthly weeding. Flowering induction was made twelve month after planting, with diluted calcium carbide. For this purpose, 1 kg of carbide was diluted in a 200-liter drum. Each plant received 50 cm^3 of acetylene carbide during the cooler hours of the day (between 6 AM and 8 AM). The harvest took place five months after flowering induction.

2.3. Fruit Weight Attributes and Fresh Juice Physico-Chemical Measurements

The fruits were harvested in each experimental unit at C3 stage (yellow/orange on two thirds of the fruit surface) [24]. The weight of each fruit and crown were assessed using a brand (DH2-000050, ±0.0001, Zawiera, Tianjin, China). The measurements of median diameter, total fruit length, infructescence length and crown length were taken for each fruit with a tape and a ruler. Each fruit was then peeled and crushed. The obtained product was pressed, the fresh juice filtered, and the juice volume weighed. The total soluble solids (TSS) content of the juice was determined with the refractometer (HI96801, HANNA instruments, Bucharest, Romania) and the pH with the pH meter (HI96107, ±0.1 pH, HANNA instruments, Villafranca Padovana, Italy).

For each fruit, the crown length: fruit length ratio was calculated and associated with fruit weight and TSS for the determination of percentage of exportable pineapple fruits per treatment. Minimum quality criteria for fruits meeting European export standards include: the fruit weight should be between 0.70 and 2.75 kg, the crown length: fruit length ratio should be between 50% and 150% and the TSS should be at least 12° Brix [21].

The concentration of various sugars (sucrose, glucose, fructose, raffinose) and organic acids (oxalate, citrate, malate, propionate, lactic acid and formic acid) present in the fresh juice was determined using high performance liquid chromatography (HPLC) [25] on fruits harvested the same day for each experimental unit. The juice from each fruit was deducted with a syringe and filtered through a sterile filter (Sartorius Minisart) of 0.20 μm. The filtrate was collected in an Eppendorf tube of 2 mL. Solubles, sugars, and acids present in the juice were separated by liquid chromatography on a column SUPELCOGEL H of dimensions 30 cm × 7.8 mm. Twenty (20) microliters of juice was injected per sample into the system. The eluent used was sulfuric acid (5 mM). Each compound was identified from its retention time and quantified.

2.4. Panel of Tasters' Selection and Fruit Sensory Characteristics Measurements

The evaluation of sensory characteristics of fresh pineapple pulp was made according to [26]. Initially, a group of 25 students from the University of Abomey-Calavi was selected and filtered based on their sensory acuity on elemental flavors. The examination consisted of asking them to identify the elemental flavors and the smells of the basic flavors, on coded samples of sugar solutions (sweet flavor), salt (salt flavor), lemon (acid flavor) and caffeine (bitter flavor). Thus, the sugar solution was prepared with sugar 10 g·kg^{-1}; the saline solution with salt 2 g·kg^{-1}; the acid solution with citric acid 0.4 g·kg^{-1}; and the bitter solution with caffeine 0.5 g·kg^{-1}. At the end of this step, 20 people were able to identify each of these solutions and were therefore selected.

After this first selection, the 20 tasters were subjected to an intensity notation test realized on five coded pineapple samples (purchased in the market) whose characteristics were known and different. The purpose of this step was to judge the sensitivity of each taster to different characteristics of pineapple fruits. They were asked to assess each sample by indicating the intensity of its sweet taste, its acid taste, its aroma, its fiber content and its acceptability, according to a scale with five categories (not at all, slightly, moderately, strongly, extremely). At the end of this test, 16 tasters had demonstrated superior performance and were selected for the test on the pineapple samples from the experimental test.

Fruits were harvested by experimental unit and their sensorial characteristics were evaluated by the selected and trained panel of tasters. Thus, a total of 16 samples, 15 corresponding to the 15 treatments of the experiment and one control (T0) (fruit purchased in the market), were evaluated.

Each fruit was peeled and cut into pieces. The samples were coded with three random digits. All samples were presented simultaneously to the tasters, in a random order. Each taster could taste a sample several times, and rinse his mouth with water before moving to another sample. The sensorial characteristics such as the sweetness, acidic taste, aroma, and fiber content were evaluated by an intensity notation test, using a scale with five categories (not at all, slightly, moderately, strongly, extremely). The overall acceptability was assessed with a nine-point hedonic scale, with nine representing the most acceptable. Two sections of the test were carried out on different days.

2.5. Statistical Analysis

A two-way ANOVA test for the split plot was performed on chemical characteristics of fresh juice quality and fruit weight attributes (juice yield, TSS, pH, sucrose, glucose, fructose, raffinose, oxalate, citrate, malate, propionate, formic acid, malic acid) using package agricolae in R software (version R.3.1.0, 2014, R Core Team, Vienna, Australia) [27]. The percentages of fruits meeting European standards was transformed into arcsine square root [28]. Proportions equal to 0 and 1 were replaced by $(1/4n)$ and $(1-(1/4n))$ respectively, where n is the total number of fruits per net plot [28]. The normality [29] and homoscedasticity [30] conditions of model residues were checked for validation. Means were separated using the LSD test, with different LSD values being necessary for comparisons [31]. To evaluate fruit acceptability for fresh local consumption, a principal components analysis followed by a hierarchical classification was performed with the package FactoMineR on the sensory and physico-chemical characteristics of fruits. The chemical parameters were considered as supplementary quantitative variables, because they are directly related to sensory characteristics [32,33]. For each sensory and physico-chemical characteristics class, we measured the difference between the values for class and overall values. These statistics can be converted into a criterion called value-test, used to select the most characteristic variables [33,34]. The most characteristic variables of a class are those whose associated values are greater in absolute value than 2. Moreover, if this value test is positive for a variable, it has a high value in the class under consideration. In contrast, if the value is negative, the variable has a low value for the class. Linear regressions were performed between fruit weight attributes and physico-chemical characteristics. The conditions of normality [29] and homoscedasticity [35] of regression residues, residues independence [36] and linearity [37] were checked for the model validation.

3. Results

3.1. Influence of Mulching and N-K Fertilization on Pineapple Fresh Fruit Juice Quality and Proportion of Fruit Meeting European Standards

Significant effects of mulching and/or mineral fertilizer were observed on some fruit juice attributes (Figure 1). Mulching had significant effects on total soluble solids (TSS) ($p = 0.020$). Juice from pineapples with burying residues had more sweetness ($15.5° \pm 0.87°$ Brix) than from those with no mulching ($15.1° \pm 1.2°$ Brix) and mulching ($14.9° \pm 1.03°$ Brix) (Figure 1A). N-K fertilization had a significant effect on the TSS ($p = 0.000$), glucose ($p = 0.007$) and fructose ($p = 0.009$) content of juice. Treatments T4, T3 and T2 had the higher juice TSS ($16.1° \pm 0.5°$ Brix, $15.8° \pm 0.7°$ Brix and $15.7° \pm 0.5°$ Brix, respectively) whereas the lower TSS values were obtained with treatments T5 (10 N and 1.6 K) and T1 (1.6 N and 1.6 K) (respectively, $14.2° \pm 0.5°$ Brix and $14° \pm 0.5°$ Brix) (Figure 1B). Glucose content was higher with T3 and T4 (1.6 N and 11.6 K) (14.9 ± 2.2 mg·mL^{-1} and 14.81 ± 3.0 mg·mL^{-1}, respectively), while the lower value (11.1 ± 2.8 mg·mL^{-1}) was obtained with treatment T5 (Figure 1C). The highest and lowest values of fructose content were obtained respectively with treatments T4 (23 ± 7.3 mg·mL^{-1}) and T5 (15.3 ± 3.4 mg·mL^{-1}) (Figure 1D).

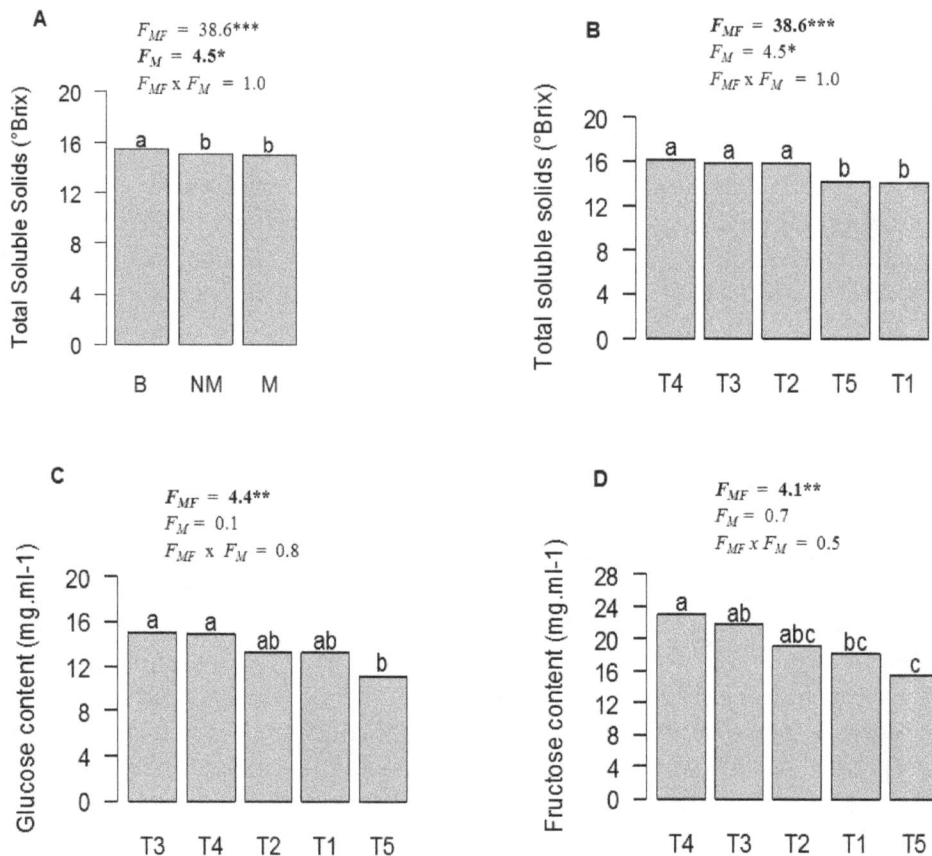

Figure 1. Influence of mulching and N-K fertilization on TSS (**A**,**B**) and organic sugar (**C**,**D**) content of pineapple fresh fruit juice. * = significant, ** = highly significant, *** = very highly significant. F_{MF} = Fisher value for mineral fertilizer, F_M = Fisher value for mulching, $F_{MF} \times F_M$ = Fisher value for mineral fertilizer and mulching interaction. T1 = 1.6 N and 1.6 K, T2 = 5.8 N and 6.6 K, T3 = 10 N and 11.6 K, T4 = 1.6 N and 11.6 K, T5 = 10 N and 1.6 K (g·plant^{-1}). B = Burying, M = Mulching, NM = No mulching. Means that do not share a letter are significantly different.

In terms of juice pH and organic acid, significant influences were observed with mulching and/or N-K fertilization (Figure 2). Mulching associated with N-K fertilization had a significant effect on the lactic acid content of juice ($p = 0.006$). We noticed indeed that higher lactic acid content was observed with the T5NM (treatment T5 + no mulching) (1.42 ± 0.28 mg·mL^{-1}) and lower content with T1NM (treatment T1 + no mulching) (0.12 ± 0.12 mg·mL^{-1}) (Figure 2A). Mulching had significantly influenced juice citrate content ($p = 0.05$). The citrate content was higher with the influence of no mulching (86.7 ± 16.5 mg·mL^{-1}), followed respectively by mulching (78.9 ± 16.7 mg·mL^{-1}) and burying residues (73.2 ± 16.5 mg·mL^{-1}) (Figure 2B). N-K fertilization had a significant effect on pH ($p = 0.011$) and the malate content of juice ($p = 0.040$). Juice was less acid with T3 treatment (pH = 5.18 ± 0.63), than treatments T2 (pH = 5.0 ± 0.6), T1 (pH = 4.98 ± 0.51), T4 (pH = 4.95 ± 0.55) and T5 (pH = 4.93 ± 0.63) (Figure 2C). The malate content was higher with the treatments T4 (43.5 ± 8.9 mg·mL^{-1}) and T3 (40.8 ± 10.4 mg·mL^{-1}), followed by treatments T1 (37.1 ± 6.9 mg·mL^{-1}) and T2 (36.4 ± 4.8 mg·mL^{-1}) (Figure 2D). Moreover, mulching and/or mineral fertilizer were not found to have significant effects ($p > 0.05$) on sucrose, raffinose, oxalate, proprionate and formic acid.

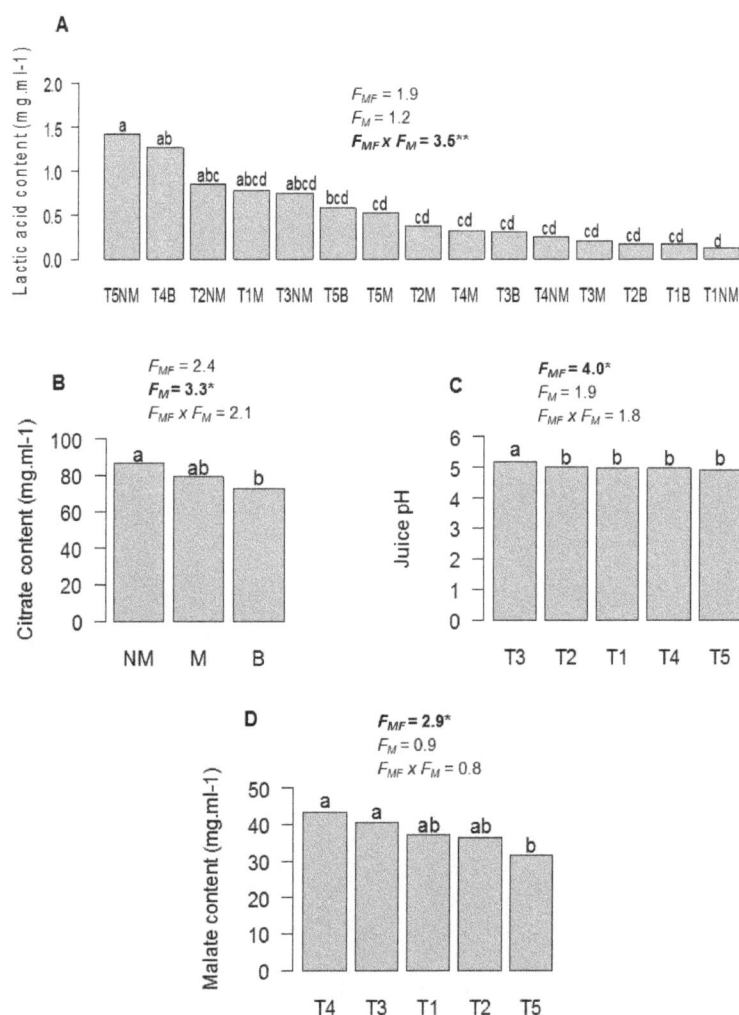

Figure 2. Influence of mulching and N-K fertilization on pH (**C**) and organic acid (**A,B,D**) content of pineapple fresh fruit juice. * = significant, ** = highly significant, F_{MF} = Fisher value for mineral fertilizer, F_M = Fisher value for mulching, $F_{MF} \times F_M$ = Fisher value for mineral fertilizer and mulching interaction. T1 = 1.6 N and 1.6 K, T2 = 5.8 N and 6.6 K, T3 = 10 N and 11.6 K, T4 = 1.6 N and 11.6 K, T5 = 10 N and 1.6 K (g·plant^{-1}). B = Burying, M = Mulching, NM = No mulching. Means that do not share a letter are significantly different.

Mulching and/or mineral fertilizer significantly influenced fruit quality attributes and proportion of fruit for exportation (Figures 3 and 4). Mulching had significant effects on infructescence length ($p = 0.008$) and fruit weight ($p = 0.002$). Infructescence length was higher with burying (15.81 ± 0.51 cm), followed respectively by mulching (14.95 ± 0.63 cm) and no mulching (13.96 ± 0.58 cm) (Figure 3A). Meanwhile, fruit weight was higher with burying (1.36 ± 0.07 kg·fruit^{-1}) and mulching (1.25 ± 0.09 kg·fruit^{-1}), followed by no mulching (1.11 ± 0.08 kg·fruit^{-1}) (Figure 3B). Infructescence length, fruit weight and the ratio of crown length: fruit length was significantly influenced by N-K fertilization ($p = 0.000$, $p = 0.000$ and $p = 0.000$ respectively). The highest infructescence length and fruit weight were obtained with T3 treatment (17.03 ± 0.52 cm and 1.57 ± 0.08 kg·plant^{-1} respectively), followed first by T4 (14.97 ± 0.64 cm and 1.28 ± 0.1 kg·plant^{-1}) and T2 (14.93 ± 0.44 cm and 1.25 ± 0.08 kg·fruit^{-1}), and then T5 (14.48 ± 0.92 and 1.11 ± 0.1 kg·fruit^{-1}) and T1 (13.11 ± 0.76 cm and 0.99 ± 0.01 kg·fruit^{-1}) (Figure 3C,D). The ratio of crown length: fruit length was higher with treatment T1 (0.61 ± 0.01), followed respectively by treatments T5 (0.59 ± 0.02), T2 (0.56 ± 0.01), T4 (0.55 ± 0.01) and T3 (0.53 ± 0.01) (Figure 3E).

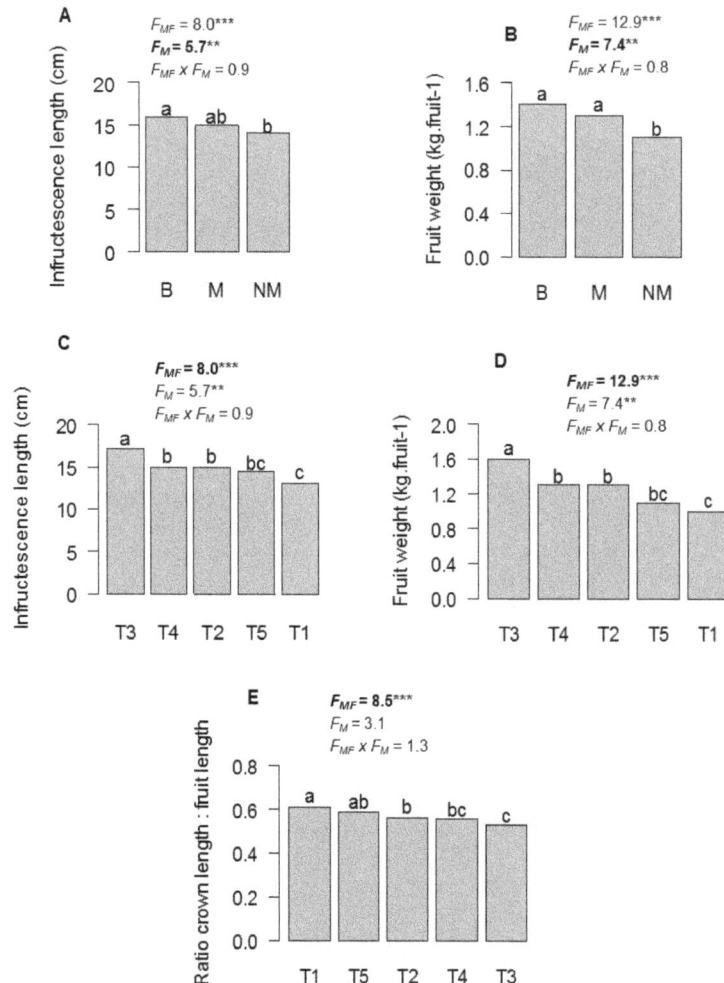

Figure 3. Influence of mulching and N-K fertilization on infructescence length (**A,C**), fruit weight (**B,D**) and ratio crown length: fruit length (**E**) of pineapple. * = significant, *** = very highly significant. F_{MF} = Fisher value for mineral fertilizer, F_M = Fisher value for mulching, $F_{MF} \times F_M$ = Fisher value for mineral fertilizer and mulching interaction. T1 = 1.6 N and 1.6 K, T2 = 5.8 N and 6.6 K, T3 = 10 N and 11.6 K, T4 = 1.6 N and 11.6 K, T5 = 10 N and 1.6 K (g·plant^{-1}). B = Burying, M = Mulching, NM = No mulching. Means that do not share a letter are significantly different.

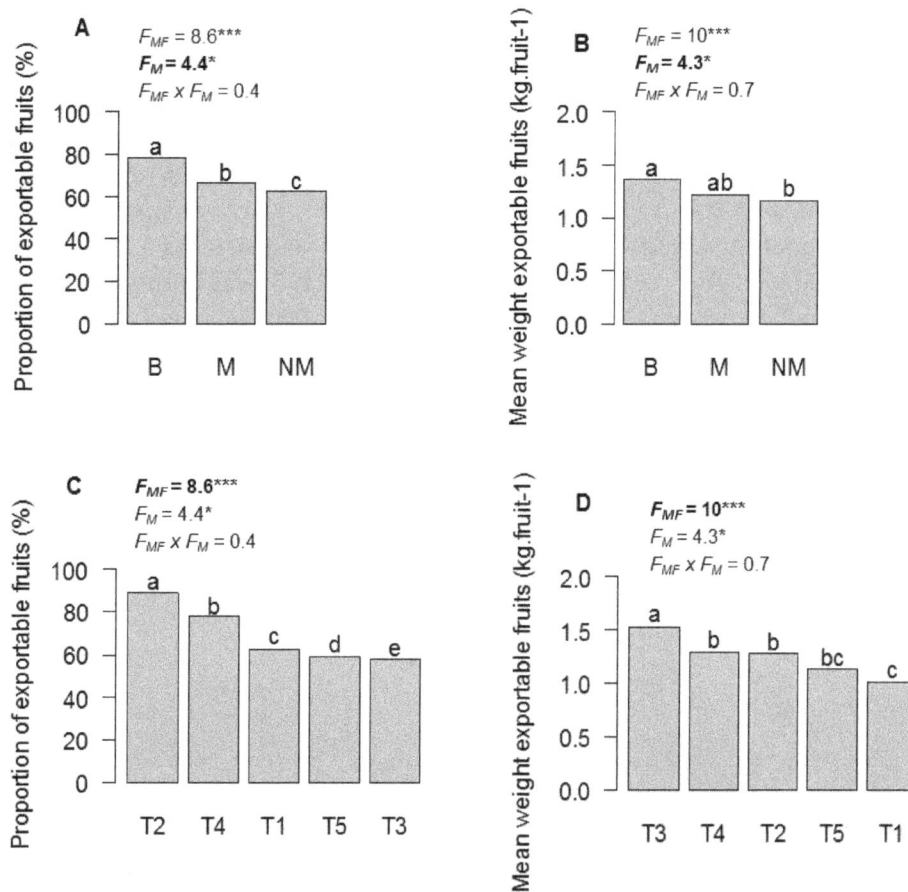

Figure 4. Influence of mulching and N-K fertilization on proportion of exportable fruits (**A,C**) and mean weight exportable fruits of pineapple (**B,D**). * = significant, *** = very highly significant. F_{MF} = Fisher value for mineral fertilizer, F_M = Fisher value for mulching, $F_{MF} \times F_M$ = Fisher value for mineral fertilizer and mulching interaction. T1 = 1.6 N and 1.6 K, T2 = 5.8 N and 6.6 K, T3 = 10 N and 11.6 K, T4 = 1.6 N and 11.6 K, T5 = 10 N and 1.6 K (g·plant^{-1}). B = Burying, M = Mulching, NM = No mulching. Means that do not share a letter are significantly different.

Mulching had a significant effect on the proportion of exportable fruits ($p = 0.02$) and the mean weight of exportable fruits ($p = 0.02$). The highest proportion of exportable fruits and the highest mean weight of exportable fruits were obtained with burying (78.67% $\pm$ 4.99% and 1.35 $\pm$ 0.07 kg·fruit^{-1}), followed respectively by mulching (66.67% $\pm$ 4.88% and 1.23 $\pm$ 0.08 kg·fruit^{-1}) and no mulching (62.22% $\pm$ 5.26% and 1.16 $\pm$ 0.07 kg·fruit^{-1}) (Figure 4A,B). N-K fertilization had significantly influenced the proportion of exportable fruits ($p = 0.000$) and the mean weight of exportable fruits ($p = 0.000$). The proportion of exportable fruits was higher with T2 treatment (88.88% $\pm$ 4.81%), followed respectively by T4 (77.77% $\pm$ 6.21%), T1 (62.59% $\pm$ 6.54%), T5 (59.26% $\pm$ 4.90%) and T3 (57.41% $\pm$ 5.63%) (Figure 4C). Treatment T3 (1.53 $\pm$ 0.06 kg·fruit^{-1}) induced the highest mean weight of exportable fruits, followed first by T4 (1.29 $\pm$ 0.09 kg·fruit^{-1}) and T2 (1.27 $\pm$ 0.09 kg·fruit^{-1}), and then T5 (1.13 $\pm$ 0.09 kg·fruit^{-1}) and T1 (1.02 $\pm$ 0.07 kg·fruit^{-1}) (Figure 4D). Finally, crown length had not been significantly ($p > 0.05$) influenced by mulching and/or mineral fertilizer. Therefore, the treatments T2 or T4 containing a lower quantity of fertilizer associated with buried residues could be advocated to farmers.

3.2. Influence of Mulching and N-K Fertilization on Pineapple Fruit Quality for Local Consumption

The principal component analysis performed on the sensorial and physico-chemical characteristics of fruits showed that the first two axes explained 62.7% of the total inertia. The first dimension was characterized by the variables juice yield ($r = 0.85$, $p = 0.000$), fruit weight ($r = 0.83$, $p = 0.000$), sweet taste ($r = 0.71$, $p = 0.003$), glucose ($r = 0.70$, $p = 0.003$), TSS ($r = 0.66$, $p = 0.008$), pH ($r = 0.65$, $p = 0.008$), acceptability ($r = 0.64$, $p = 0.009$), and fructose ($r = 0.51$, $p = 0.049$), which were positively correlated to it; meanwhile the variables raffinose ($r = -0.58$, $p = 0.024$) and acid taste ($r = -0.76$, $p = 0.000$) were negatively correlated to it. The variable fructose ($r = 0.62$, $p = 0.013$) was positively correlated to the second dimension, while the variable aroma ($r = -0.79$, $p = 0.000$) was negatively correlated to it (Figure 5).

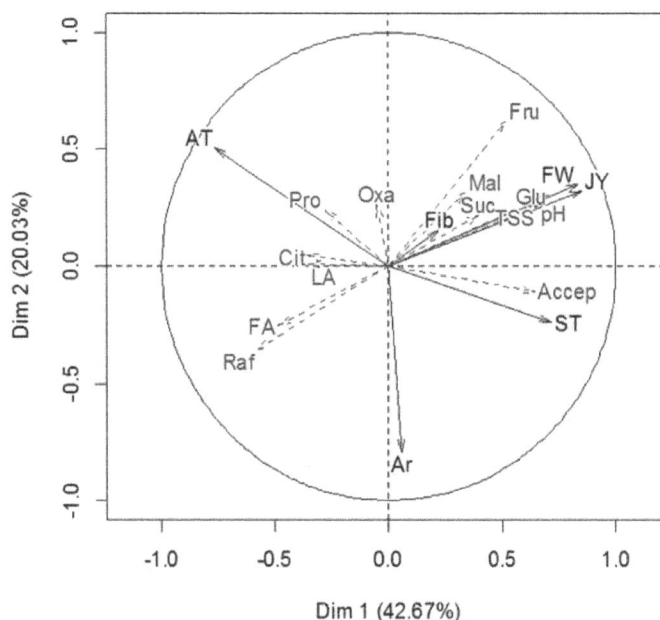

Figure 5. Correlation between physico-chemical and organoleptic characteristics of pineapple fruit and the PCA of the first two dimensions. Raf = Raffinose, FA = Formic Acid, Latic Acid, Pro = Proprionate, Oxa = Oxalate, Fib = Fiber, Fru = Fructose, Mal = Malate, Suc = Sucrose, Glu = Glucose, FW = Fruit Weight, TSS = Total Soluble Solids, Accep = Acceptability, ST = Sweet taste, Ar = Aroma.

The fertilizer treatments were grouped into four classes (Figure 6). The first class was represented by T2B, T3B, T3M, T3NM and T4B, of which fruits were heavier (V.test > 2, $p < 0.01$) and high yielding in juice (V.test > 2, $p < 0.01$), with high pH (V.test > 2, $p < 0.05$), glucose (V.test > 2, $p < 0.05$), TSS (V.test > 2, $p < 0.05$) and low raffinose (V.test < −2, $p < 0.05$). Among these interesting treatments, T2B or T4B, containing lower fertilizer quantities and buried residues, could be used for local-oriented production. The second class, represented by treatments T1NM, T1B, T2NM, T2M, T4NM and T5M, was defined by fruits having a very high content of raffinose (V.test > 2, $p < 0.05$) and very low juice yield (V.test < −2, $p < 0.05$). The third class included T4M, whose fruits were very little flavored (V.test < −2, $p < 0.01$). The treatments T1M, T5NM and T5B belonged to the fourth class, where fruits had a very acidic taste (V.test > 2, $p < 0.05$), were very weakly sweet (V.test < −2, $p < 0.01$), and contained low sucrose levels (V.test < −2, $p < 0.05$) and low TSS (V.test < −2, $p < 0.05$) (Table 1).

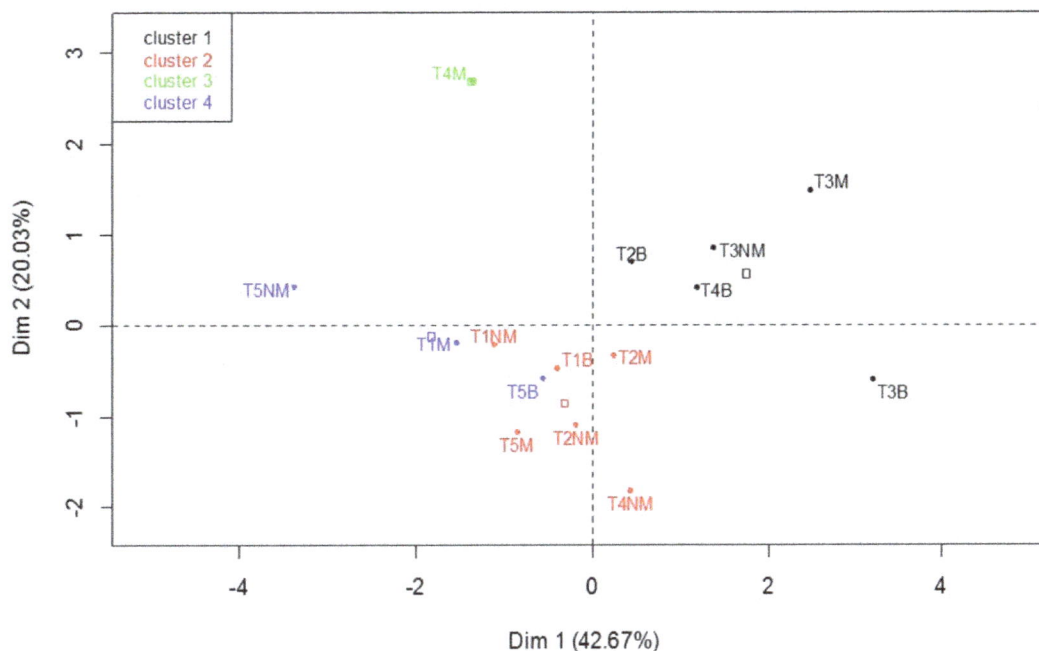

Figure 6. Projection of treatment classes according to physico-chemical and organoleptic parameters of pineapple fruit in the PCA factorial design. T1B = 1.6 N + 1.6 K + Burying, T1M = 1.6 N + 1.6 K + Mulching, T1NM = 1.6 N + 1.6 K + No mulching; T2B = 5.8 N + 6.6 K + Burying, T2M = 5.8 N + 6.6 K + Mulching, T2NM = 5.8 N + 6.6 K + No mulching; T3B = 10 N + 11.6 K + Burying, T3M = 10 N + 11.6 K + Mulching, T3NM = 10 N + 11.6 K + No mulching; T4B = 1.6 N + 11.6 K + Burying, T4M = 1.6 N + 11.6 K + Mulching, T4NM = 1.6 N + 11.6 K + No mulching; T5B = 10 N + 1.6 K + Burying, T5M = 10 N + 1.6 K + Mulching, T5NM = 10 N + 1.6 K + No mulching.

Table 1. Discriminant variables of treatment classes for physico-chemical and organoleptic quality of pineapple fruit.

Characteristics	Classes	V. test	Mean	Probability
Juice yield		3.15	0.69 ± 0.09	0.002
Fruit Weight		3	1.43 ± 0.17	0.003
pH	1	2.42	4.27 ± 0.08	0.015
Glucose		2.36	14.91 ± 1.45	0.018
TSS		2.08	16.29 ± 0.53	0.037
Raffinose		-2.29	0.2 ± 0.01	0.022
Raffinose	2	2.35	0.27 ± 0.04	0.02
Juice volume		-2.06	0.46 ± 0.06	0.04
Aroma	3	-2.77	1.64 ± 0.00	0.00
Acid taste		2.05	1.61 ± 0.13	0.04
Sucrose	4	-2.14	74.81 ± 6.04	0.03
TSS		-2.39	14.33 ± 0.75	0.01
Sweet taste		-3.00	3.09 ± 0.27	0.00

V.test = V - computed from the V.test in comparison of the mean value of the sensorial parameter for a given variable to the overall mean. A V - computed greater than 2 means the class had significantly ($p < 0.05$) greater value for the given variable. But V - computed lower than 2 implies that the class had significantly ($p < 0.05$) lower value for the given variable.

3.3. Relation Between Fruit Morphological and Physico-Chemical Properties

Linear regressions between morphological and physico-chemical characteristics of fruit revealed significant relationships. It was shown that pH, total soluble solids (TSS) ($p < 0.05$) and fruit weight ($p < 0.001$) decreased linearly as the ratios of crown length: fruit length, crown length: infructescence length and crown length: median diameter increased (Figure 7). This implies that the higher these ratios are, the weaker the fruit quality becomes. Therefore, these ratios have been revealed to all be useful for selecting fruit for exportation. An increase of fruit weight was significantly associated with a linear increase of fructose, glucose, sucrose ($p < 0.01$), TSS ($p < 0.01$), and pH ($p < 0.01$). Conversely, raffinose ($p < 0.01$) content decreased linearly as fruit weight increased (Figure 8).

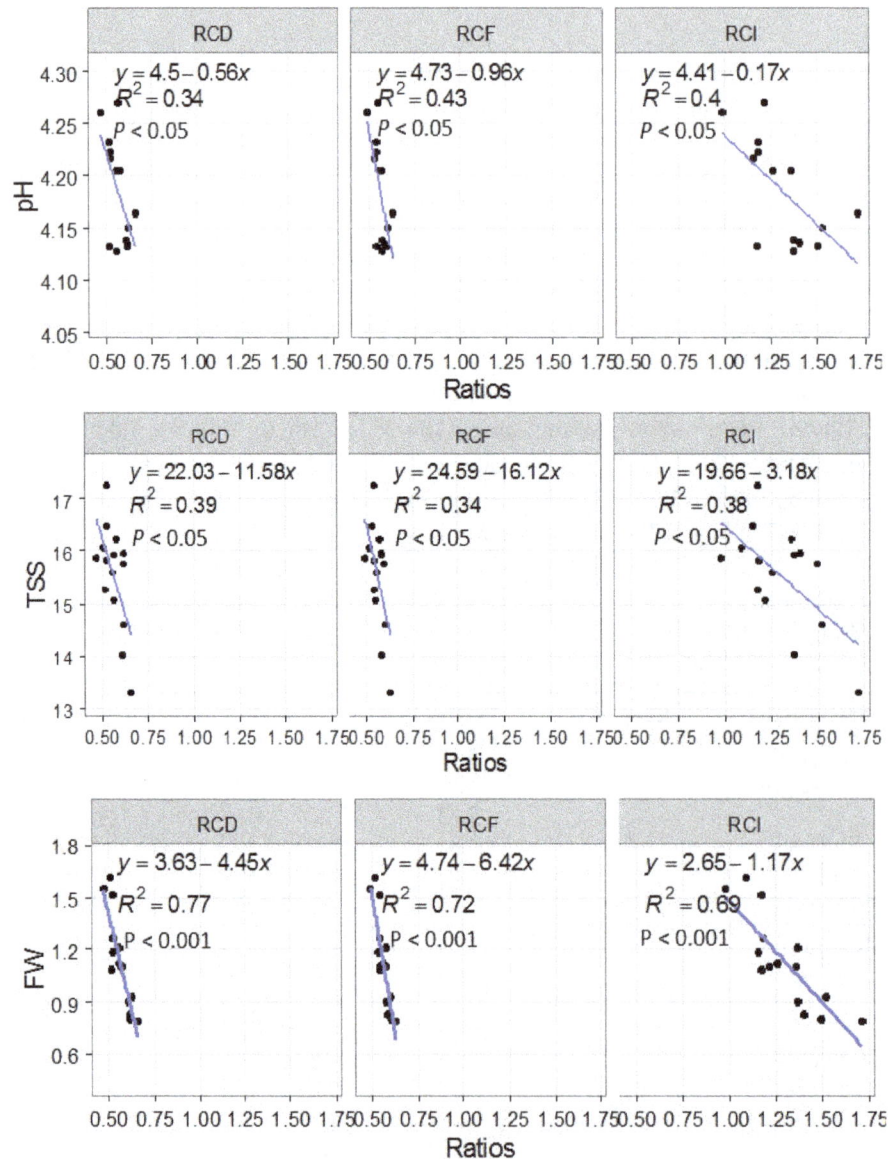

Figure 7. Relationship between pH, TSS, fruit weight and fruit characteristics ratios. RCD = Ration crown length/median diameter, RCF = Ration crown length/fruit length, RCI = Ration crown length/infructescence, TSS = total soluble solids ($°$ Brix), FW = fruit Weight (kg·fruit^{-1}), P = Probability, R^2 = Coefficient of determination.

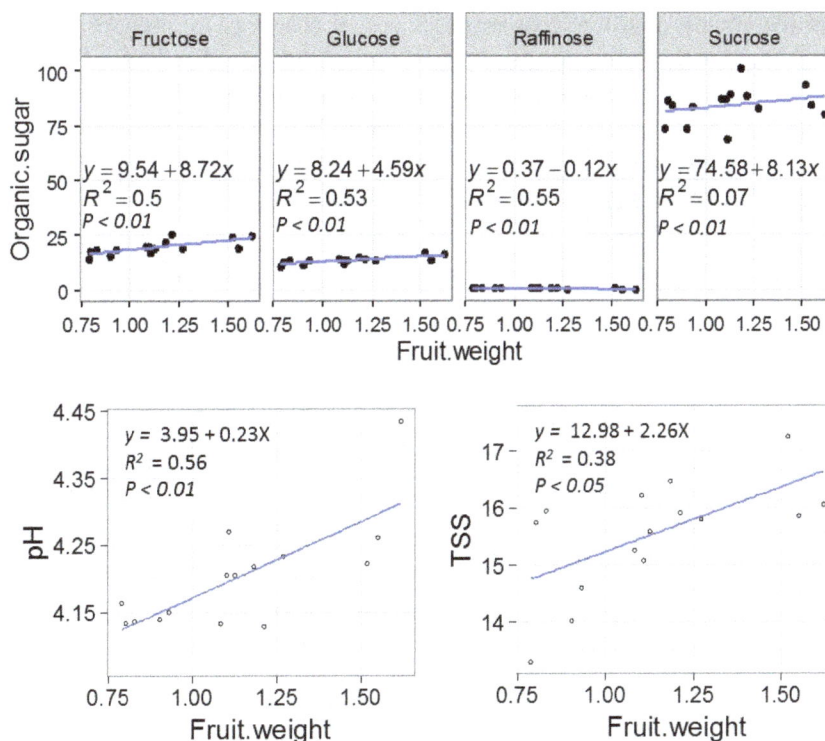

Figure 8. Relationship between fruit weight and physico-chemical characteristics. TSS = total soluble solids (° Brix), P = Probability, R^2 = Coefficient of determination.

4. Discussion

4.1. Influence of Mulching and N-K Fertilization on Pineapple Fresh Fruit Juice Quality and Proportion of Fruit Meeting European Standards and Local Acceptability

Mulching associated with N-K fertilization had influenced positively the quality of sugarloaf pineapple and, as a result, had improved the percentage of fruit meeting European standards and standards for local consumption. Mineral and/or organic fertilizer effects on fruit quality attributes has been reported by some authors [12,14–17,32,38], but the impact of pineapple residues on fruit quality was not explored. Likewise, information about fertilization influence on organic sugars and acids had not yet been reported. Plant vigor at flowering induction time ensures fruit quality (fruit weight and length attributes) [39]. However, cultural practices that guarantee a plant vigor before flowering induction need to receive considerable attention.

Our results suggested that burying or mulching pineapple residues enhanced TSS, fruit weight, infructescence length, proportion of exportable fruits, and mean weight of exportable fruits while decreasing citrate content. These findings were consistent with the works of [17] and [12], showing the highest TSS and lowest titrable acidity respectively with application of M. puriens green manure and dry grass of panicum maximum. The residues effects were improved with burying. Buried residues associated with N-K fertilization had increased fruit characteristics for local consumption. Pineapple residues burying thus remains an alternative cultivation practice that could reduce N and K chemical fertilizer use. This can be explained by the rapid decomposition of residues in burying [40], which release soil nutrients including potassium and nitrogen [41]. An increase of 39.6% of total sugar content of pineapple fruit by the addition of compost pineapple residue to topsoil, compared to no application of compost, was observed [16].

N-K fertilization also enhanced TSS, glucose, fructose and malate contents, as well as pH, fruit weight, infructescence length, proportion of exportable fruits and mean weight of exportable fruit. High values were globally obtained with treatments which associated higher nitrogen and potassium doses. These results are comparable to those of [32], who observed an increase in pH of pineapple "Smooth Cayenne" under the influence of nitrogen and potassium. Similarly, [11] and [38] noted a reduction in juice acidity under the influence of increasing nitrogen rates. However, [42] reported that potassium improves fruit quality through increasing the acidity and flesh firmness, increasing sugar content and flavor of fruit with the expansion of stalk diameter.

Our findings suggested that the best mulching and N-K fertilization for satisfying fruit characteristics for local consumption did not also lead to all fruit meeting exportation standards. Consequently, advocating optimum rates of mulching and N-K fertilization to farmers should be done with respect to the destination of the production.

4.2. Fruit Morphological Characteristics Were Correlated with Physico-Chemical Properties

Interestingly, our results suggested that the ratios of RCF (ratio crown length: fruit length), RCI (ratio crown length: infructescence length) and RCD (ratio crown length: median diameter) were significantly associated with fruit quality, which had not been previously reported. The optimum range for RCF for pineapple meeting export standards is 50%–150% [21]. RCI and RCD also remain relevant to estimate fruit quality attributes. According to these ratios, fruits with too long of crowns cannot be of good physico-chemical quality. Longer crowns can accumulate more nutrients than necessary for infructescence growth and quality. Dry-matter accumulation in the crown was 1.2%–2.5% and the accumulation of macronutrients ranged from 0.30%–1.29% for nitrogen, 0.06%–0.22% for phosphorus and 0.45%–2.68% for potassium, depending on cultivars [43]. In this work, heavier fruit tended to have interesting organic sugars content, with higher pH and total solids. Therefore, fruit weight is an indicator of fruit quality while it is grown under mulching and N-K fertilization.

5. Conclusions

The study revealed that mulching and mineral fertilization positively influenced fruit physico-chemical and organoleptic characteristics for sugarloaf pineapple, and consequently enhanced the percentage of fruit meeting European standards and the standards of acceptability for local consumption. Buried pineapple residues or surface mulching enhanced TSS, fruit weight, infructescence length, proportion of exportable fruits and mean weight of exportable fruits, while decreasing citrate content. N-K fertilization also enhanced TSS, glucose, fructose and malate contents, pH, fruit weight, infructescence length, proportion of exportable fruits and mean weight of exportable fruit. Interesting treatments such as T2B or T4B, containing a lower quantity of fertilizer associated with buried residues, could be advocated to farmers for exportation and local-oriented production. Finally, the ratios RCF (ratio crown length: fruit length), RCI (ratio crown length: infructescence length), and RCD (ratio crown length: median diameter) were correlated to fruit quality, which has not been previously reported.

Acknowledgments: This experiment was supported by the Competitive Fund Program for Researches of Abomey-Calavi University (PFCR/UAC).

Author Contributions: Elvire L. Sossa, Codjo E. Agbangba and Sènan G. G. S. Accalogoun performed the research design, data collection, data analysis and wrote the manuscript. Guillaume L. Amadji, Kossi E. Agbossou and Djidjoho J. Hounhouigan performed the whole write up and contributed significantly to manuscript improvement.

Conflicts of Interest: The authors declare no conflict of interest.

References

1.   Vagneron, I.; Faure, G.; Loeillet, D. Is there a pilot in the chain? Identifying the key drivers of change in the fresh pineapple sector. *Food Policy* **2009**, *34*, 437–446. [CrossRef]

2.   Calderon, M.M.; Graü, M.A.R.; Belloso, O.M. Mechanical and chemical properties of Gold cultivar pineapple flesh (*Ananas comosus*). *Eur. Food Res. Technol.* **2009**, *230*, 675–686. [CrossRef]

3.	Hounhouigan, M.H.; Lineman, A.R.; Ingenbleek, P.T.M.; Soumanou, M.M.; van Trijp, H.C.M.; van Boekel, M.A.J.S. Effect of physical damage and storage of pineapple fruits on their suitability for juice production. *J. Food Qual.* **2014**, *37*, 268–273. [CrossRef]

4.	INSAE (Institut National de la Statistique et de l'Analyse Economique). *Dynamique des Filières D'exportation au Bénin*; Ministère du Plan: Cotonou, Benin, 2009; p. 50.

5.	Fassinou, H.V.N.; Lommen, W.; van der Vorst, J.; Agbossou, K.E.; Struik, P. Analysis of pineapple production systems in Benin. *Acta Hortic.* **2012**, *928*, 47–58. [CrossRef]

6.	Fassinou, H.V.N.; Lommen, W.J.M.; Agbossou, K.E.; Struik, P.C. Influence of weight and type of planting material on fruit quality and its heterogeneity in pineapple (*Ananas comosus* (L.) Merrill). *Front. Plant Sci.* **2015**, *5*, 1–16. [CrossRef] [PubMed]

7.	Djalalou-Dine, A.A.A. Marketing channel selection by smallholder farmers in pineapple supply chain in Benin. *J. Food Prod. Mark.* **2015**, *21*, 337–357.

8.	Agbangba, C.E.; Olodo, G.P.; Dagbenonbakin, D.G.; Akpo, L.E.; Sokpon, N. Preliminary DRIS Model parametrization to access pineapple variety 'Perola' nutrient status in Benin (West Africa). *Afr. J. Agric. Res.* **2011**, *6*, 5841–5847.

9.	Malezieux, E.; Bartholomew, D.P. Plant nutrition. In *The Pineapple: Botany, Production and Uses*; Bartholomew, D.P., Paull, R.E., Rohrbach, K.G., Eds.; CABI Publishing: New York, NY, USA, 2003; pp. 143–165.

10.	Soares, A.G.; Trugo, L.C.; Botrel, N.; Sousa, L.F.S. Reduction of internal browning of pineapple fruit (*Ananas comusus* L.) by preharvest soil application of potassium Postharvest. *Biol. Technol.* **2005**, *35*, 201–207. [CrossRef]

11.	Spironello, A.; Quaggio, J.A.; Teixeira, L.A.J.; Furlani, P.R.; Sigrist, J.M.M. Pineapple Yield and Fruit Quality Affected By NPK Fertilization in a Tropical Soil. *Rev. Bras. Frutic.* **2004**, *26*, 155–159. [CrossRef]

12.	Darnaudery, M.; Fournier, P.; Léchaudel, M. Low-input pineapple crops with high quality fruit: Promising impacts of locally integrated and organic fertilization compared to chemical fertilizers. *Exp. Agric.* **2016**, *52*, 1–17. [CrossRef]

13.	Zougmore, R.; Ouattara, K.; Mando, A.; Ouattara, B. Rôle des nutriments dans le succès des techniques de conservation des eaux et des sols (cordons pierreux, bandes enherbées, zaï et demi-lune) au Burkina Faso. *Sécheresse* **2004**, *15*, 41–48.

14.	Amorim, A.V.; Garruti, D.S.; de Lacerda, C.F.; Moura, C.F.H.; Filho, E.G. Postharvest and sensory quality of pineapples grown under micronutrients doses and two types of mulching. *Afr. J. Agric. Res.* **2013**, *8*, 2240–2248.

15.	Liu, C.; Liu, Y.; Yi, G. Effects of Film Mulching on Aroma Components of Pineapple Fruits. *J. Agric. Sci.* **2011**, *3*, 196–201. [CrossRef]

16.	Liu, C.H.; Liu, Y.; Fan, C.; Kuang, S.Z. The effects of composted pineapple residue return on soil properties and the growth and yield of pineapple. *J. Soil Sci. Plant Nutr.* **2013**, *13*, 433–444. [CrossRef]

17.	Norman, J.C. Effects of mulching and nitrogen fertilization on "sugarloaf" pineapple, *Ananas comosus* (L.) Merr. *Den Trop. Subtrop.* **1986**, *87*, 47–53.

18.	Ahmed, O.H.; Husni, M.H.A. Exploring the nature of the relationships among total, extractable and solution phosphorus in cultivated organic soils. *Int. J. Agric. Res.* **2010**, *5*, 746–756.

19.	Heard, J.; Cavers, C.; Adrian, G. Up in Smoke—Nutrient Loss with Straw Burning. *Better Crops* **2006**, *90*, 10–11.

20.	Sossa, E.L.; Amadji, G.L.; Vissoh, P.V.; Hounsou, B.M.; Agbossou, K.E.; Hounhouigan, D.J. Caractérisation des systèmes de culture d'ananas (*Ananas comosus* (L.) Merrill) sur le plateau d'Allada au Sud-Bénin. *Int. J. Chem. Biol. Sci.* **2014**, *8*, 1030–1038. [CrossRef]

21.	Codex Alimentarius. *Codex Standard for Pineapples*; CODEX STAN 182-1993 2005; FAO-WHO: Rome, Italy; p. 5.

22.	Sossa, E.L.; Amadji, G.L.; Aholoukpè, N.S.H.; Hounsou, B.M.; Agbossou, K.E.; Hounhouigan, D.J. Change in a ferralsol physico-chemical properties under pineapple cropping system in southern of Benin. *J. Appl. Biosci.* **2015**, *91*, 8559–8569. [CrossRef]

23.	Scohier, P.; Texido, R.A. *Agriculture en Afrique Tropical*; Raemaekers, H.R., Ed.; Direction Générale de la Coopération Internationale: Ministère des Affaires Etrangères, du Commerce Exterieur et de la Cooperation Internationale: Brussels, Belgique, 2001; p. 1634.

24. Soler, A. *L'ananas: Critères de Qualité*; CIRAD-IRFA: Paris, France, 1992; p. 48.

25. Mestres, C.; Rouau, X. Influence of natural fermentation and drying conditions on the physicochemical characteristics of cassava starch. *J. Sci. Food Agric.* **1997**, *74*, 147–155. [CrossRef]

26. Watts, B.M.; Ylimaki, G.L.; Jeffery, L.E.; Elias, L.G. *Méthodes de Base pour l'Evaluation Sensorielle des Aliments*; CRDI: Ottawa, ON, Canada, 1991; p. 159.

27. R Core Team. *A Language and Environment for Statistical Computing*; R Foundation for Statistical Computing: Vienna, Austria, 2015. Available online: https://www.R-project.org/ (accessed on 15 January 2015).

28. Fernandez, G.C. Residual analysis and data transformations: Important tools in statistical analysis. *HortScience* **1992**, *27*, 297–300.

29. Shapiro, S.S.; Wilk, M.B. An analysis of variance test for normality (complete samples). *Biometrika* **1965**, *52*, 591–611. [CrossRef]

30. Snedecor, G.W.; Cochran, W.G. *Statistical Methods Applied to Experiments in Agriculture and Biology*; The Iowa State College Press: Ames, LA, USA, 1956; p. 250.

31. Gomez, K.A.; Gomez, A.A. *Statistical Procedures for Agricultural Research*, 2nd ed.; John Wiley and Sons: New York, NY, USA, 1984; p. 300.

32. Agbangba, C.E.; Dagbenonbakin, G.D.; Djogbénou, C.P.; Houssou, P.; Assea, D.E.; Sossa, E.L.; Kotomalè, U.A.; Ahotonou, P.; Ndiaga, C.; Akpo, L.E. Influence de la fertilisation minérale sur la qualité physico-chimique et organoleptique du jus d'ananas Cayenne lisse au Bénin. *Int. J. Biol. Chem. Sci.* **2015**, *9*, 1277–1288. [CrossRef]

33. Husson, F.; Lê, S.; Pagês, G. *Exploratory Multivariate Analysis by Exemple Using R*; CRC Press, Taylor and Françis Group: New York, NY, USA, 2010; p. 205.

34. Morineau, A. Note sur la Caractérisation Statistique d'une Classe et les Valeurs-tests. *Bull. Tech. Cent. Stat. d'Inform. Appl.* **1984**, *2*, 20–27.

35. Breusch, T.; Pagan, A. Simple test for heteroscedasticity and random coefficient variation. *Econom. Soc.* **1979**, *47*, 1287–1294. [CrossRef]

36. Durbin, J.; Watson, G.S. Testing for Serial Correlation in Least Squares Regression. *Biometrika* **1950**, *37*, 409–428. [CrossRef] [PubMed]

37. Ramsey, J.B. Tests for Specification Errors in Classical Linear Least Squares Regression Analysis. *J. R. Stat. Soc. Ser. B* **1969**, *31*, 350–371.

38. Omotoso, S.O.; Akinrinde, E.A. Effect of nitrogen fertilizer on some growth, yield and fruit quality parameters in pineapple (*Ananas comosus* L. Merr.) plant at Ado-Ekiti Southwestern, Nigeria. *Int. Res. J. Agric. Sci. Soil Sci.* **2013**, *3*, 11–16.

39. Fassinou, H.V.N.; Lommen, W.J.M.; Agbossou, E.K.; Struik, P.C. Heterogeneity in pineapple fruit quality within crops results from plant heterogeneity at flower induction. *Front. Plant Sci.* **2014**, *5*, 670.

40. Eusufzai, M.K.; Fujii, K.; Iiyama, I. Decomposition of Surface Applied and Buried Residue Biomass. *Geol. Soc. Am. Abstr. Programs* **2008**, *40*, 358.

41. Ching, H.Y.; Ahmed, O.H.; Kassim, S.; Ab Majid, N.M. Co-composting of pineapple leaves and chicken manure slurry. *Int. J. Recycl. Org. Waste Agric.* **2013**, *2*, 23. [CrossRef]

42. Teixeira, L.; Quaggio, J.; Cantarella, H.; Mellis, E. Potassium fertilization for pineapple: Effects on plant growth and fruit yield. *Rev. Bras. Frutic. Jaboticabal-SP* **2011**, *33*, 618–626. [CrossRef]

43. Hanafi, M.M.; Selamat, M.M.; Husni, M.H.A.; Adzemi, M.A. Dry matter and nutrient partitioning of selected pineapple cultivars grown on mineral and tropical peat soils. *Commun. Soil Sci. Plant Anal.* **2009**, *40*, 3263–3280. [CrossRef]

9

Development of a Statistical Crop Model to Explain the Relationship between Seed Yield and Phenotypic Diversity within the *Brassica napus* Genepool

Emma J. Bennett [1,†], Christopher J. Brignell [2,†], Pierre W. C. Carion [3], Samantha M. Cook [3], Peter J. Eastmond [3], Graham R. Teakle [4], John P. Hammond [5], Clare Love [6], Graham J. King [7], Jeremy A. Roberts [8] and Carol Wagstaff [1,*]

[1] Department of Food and Nutritional Sciences and Centre for Food Security, University of Reading, Whiteknights, PO Box 226, Reading, Berkshire RG6 6AP, UK; e.j.bennett@reading.ac.uk
[2] School of Mathematical Sciences, University of Nottingham, University Park, Nottingham NG7 2RD, UK; Chris.Brignell@nottingham.ac.uk
[3] Rothamsted Research, Harpenden, Hertfordshire AL5 2JQ, UK; pierre.carion@rothamsted.ac.uk (P.W.C.C.); sam.cook@rothamsted.ac.uk (S.M.C.); peter.eastmond@rothamsted.ac.uk (P.J.E.)
[4] Warwick Crop Centre, University of Warwick, Wellesbourne CV35 9EF, UK; graham.teakle@warwick.ac.uk
[5] School of Agriculture, Policy and Development and Centre for Food Security, University of Reading, Earley Gate, Whiteknights Road, PO Box 237, Reading RG6 6AR, UK; j.p.hammond@reading.ac.uk
[6] Rothamsted Research, Harpenden, Hertfordshire AL5 2JQ, UK; clare.love@mcri.edu.au
[7] Southern Cross Plant Science, Southern Cross University, PO Box 157, Lismore, NSW 2480, Australia; Graham.King@scu.edu.au
[8] Office of the Vice-Chancellor, 18 Portland Villas, University of Plymouth, Plymouth, Devon PL4 8AA, UK; jerry.roberts@plymouth.ac.uk
*   Correspondence: c.wagstaff@reading.ac.uk
†   These authors contributed equally to the work.

Academic Editor: Karin Krupinska

Abstract: Plants are extremely versatile organisms that respond to the environment in which they find themselves, but a large part of their development is under genetic regulation. The links between developmental parameters and yield are poorly understood in oilseed rape; understanding this relationship will help growers to predict their yields more accurately and breeders to focus on traits that may lead to yield improvements. To determine the relationship between seed yield and other agronomic traits, we investigated the natural variation that already exists with regards to resource allocation in 37 lines of the crop species *Brassica napus*. Over 130 different traits were assessed; they included seed yield parameters, seed composition, leaf mineral analysis, rates of pod and leaf senescence and plant architecture traits. A stepwise regression analysis was used to model statistically the measured traits with seed yield per plant. Above-ground biomass and protein content together accounted for 94.36% of the recorded variation. The primary raceme area, which was highly correlated with yield parameters (0.65), provides an early indicator of potential yield. The pod and leaf photosynthetic and senescence parameters measured had only a limited influence on seed yield and were not correlated with each other, indicating that reproductive development is not necessarily driving the senescence process within field-grown *B. napus*. Assessing the diversity that exists within the *B. napus* gene pool has highlighted architectural, seed and mineral composition traits that should be targeted in breeding programmes through the development of linked markers to improve crop yields.

Keywords: oilseed rape; statistical crop model; plant development; yield; seed composition

1. Introduction

Crop yield is a complex trait determined by a number of contributing environmental and genetic factors. Mathematical modelling approaches are one way of synthesising information and simplifying the complex interactions that exist throughout a plant's life cycle to gain information about the most appropriate target traits that could increase yields. The rationale is that, if used as early predictors of final yield, models have the potential to inform growers of alterations in farming practices that, if implemented early in the season, could increase yields. Models exist for the three major cereal crops grown worldwide (rice, wheat and maize) [1–3], and all aim to predict yield in the face of environmental or genetic variation. Crops such as oilseed rape (canola, rapeseed, colza) are of increasing economic importance, yet they have an indeterminate growth habit compared to cereals, having been domesticated for only 4000 years compared to the 10,000 years for wheat, and therefore, require a dedicated approach in order to determine yield components that are of use to growers and breeders. Several models of yield prediction in oilseed rape have been generated previously [4–8]. As with all models, they have limitations; for instance, only applying data from optimum growth conditions, using just one variety of oilseed rape and/or being solely based on parameters derived from the literature. Hence, whilst modelling oilseed rape growth provides a useful tool to evaluate the parameters affecting yield, to date, this technique is still in its infancy and has not had the power of the diversity trial used in the present study to inform the modelling approach.

Oilseed rape (*Brassica napus*) is a crop species primarily harvested for its oil-containing seeds within temperate regions and is globally ranked as the third leading source of plant oil after soybean (*Glycine max*) and oil palm (*Elaeis guineensis*) [9]. The oil is used in the food industry, but the seed and remaining biomass material also have a number of other roles, including use as a protein meal for the feed industry and as a feedstock for biofuels. Its success as an oilseed can be attributed to the fact that its seeds are composed of ~25% protein and ~50% oil (w/w); the oil having a desirable composition of almost entirely unsaturated fatty acids [10], of which the primary components are the unsaturated fatty acids oleic acid (C18:1), linoleic acid (C18:2) and linolenic acid (C18:3; [11]. *Brassica napus* (AACC) is an amphidiploid species arising from a spontaneous hybridisation between *B. rapa* (AA) and *B. oleracea* (CC) [12]. Compared to other crops, such as wheat, oilseed rape has only recently undergone a domestication event [13]; this is highlighted by the increase in its cultivation as a crop between 1961 and 2013 when there was a 481% global increase in the area of oilseed rape harvested, contrasting with a 7% increase in wheat over the same period [14]. This observation could be partly attributed to the fact that oilseed rape is often used as a break crop within a wheat rotation, yet within these few decades, seed quality has already been improved by reducing the concentration of erucic acid and glucosinolates (GSLs), which were believed to be anti-nutritional seed components [15]. This may have been achieved at a cost, as one study linked Quantitative Trait Loci (QTL) for seed yield in winter oilseed rape to high Glucosinolate (GSL) concentration, revealing that reducing the latter has the potential to negatively impact seed yield [16].

As crops become domesticated, there is strong selection against the natural variation in plant development and architecture within the population, leading to increasingly uniform monocultures that are suited to mechanised harvest at a single time point. In general, the growth habit, yield, pest/pathogen defence system and response to environmental conditions throughout the growing season largely depend on the genotype of the accession sown [17].

In order to maximise crop yield, a plant within a field-based canopy has to balance the extent of vegetative growth, the length of the photosynthetic period, the number of reproductive structures in which it invests and the amount of resource (protein, lipid, carbohydrate) that it imports into each seed [18]. It is also important that the plants within a canopy retain the adaptive capacity (plasticity) to respond to environmental changes and exploit any extra resources that may accrue during the growing season [19]. The optimal idiotype of oilseed rape is yet to be defined; the crop currently produces extensive above-ground vegetative biomass and hence has a low harvest index (the ratio between harvestable yield and total plant biomass). Our hypothesis is that plants with high yield in terms of seed mass per plant achieve this as a consequence of more pods per plant rather than a positive increase in resource accumulation within individual seeds or pods. Whilst the vegetative structure of the plant contributes towards increased pod production by providing more sites at which pods can be formed, it can also reduce the amount of incident light reaching photosynthetic components lower down in the canopy. Even the relatively small petals are believed to block out ~60% of the photosynthetically-active radiation (PAR) [20].

The timing of leaf senescence is not predetermined, but is influenced by both environmental [21] and genotypic factors [22,23], which will ultimately affect both the number of fruits that develop and the extent to which the seed will fill. However, the relationship between senescence and yield is a complex one and not fully understood [24], especially in oilseed rape. The role of the pod wall extends far beyond that of a protective organ [18,25], as the pods are themselves photosynthetic units that senesce, contributing 50%–60% of the final dry mass of the mature plant [26]. It has been noted that *Arabidopsis* accessions that naturally senesce early have a greater number of pods per plant compared to late-senescing accessions (although the nutritional composition and weight of these seeds were not stated), indicating a greater investment in reproductive as opposed to vegetative development [22,23]. Some *Arabidopsis* mutants that exhibit a delayed senescence phenotype, such as the abscisic acid insensitive (*abi3*) mutant, develop more pods per plant compared to the wild-type [27]. This delay was predicted to enhance yields by providing a longer photosynthetic period in which to synthesise photoassimilates for subsequent re-allocation into reproductive structures. Whilst this might increase carbon assimilation in the seed, nitrogen (N) remobilisation becomes delayed in late senescing *Arabidopsis* phenotypes [28], so while delaying senescence might result in higher yields, the trade-off is that the seeds contain less protein [17]. Hence, when choosing ideal traits for crops, there is confusion over whether maximum yield will be achieved from fast or slow senescing lines [29].

The diversity trials performed as part of the Oilseed RapE Genetic Improvement Network (OREGIN) project [30,31] sampled in the current study provided a robust basis to begin modelling, as they represent the genetic diversity that exists across the domesticated *B. napus* gene pool, therefore providing a useful insight into direct future breeding programmes. The present study developed statistical models of yield across two years that were able to elucidate the most important architectural and quality traits related to yield and that provide a predictive model of yield based on parameters that could be measured prior to pod formation and then acted on to maximize yield.

2. Results

2.1. Development of an Efficient Statistical Model to Explain Yield

Using data for 45 architectural and physiological traits measured in Year 1 of the field trial, forward selection was used to develop a deterministic statistical model to explain log[yield]. Data from only the varieties present in both years of the trial were included. Yield was defined as the weight of seeds harvested per plant. The varieties were grown under two nitrogen treatments in Year 1, but only in residual soil nitrogen in Year 2; residual N was found to be 18–22 kg ha^{-1} in both years, and high N in Year 1 was determined as 148 kg ha^{-1}. A two-factor ANOVA from Year 1 data to compare the significance of variety and nitrogen on yield showed that both terms are significant factors in predicting yield, with variety explaining 44.5% of the variance and nitrogen explaining 23.1% of the variance, respectively (Table S1). A comparison by ANOVA using variety and year terms showed that variety was significant (explained 37% of the variance), but year was not (explained 0.7% of the variance), thus validating our approach of comparing data across years. Variety was excluded as a significant term from the analysis in order to examine which traits consistently contribute to high or low yield across different accessions (Table 1). The model explained over 50% of the yield variance with the inclusion of five trait terms: chlorophyll a from canopy leaves, seed linolenic acid content, number of seeds per pod, the rate of N applied to the plot, glucosinolate content of the seeds. The effect of N was positive overall and confirmed that addition of N increases yield. However, N treatment (high/low) did not segregate the dataset into two distinct groups according to yield; some varieties grown on low N had higher yield than others grown on high nitrogen and vice versa (Figure 1a). A two-way ANOVA of log[yield] against variety and N showed that the size of the effect differs between varieties, and for most varieties in the present study, the effect of N on seed yield per plant is not significant (Figure 1b), although if more years/blocks were available, other varieties in the middle of the graph may also show a significant response.

Table 1. Statistical model to explain log[yield].

Coefficients	Estimate	Standard Error	t-Value	Pr (>|t|)	Significance	Cumulative % Variance Explained
(Intercept)	2.472	0.285	8.672	1.42×10^{-14}	***	-
Log(canopy leaf Chl a)	0.491	0.179	2.740	0.0070	**	28.81
Seed linolenic acid	−0.058	0.016	−3.635	0.0004	***	44.05
Seeds per pod	0.019	0.008	2.287	0.0238	*	48.70
Nitrogen applied to plot	0.329	0.088	3.722	0.0003	***	52.91
Log(seed glucosinolate)	−0.067	0.031	−2.200	0.0296	*	54.59

Forty-five traits were considered as candidates for terms in the model. Starting with no terms in the model, terms were added using forward selection (see the Materials and Methods). Traits that do not appear in the model did not significantly improve the estimation of log[yield]. The data modelled represented 35 varieties grown at residual N (~18 kg ha^{-1}) and high N (148 kg ha^{-1}) concentrations in duplicate plots in a randomised block design. Residual standard error: 0.3629 on 131 degrees of freedom; multiple R-squared: 0.5459; adjusted R-squared: 0.5285; significance codes: 0.001 '***'; 0.01 '**'; 0.05 '*'.

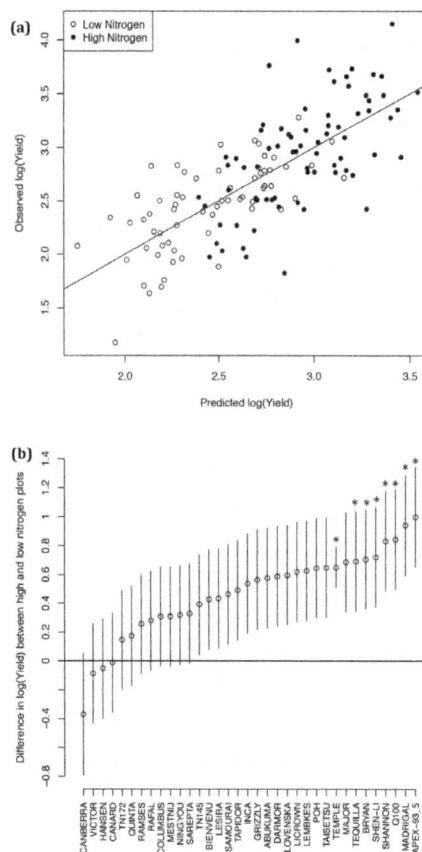

Figure 1. Responses of 35 varieties of *Brassica napus* to nitrogen fertilizers in Year 1. (**a**) Comparison of observed yield to that predicted by the statistical model for Year 1 data from low (18 kg ha^{-1}; open circle) and high (148 kg ha^{-1}; closed circle) nitrogen plots (R^2 = 0.5404). (**b**) Size of the nitrogen effect for each variety. Bars = SEM. Varieties for which the addition of nitrogen had a significant impact on yield are marked with *.

2.2. Genetic Factors Are Stronger Predictors of Yield than Environmental Factors within Year

In order to explore models that could predict yield independently of N application, and in preparation for characterising Year 2 data where only one N treatment was included in the trial, statistical models were developed independently for low and high N treatments (Figure 2 and Table 2). Both models performed slightly less well than the combined model (Figure 1a). Each shared some covariates with the combined model, with seeds per pod and seed linolenic acid common between the combined and low N models and log[seed glucosinolate] common between the combined and high N models. There were no common covariates between the low and high N models. Having seed compositional traits as core terms in both the combined and separated models, as opposed to the terms being entirely of growth and development traits, suggests that the combined model is applicable to a range of growing environments. The six terms in the low N model accounted for 52.67% of the variance in yield. The low N treatment demonstrated the importance of senescence when resources are limited; a higher percentage of green pods at the stages examined showed a positive and significant ($p < 0.05$) relationship with yield (Table 2). Variety was excluded as a candidate for this model; however, in a separate model, where only variety was considered, it explained 44.50% of the variation in yield (across all plots), rising to 60.92% in high N plots and 78.43% in low N plots. Therefore, although N does have an impact on yield, the genetic basis of yield (i.e., variety) is a much more powerful predictor.

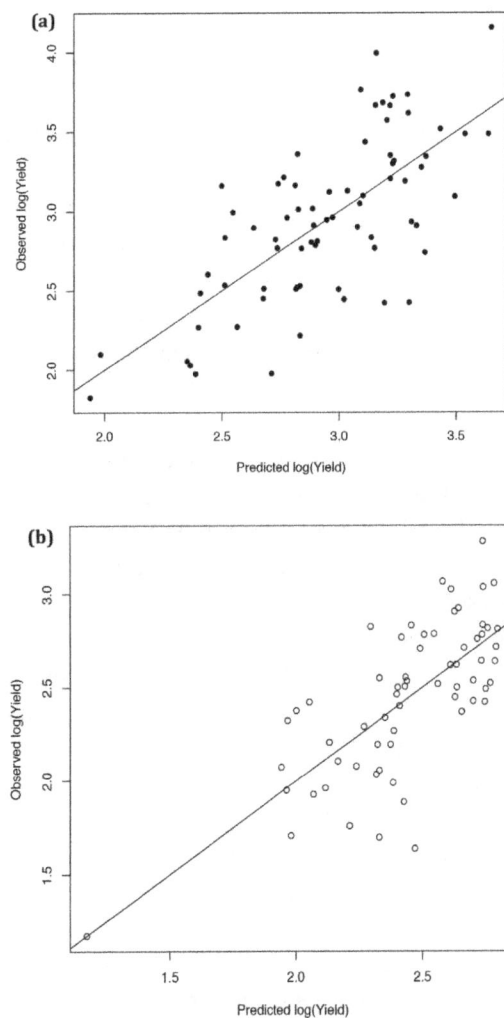

Figure 2. Comparison of observed yield of 35 varieties of *Brassica napus* to that predicted by the statistical model for Year 1 data using data from (**a**) high nitrogen (148 kg ha^{-1}) plots ($R^2 = 0.4922$) and (**b**) low nitrogen (18 kg ha^{-1}) plots ($R^2 = 0.5221$).

Table 2. Statistical model to explain log[yield] under high and low nitrogen.

High Nitrogen Coefficients	Estimate	Standard Error	t-Value	Pr (> |t|)	Significance	Cumulative % Variance Explained
(Intercept)	1.649	0.833	1.980	0.0518	.	-
Seed protein content	−0.051	0.018	−2.767	0.0073	**	33.77
Log[early leaf Mn]	0.632	0.223	2.836	0.0060	**	40.51
Pod dehiscence Low	0.672	0.282	2.379	0.0202	*	-
Pod dehiscence Medium	0.731	0.278	0.268	0.0107	*	46.17
Log[seed glucosinolate]	−0.108	0.054	−2.019	0.0475	*	49.26
Low Nitrogen Coefficients						
(Intercept)	2.113	0.342	6.180	7.28×10^{-8}	***	-
Seed linolenic acid	−0.033	0.021	−1.597	0.1159	-	35.59
Seeds per pod	0.028	0.009	2.995	0.0041	**	41.66
Pod development: Yellow stage	−0.276	0.243	−1.137	0.2602	-	-
Pod development: Yellow-green stage	−0.048	0.216	−0.222	0.8253	-	52.67

Forty-five traits were considered as candidates for terms in the model. Starting with no terms in the model, terms were added using forward selection (see Methods). Traits that do not appear in the model did not significantly improve the estimation of log[yield]. The data modelled represented 35 varieties grown at high N (148 kg ha^{-1}) and low (18 kg ha^{-1}) concentrations in duplicate plots in a randomised block design. Residual standard error: 0.3795 on 67° of freedom; multiple R-squared: 0.4926; adjusted R-squared: 0.4547; residual standard error: 0.2931 on 57° of freedom; multiple R-squared: 0.5267; adjusted R-squared: 0.4769; significance codes: 0.001 '***'; 0.01 '**'; 0.05 '*'; 0.1 '.'.

2.3. Developing an Improved Model to Explain Yield

Application of the Year 1 low N model to Year 2 data (the crop was only grown under low N in Year 2) showed that the model was poor at explaining yield, being able to only account for 25.48% of yield variation (Table 3, Figure 3a). This is half of the yield variation explained by the same model in Year 1 (52.67%) and demonstrates that within-year variation in yield is dominated by genetic factors, whilst between-year variation is strongly influenced by the environmental conditions to which the plants were exposed. However, although the estimates of variation for the terms used in the model are different between years, we see the same directional effects, i.e., there is a negative estimate of variation for seed linolenic acid and a positive estimate of variation for seeds per pod.

To develop the model further, a second forward selection process was performed to determine which traits best explained log[yield] in Year 2. Plants in Year 2 were grown on residual N only, and additional traits were recorded. As predicted, variety was highly significant ($p = 0.009$) in predicting log[yield], but as with Year 1, this was not included as a term in the following analyses in order to determine traits that contribute to the yield difference observed between varieties. Seed compositional traits, such as linolenic acid, total seed oil and seed protein content, were important factors; however, total vegetative mass (excluding seeds) was the most significant term, explaining 73.70% of log[yield] (Table 3, Figure 3b). Seed oil was shown to be a positive term, whereas protein content had a negative impact on yield, demonstrating that the oil:protein ratio is important in determining total yield. In total, the forward selection model accounted for 95.12% of the variation contributing to log[yield].

Given that total above-ground biomass at harvest was such an important predictor of log[yield], we investigated this trait further and used forward selection to identify the traits and variables that explained above-ground vegetative biomass at harvest (Table 3, Figure 3c). A consequence of including a trait in the model is that other traits correlated with it are less likely to be significant themselves and are not included in the model; hence, none of the terms explaining biomass appear in the yield model, as the biomass terms acts as a proxy for them all. The first three predictors of vegetative biomass were all architectural traits with positive effects, but these three traits only explain 77.74% of the vegetative biomass, suggesting that other traits listed in both models make a small, but significant contribution to yield. The first four terms in the biomass model (stem area, number of branches on the primary stem, plant height and pod carotenoid content at Week 41) were all measured pre-harvest, thus providing a means of predicting yield.

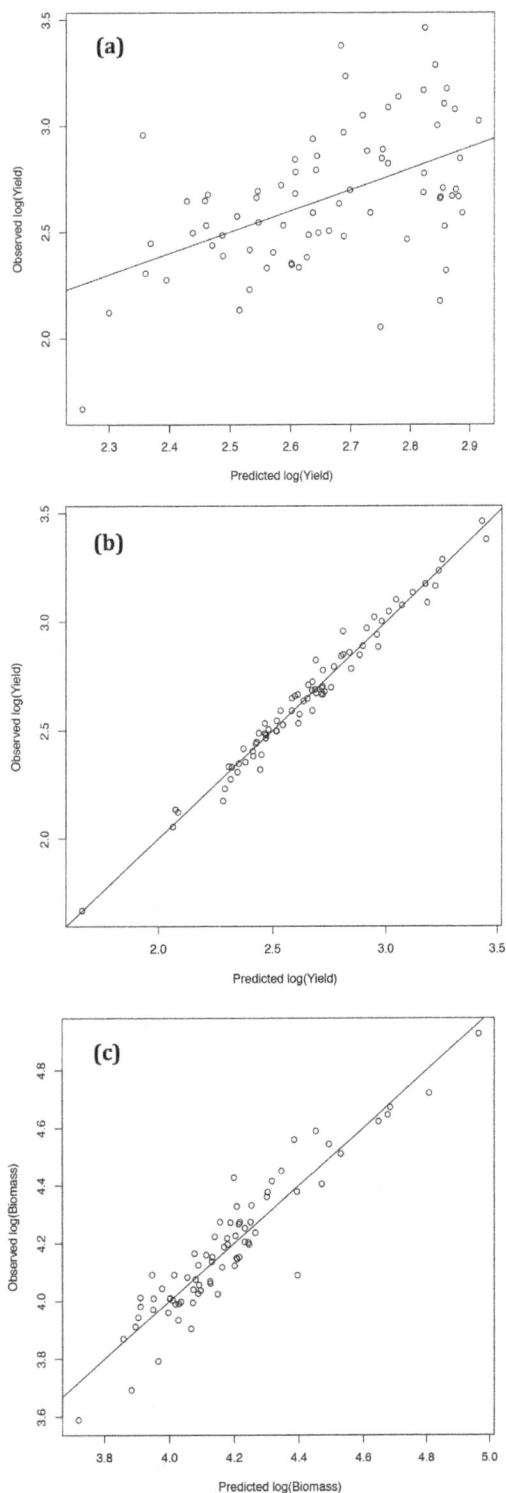

Figure 3. Models: (**a**) Comparing observed yield of *Brassica napus* in Year 2 to that predicted by the statistical model developed on Year 1 data using plants grown on low nitrogen (18 kg ha^{-1}) plots. $R^2 = 0.2548$. (**b**) Explaining yield in Year 2 using forward selection from 133 possible terms. The final model used nine terms to explain 95.12% of the yield variance using plants grown on low nitrogen (18 kg ha^{-1}) plots in a randomised block design. $R^2 = 0.951$. (**c**) Explaining vegetative biomass in Year 2 using forward selection from 133 possible terms. The final model used four terms to explain 80.04% of the biomass variance using plants grown on low nitrogen (18 kg ha^{-1}) plots in a randomised block design. $R^2 = 0.8004$.

Table 3. Statistical model to explain log[yield] using the Year 1 model on Year 2 data, the forward selection statistical model on Year 2 data and the forward selection statistical model to explain the log[biomass] term from Year 2 data.

Coefficients	Estimate	Standard Error	*t*-Value	Pr (> | *t* |)	Significance	Cumulative % Variance Explained
Year 1 Model On Year 2 Data						
(Intercept)	2.186	0.325	6.730	3.65×10^{-9}	***	-
Linolenic acid	−0.008	0.023	−0.357	0.7224	-	-
Seeds per pod	0.012	0.007	1.777	0.0798	.	-
Pod development: Yellow stage	0.192	0.103	1.863	0.0666	.	-
Pod development: Yellow-green stage	0.332	0.103	3.219	0.0019	**	-
Year 2 Model on Year 2 Data						
(Intercept)	−2.844	0.494	−5.757	2.42×10^{-7}	***	-
Log[Vegmass]	1.262	0.042	30.377	$<2 \times 10^{-16}$	***	73.70%
Total Protein	−0.019	0.010	−1.952	0.0551	.	89.16%
Canopy leaf Chlorophyll b	0.337	0.072	4.666	1.55×10^{-5}	***	91.41%
Total Lipid	0.015	0.006	2.736	0.0204	*	92.04%
Seed weight per pod	0.001	0.000	4.036	0.0001	***	92.80%
No. pods on primary raceme	0.007	0.001	4.427	3.67×10^{-5}	***	93.50%
Canopy leaf calcium	−0.125	0.029	−4.336	5.07×10^{-5}	***	94.40%
Linolenic acid	−0.021	0.008	−2.517	0.0143	*	94.81%
log(Pod Chlorophyll a content Wk43)	−0.024	0.012	−2.046	0.0448	*	95.12%
Log[Vegmass] Model on Year 2 Data						
(Intercept)	2.423	0.108	22.435	$<2 \times 10^{-16}$	***	-
Stem area	0.001	0.000	3.407	0.0011	**	47.16%
Log[No. branches on primary stem]	0.029	0.003	9.574	2.06×10^{-14}	***	59.93%
Height	0.005	0.001	7.373	2.42×10^{-10}	***	77.74%
Total branch No.	0.034	0.004	9.793	2.04×10^{-14}	***	75.63%
Pod carotenoid content (Wk41)	0.021	0.007	2.857	0.005	**	80.04%

The data modelled represented 35 varieties grown at residual N (~18 kg ha^{-1}) in duplicate plots in a randomised block design. Residual standard error: 0.2893 on 71° of freedom; multiple R-squared: 0.2548; adjusted R-squared: 0.2128; residual standard error: 0.07675 on 66° of freedom; multiple R-squared: 0.9512; adjusted R-squared: 0.9446; residual standard error: 0.102 on 71° of freedom; multiple R-squared: 0.8004; adjusted R-squared: 0.7891; significance codes: 0.001 '***'; 0.01 '**'; 0.05 '*'; 0.1 '.'.

2.4. Ten-Fold Cross-Validation of the Year 2 Models

In order to investigate how robust these models are to new data, we divided the plots from Year 2 at random into ten groups (folds). We then fitted the models (estimated the parameters) using nine of the folds and then applied the model to the remaining fold and measured the percentage of variance explained. We repeated this procedure for each of the ten folds in turn. Over the 10 folds, the percentage of variance explained by the yield model ranged from 83.28%–97.90%, with an average of 93.61%. The same approach was applied to the "Vegmass" model, where the average percentage of variance explained was 67.57%. As expected, there is a slight drop in the percentage of variation explained compared to the original model that used all of the available data, but the decrease is not substantial, suggesting that the models are robust to new data obtained from plants grown under similar environmental conditions.

2.5. Seed Oil Content Rather Than Protein Drives Seed Yield

A correlation matrix was used to visualise positive and negative relationships between different traits (Figure 4). The matrix showed (factors mentioned in the text below are boxed in blue in Figure 4) that the oil:protein ratio was correlated positively with seed number per pod (0.756) and negatively with seed packing density per pod (−0.702). This appeared to be driven by the lipid component, since there was a negative relationship between protein content and seed weight per pod (−0.627). This suggests that seed yield gain for an individual plant was driven by increased oil content. There was also a strong negative relationship between pigment (chlorophyll and carotenoid) content of the pod and protein/sugars (range from −0.239——0.579). This implies that senescence may drive the remobilisation of sugars and proteins into the developing seed and that retention of pigment in the pod during development does not necessarily imply that resources will be translocated into seeds at a later stage. Flowering window has a non-significant relationship with yield and biomass. The relationship between the majority of the leaf minerals analysed and yield was also non-significant, the exceptions being canopy leaf potassium, which was strongly negatively correlated with yield (-0.432), total biomass (-0.403) and vegetative biomass (-0.375) and early leaf phosphorus, which was positively correlated with yield (0.281), total biomass (0.384) and vegetative biomass (0.408).

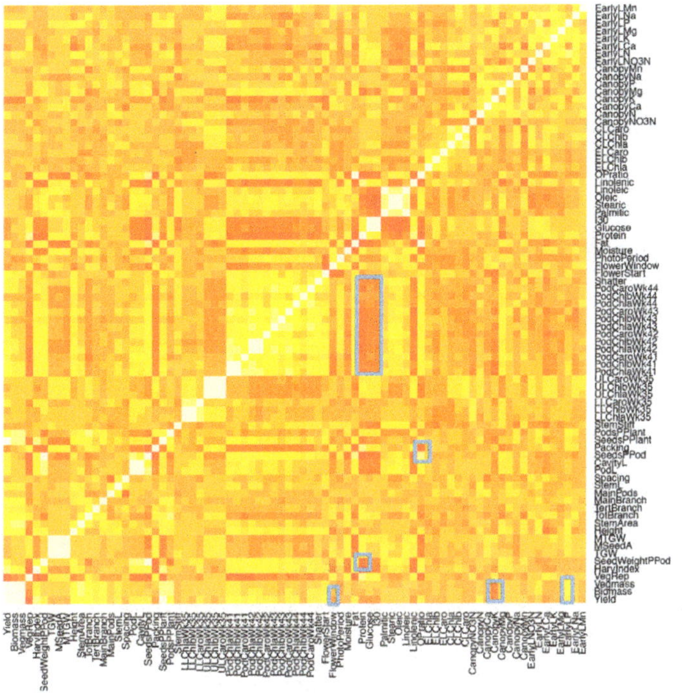

Figure 4. Correlation matrix showing the relationship between all traits measured from *Brassica napus* plants in Year 2. White squares = positive correlation of +1; red squares indicate a strong negative correlation. Note that where a measurement was taken on multiple occasions, some time points are not shown for clarity. Blue squares indicate regions of the matrix that are discussed in the main text.

Multi-Dimensional Scaling (MDS) was used to project the distances between traits in a schematic diagram (Figure 5). The correlation between each pair of traits was converted into a distance, where a correlation of ± 1 was given a distance of zero and a correlation of zero was given a distance of one. We then used MDS to show the best two-dimensional projection of these distances; traits that are positioned close together are highly correlated. Traits that the models (Table 3) deem useful for predicting yield and/or biomass are circled, and colour is used to group traits according to their type: yield, architecture, leaf chlorophyll, pod chlorophyll, progression of plant development, oil/protein content and minerals. Biomass, the most important predictor of yield, was located relatively close to the yield term, and biomass itself is positioned closely to the architectural traits nearby. The next two predictors from the linear yield model, seed protein and canopy leaf chlorophyll b content, are located further away from the yield term. Leaf mineral analysis makes no contribution to the yield models. Perhaps surprisingly, traits relating to floral development are not important for the yield model, so the date of first flower opening, photoperiod, flowering duration and time to pod shatter had no significant impact on yield.

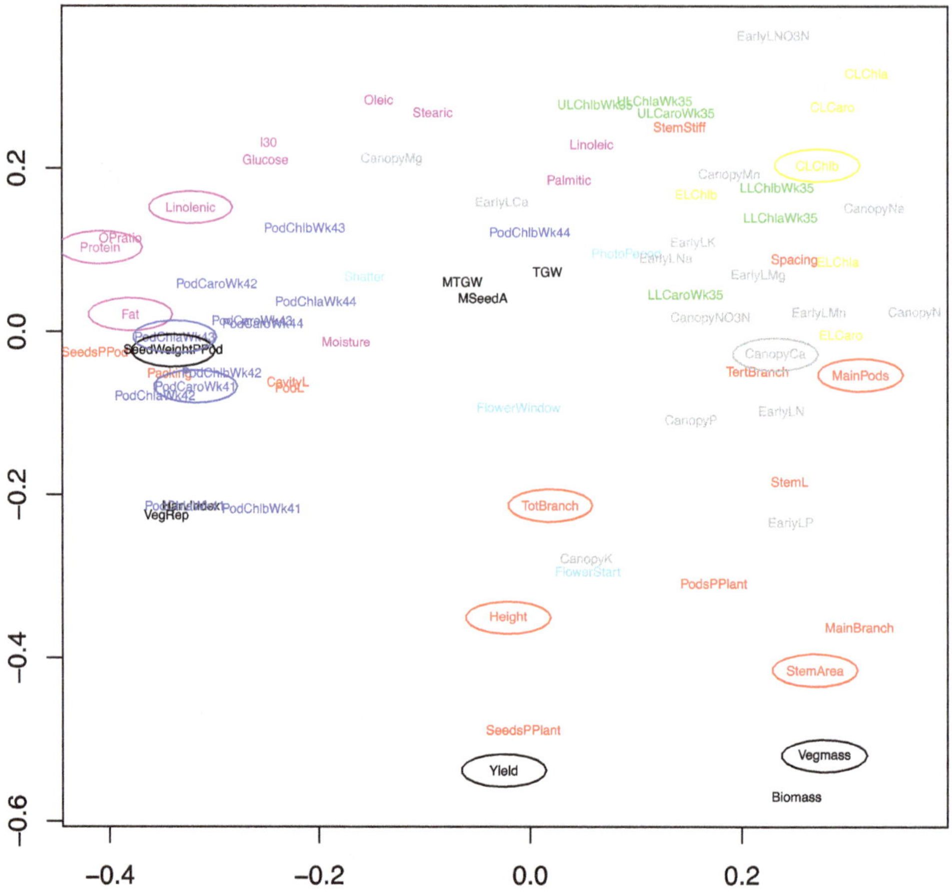

Figure 5. Multi-dimensional scaling of all traits measured from *Brassica napus*. The correlation between each pair of traits was converted into a distance, where a correlation of ±1 was given a distance of zero and a correlation of zero was given a distance of one; a two-dimensional projection of these distances is shown in the figure. Traits that are close together in this space are highly correlated. Traits are coloured according to the type of trait: measures of yield (black), architecture (red), lower leaf and upper leaf chlorophyll (green), pod chlorophyll (blue), timing of plant development (cyan), NIRS data (pink), early leaf and canopy leaf chlorophyll (yellow) and early leaf and canopy leaf minerals (grey). Traits used in the models for predicting yield and/or vegetative biomass are circled.

2.6. Analysis of Varieties to Determine Those Produce High Yields and Highlight Gaps in Breeding Potential

Having established which traits are important for predicting yield (Table 3), next we investigated which varieties displayed the traits contributing to high yield. For this analysis, we used the mean of both blocks for each variety for each trait contributing to the models for yield and above-ground vegetative biomass at harvest (Table 3). Each of the traits was scaled to have a mean = 0 and variance = 1. For the cluster analysis, we took the Euclidean distance between each pair of varieties and combined varieties using Ward's method, creating three groups from the whole dataset (Figure 6). A combined approach was then taken to include the other traits shown to be important in our earlier models and to cluster them all against the varieties studied (Figure 6b). Varieties are ordered according to the cluster analysis shown in Figure 6a. The top cluster contains all of the highest yielding varieties and the bottom cluster all the lowest yielding varieties. High seed protein, seed linolenic acid content, chlorophyll b content of the canopy leaf and early leaf calcium content were all negatively correlated with yield; however, it should be noted that canopy leaf chlorophyll b was a positive term within the statistical model for log[yield]. This is because the correlation is relatively weak with yield (0.11), biomass (0.03) and protein (0.09), so canopy leaf chlorophyll b only becomes an important term once biomass and protein are taken into account. Conversely, the architectural plant traits of raceme area and plant height were positively correlated with yield, together with lipid content and pod weight for all varieties, except Darmor, Rameses and Canard.

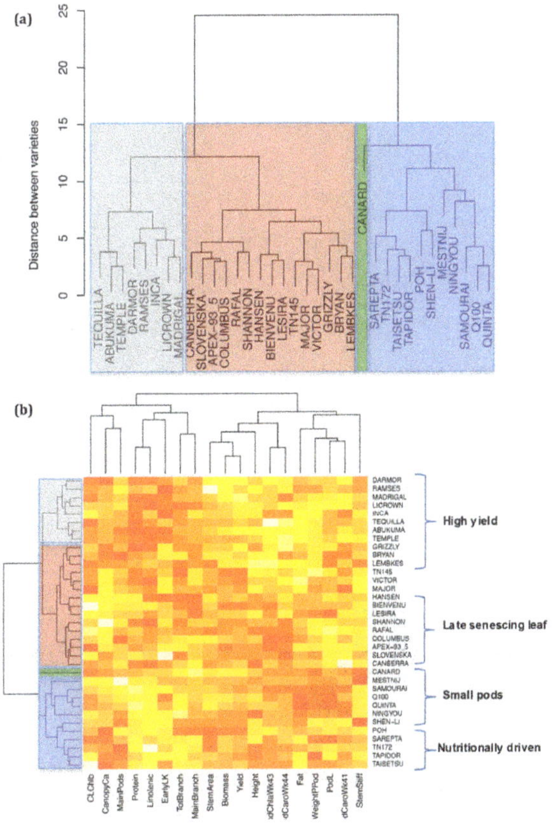

Figure 6. Cluster analysis of the *Brassica napus* traits and varieties. (**a**) Cluster dendrogram of the variety set; each cluster is blocked in green, black (grey), red or blue to correspond with Figure 6b and the colours used to circle varieties in Figure 7. The study included 35 different plant varieties. The mean of both blocks was used for each variety for each trait; each of the traits was scaled to have a mean = 0 and variance = 1, and then cluster analysis was performed by taking the Euclidean distance between each pair of varieties and combining varieties using Ward's method. The distance between varieties is defined as equal to the sum of the squares of differences across all traits analysed in Figure 5. (**b**) Analysis of the traits contributing to the yield and vegetative biomass models clustered by variety.

In the model for explaining yield, the top two terms were vegetative biomass (positive) and total protein (negative). Total seed oil also (positively) contributed to yield. Therefore, the oil:protein ratio and biomass traits were used to illustrate the clustering of different varieties in two-dimensional space (Figure 7). Canard produced the highest yield, but had a comparatively low oil:protein ratio. It also had a very poor score for raceme stiffness and has been observed to lodge readily in the field; therefore, although Canard had a high vegetative biomass, the architecture was not sufficiently robust. The varieties from the cluster coloured grey also had high yield, but show a higher oil:protein ratio than other clusters, meaning that the yield increase is achieved by assimilating more oil in the seeds. The other groups had lower vegetative biomass, lower oil:protein ratio and lower yield, with Ningyou having the poorest phenotype of all of the varieties. Yield gaps are apparent (green squares on Figure 7) where breeding and selection, or agronomy, could be used to target plants with high oil content from a smaller vegetative biomass (by changing the harvest index to put more resources into seeds), or large plants with high yields of oil-rich seeds, or large plants with low oil content that could potentially be used as biomass crops.

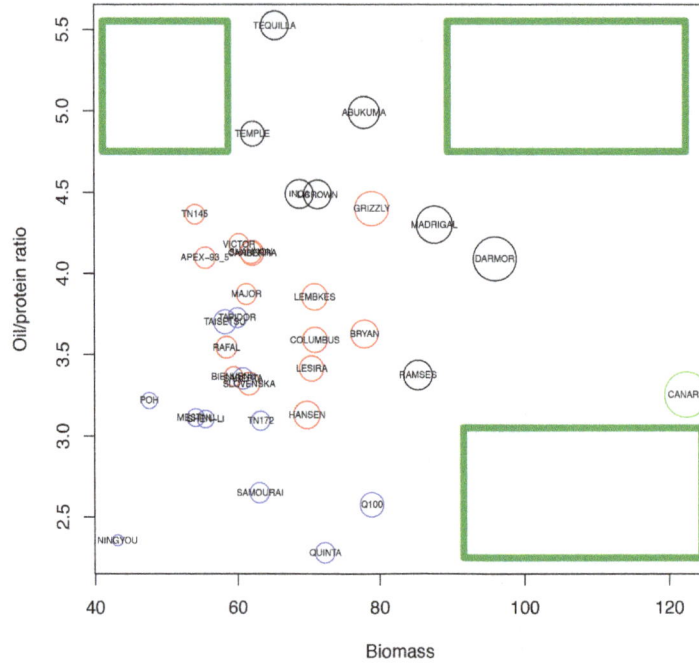

Figure 7. Varietal distribution across the *Brassica napus* diversity trial in two-dimensional space using vegetative biomass and oil:protein ratio. The yield of each variety is shown by the radius of the circle; colours are used to identify the varieties within each cluster as determined in Figure 6. Green boxes indicate yield gaps where no varieties currently exist.

2.7. Traits Measured Early in Crop Development Are Also Good Predictors of Yield

Our earlier model for explaining yield (Table 3) relied on many traits that can only be observed at, or close to, the time of harvest. While that model is useful for selecting the variety for growing in subsequent seasons, a model that forecasts yield during the growing period would also be more useful. Traditional early observations, such as date of flowering and leaf chlorophyll retention, had very poor correlation with seed yield (Figure 8).

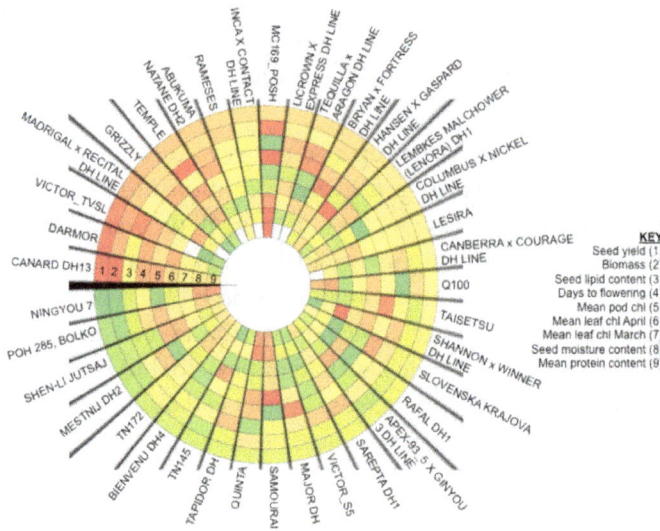

Figure 8. Scores for nine traits associated with resource allocation and yield for the thirty-five varieties used in this study, ordered by yield (outer ring) and coloured from largest (red) to smallest (green) in each ring. Averages of all replicates over both blocks were used. White spaces = missing data.

Observations in the field of the highest and lowest yielding varieties showed that there was very little visual difference in colour of pods on the main raceme from May–July (Figure 9a), although the absolute leaf chlorophyll levels in April demonstrated clear differences between the varieties (Figure 8). However, when the canopy was viewed as an entirety, there were clear differences in senescence between the high and low yielding varieties (Figure 9b) with high yielding plants having a slower senescing canopy. When the extreme lines (defined as the top ten lines for seed yield, retention of pod chlorophyll and biomass) were analysed, only two high yielding cultivars, Canard and Victor, also had high biomass and slow pod senescence. In other cases, high biomass lines overlapped with high yielding lines, but retention of pod chlorophyll did not correlate well (Figure 10a).

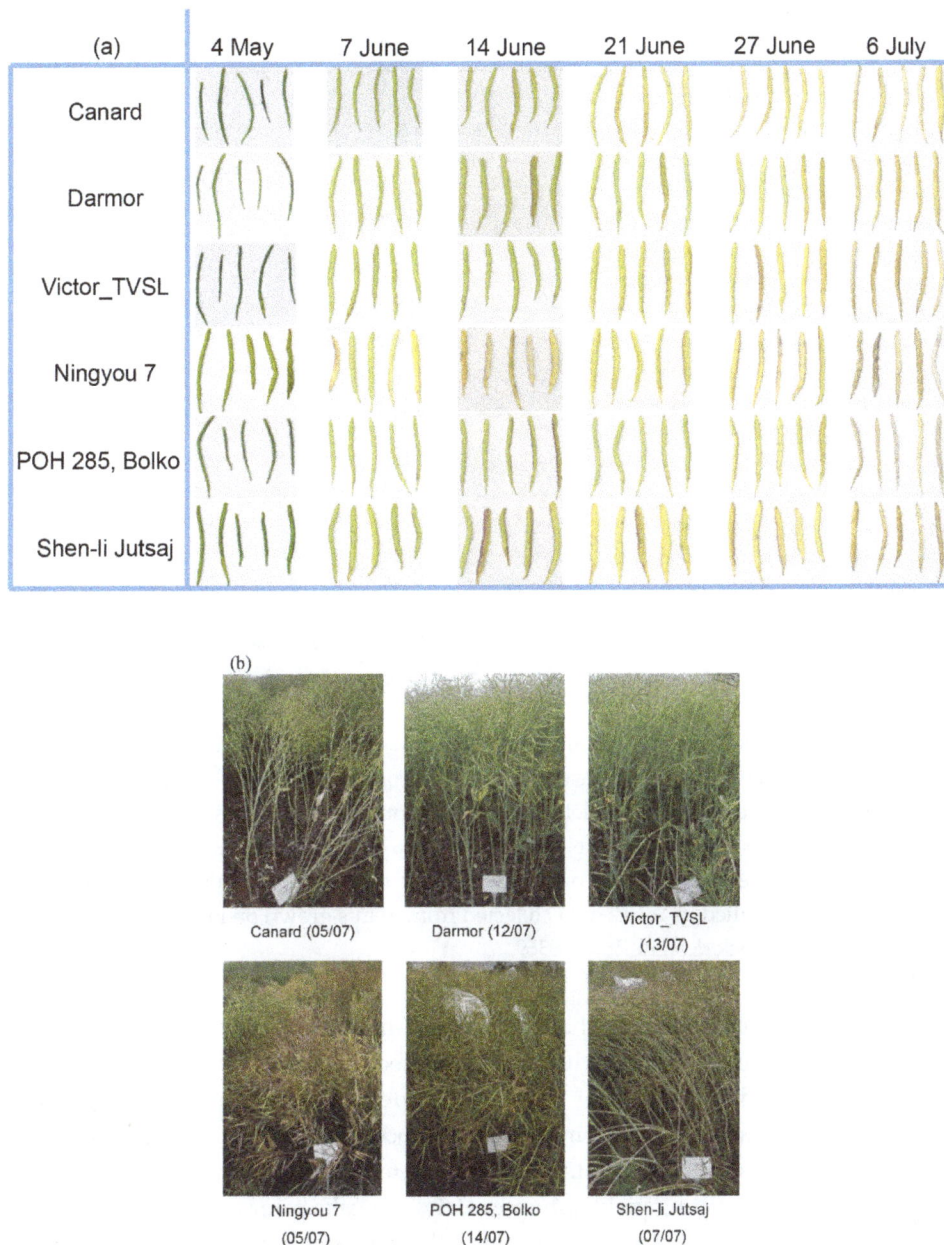

Figure 9. Contrast in uniformity of *Brassica napus*. (**a**) Pod senescence on primary stems and (**b**) Canopy senescence. Canard, Darmor and Victor TVSL are the three highest yielding lines; Ningyou7, POH285 and Shen-li Jutsai are the three lowest yielding in terms of seed mass per plant. Dates in (**a**) are the dates the images were taken. All images in (**b**) were taken on 13 June 2011, with the dates shown in brackets being the final harvest date for each line.

(a)

(b)

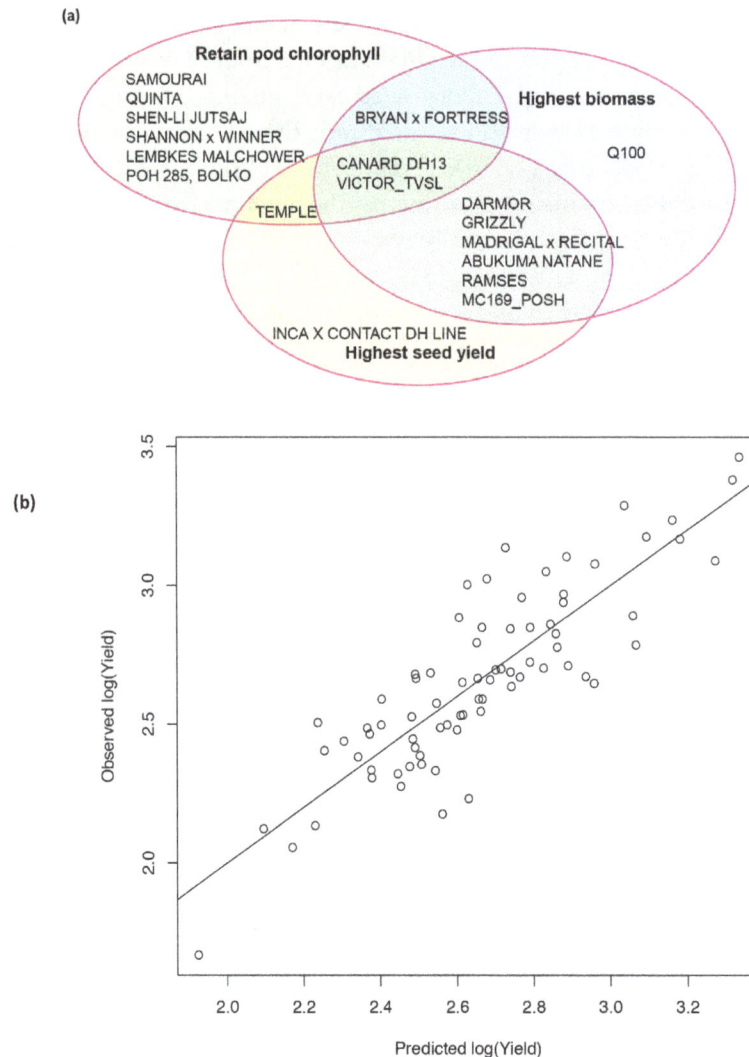

Figure 10. Plant traits as drivers for final yield in *Brassica napus*. (**a**) Top 10 lines in each category showing overlap between them. Temple is one of the most commonly-grown varieties, yet it is atypical in linking chlorophyll retention to high yield. In general, biomass and yield overlap (big plants = more seed) and varieties with a delayed senescence phenotype do not yield highly. (**b**) Model for predicting yield using traits that were observed at or before the time of flowering. The model was developed using nine terms and applied to Year 2 data collected from plants grown on low nitrogen (22 kg ha^{-1}) plots in a randomised block design. $R^2 = 0.6369$.

It was essential to develop a model that can predict final yield prior to visual indicators of yield differences. The forward selection procedure was repeated, but only allowing covariates that were unrelated to flowers or pods in the model. The fitted model (Table 4, Figure 10b) explains 63.69% of the variance in yield, compared to 95.12% in the earlier model (Table 3). This reduction was due to the omission of information regarding the later stages of plant development. However, the model still offers a good method for predicting yield approximately 15 weeks prior to harvest, i.e., approximately April of the same year. The model comprises two of the best architectural variables (stem area and numbers of branches on the primary stem) that were used to explain vegetative biomass in the earlier model. Canopy leaf chlorophyll b content, which was in the earlier model for predicting yield, was also included in the predictive model (Table 4). Interestingly, two leaf mineral concentration traits were also selected, canopy leaf potassium and canopy leaf magnesium.

Table 4. Forward selection statistical model to predict the yield from traits observable at or prior to flowering using Year 2 data. The data modelled represented 35 varieties grown at residual N ($\sim$18 kg ha^{-1}) in duplicate plots in a randomised block design.

Coefficients	Estimate	Standard Error	t-Value	Pr (>|t|)	Significance	Cumulative % Variance Explained
(Intercept)	−0.377	0.359	1.050	0.2975	-	-
Stem area	0.002	0.001	3.150	0.00241	**	36.64%
Log[No. branches on primary stem]	0.724	0.151	4.811	8.51×10^{-6}	***	44.22%
Length of flower set	0.009	0.003	3.294	0.0016	**	51.73%
Canopy leaf potassium	−0.188	0.057	−3.293	0.0016	**	56.92%
Canopy leaf chlorophyll b	0.550	0.185	2.972	0.0041	**	60.19%
Canopy leaf Mg	1.594	0.617	2.582	0.0120	*	63.69%

Residual standard error: 0.2048 on 69° of freedom; multiple R-squared: 0.6369; adjusted R-squared: 0.6054; significance codes: 0.001 '***'; 0.01 '**'; 0.05 '*'.

3. Discussion

Seed yield is a complex trait influenced by numerous interacting variables that are governed by both genetic and environmental factors. Several studies have sought to understand the contribution of individual traits to oilseed rape yield [4,6,32–35] or to model yield such that it can be predicted over future growing seasons [5,7,8,36]. However, there is some disagreement between the studies as to what are the most important traits that influence yield, be it the number of pods per plant [34], the time to maturity [6] or the duration of flowering [32]. One of the reasons for the discrepancies observed between the different studies is the limitations in the number or type of traits that were measured or the use of a limited number of cultivars, which constrained the genotypic variation that could be observed.

3.1. Greater Branching Density Drives Yield Improvement

Oilseed rape architectural traits accounted for the greatest percentage of variation in above-ground biomass at harvest. Significant positive correlations were observed between both total above-ground biomass and vegetative biomass and seed yield. Thus, although partitioning more resources into vegetative development could be seen to divert resources away from reproductive development, a larger above-ground biomass with increased branching, supported by a larger stem and taller plants, gives rise to more available positions upon which pods can develop. In this study all plots were planted at the same seed rate within year, but it is well established that lower seed rates can be used to produce plants with more branches than those in tightly-packed canopies [37].

Furthermore, an increase in seed yield per plant does not seem to be the result of heavier seeds, as judged by the lack of correlation between total seed yield and Thousand Grain Weight (TGW), a result that was also observed in *Arabidopsis* [17] and kale [38]. This also supports our hypothesis that increased yield in terms of seed mass per plant is mainly a consequence of more pods rather than a positive increase in resource accumulation within individual seeds or pods. Whilst seed yield per plant is positively correlated with plant height, this is not a trait that breeders would wish to emphasise, as tall plants are generally more susceptible to lodging [39], which can decrease seed yields by up to 16% [40]. For instance, transgenic dwarf oilseed rape lines have been found to have increased yields compared to their taller counterparts [41], and therefore, the breeding target would be to shorten the internode length whilst maintaining or increasing branch number on each plant. This is most likely to be achieved by targeting the plant hormone signalling network, which mediates the genetic regulation of responses to the environment, facilitating individual plants to change their architecture over time in response to the external conditions [42]. Cytokinins promote the emergence of axillary buds into fully-formed branches, whereas strigolactones inhibit branching. Plants are sufficiently sophisticated so that branching can be temporally or spatially regulated within an individual plant; indeed several authors regard plants as a system of competing populations of redundant organs since no individual pod, leaf or flower can be regarded as indispensable [43], and it is known that oilseed rape produces more flowers than are required to obtain maximum yield [35].

3.2. Increasing Seed Oil:Protein Ratio

The model also showed that seed oil and protein were important drivers of yield; indeed, the oil:protein ratio (derived from the seed protein and oil terms used in Model 3b) explains a further 8.1% of the variation within yield, once biomass is taken into account. Grami et al. [44] also established that seed oil content is negatively correlated with protein content; therefore, increasing the oil fraction of the seed would naturally enhance the oil:protein ratio and, according to our model, potentially help increase yields independently of changing architectural traits, although the mechanism by which this conversion occurs is presently not clear. A recent study using *Arabidopsis* populations established that seed carbon and N content were antagonistic and that % N was negatively correlated with yield [45], as it is in the present study. In *Arabidopsis*, the number of seeds and pods per plant were closely related to the concentration of seed oil and protein within the seeds [19]. It was shown for soybean that a 1-kg^{-1} increase in oil content will usually lead to a 2-kg^{-1} decrease in protein due to a negative genetic correlation between the two yield components [46]; therefore, changing the oil:protein ratio towards oil production would naturally influence seed yield when defined as the total weight of seeds per plant. None of the lines studied have both a high above-ground biomass and a high oil:protein ratio, a combination that would be predicted to produce even greater yields than those recorded. The lines that show the greatest above-ground biomass (and highest seed yields) could also have their vegetative material utilised for the production of bioethanol and biogas, easing competition with food crops for land and resources. Whilst this technology is still being developed, especially in the case of appropriate pre-treatments to extract sugars from the lignocellulosic plant cell walls, yields of 20.4 g methane 100 g^{-1} DM and 10.9 g ethanol 100 g^{-1} DM have been reported for oilseed rape [47]. Although these values are much lower than those for sugarcane (37 g ethanol 100 g^{-1} DM) [48], it is still a useful by-product of oilseed rape production and a technology under development.

In the current study, there was only limited variation between the lines in terms of seed total oil content (45%–58%), but the range of individual fatty acids was highly variable between varieties (Table S1). Many of the oilseed rape varieties grown commercially are High Oleic acid, Low Linolenic acid (HOLL) and the current population contained HOLL and non-HOLL varieties. Seed linolenic acid was consistently and negatively associated with yield; a reflection of the breeding programme that has already taken place to increase yield simultaneously with developing varieties that produce healthy, low trans-fat and stable frying oils and have good disease resistance [49]. It also demonstrates that our model is robust and a true reflection of the genetic basis of yield. Transgenic or mutation (TILLING) technologies would appear to be the most effective method of manipulating the composition of fatty acids within the seed such that it fits the purpose of use, be it for the production of lubricants in industry, or of edible vegetable oils [50], or as a way of producing omega 3-rich oils to enter the human food chain [51]. For example, expressing the yeast gene glycerol-3-phosphate dehydrogenase (*gpd1*) under the control of a seed-specific promoter increased *B. napus* seed oil content by 40% [52].

3.3. Early Leaf P Status Correlates with Final Yield

The predictive yield model showed that the traits that explain most of the variation associated with yield are architectural traits; beyond this, the concentration of minerals such as canopy leaf potassium (K), early leaf phosphorus (P) and early leaf N significantly influenced yield. Previous work surrounding the effect of P fertilisation on seed yields was inconclusive, with reports of increasing P having no effect on total seed oil and protein concentration [53,54], decreasing seed linoleic acid concentrations [55] or increasing total seed oil concentrations [56]. Therefore, more research is necessary to clarify the effect of P on yield. One hypothesis is that early leaf P reflects good access to this poorly mobile nutrient in the soil and reflects a strong root system capable of supplying resources during crop development. Given that early leaf P status is a trait that can be measured before the pods have begun to develop, it could provide a useful early indicator of seed yield. Equally, raceme width at the time of flowering was a good indicator of biomass and capable of predicting 36.64% of yield. Similarly, a study in maize found that measurements of raceme width early in the growing season also provided a good predictor of final yield [57]; these early indicators of yield could be used by growers to modify inputs and optimise yield in order to maximise profit margins.

3.4. Number of Pods Determines Yield

The model presented here used the weight of seeds per plant as a measure of yield; however, we also explored whether other parameters of yield, which are used commercially, such as TGW or the number of seeds per pod, could be used as a measure of final seed yield. From the correlation matrix, it became evident that traits previously used as a proxy for seed yield per plant, such as the number of seeds per pod [58], seed weight per pod [59], pod length [60] and even TGW, were not significantly correlated with seed yield per plant in this study (Figure 4). This finding indicates that current commercially-used measures, such as TGW, may not be the most appropriate indicator of yield for breeding and selection purposes. Only the numbers of seeds per plant and pods per plant were significantly correlated with seed yield per plant. This indicates that the pod in *B. napus* is a fairly conserved organ and that a plant will preferentially invest in the production of more pods per plant as opposed to increasing the number or mass of seeds within an individual pod. Once a seed contains the minimum amount of resources necessary to ensure germination viability, then there is no advantage for a plant in investing more resources beyond this in an individual seed, and a better survival strategy is to make more seeds in different pods, a view also reported in a study on *Arabidopsis* resource allocation [19]. Similarly, the addition of N was found to increase oilseed rape yields through the production of more pods as opposed to affecting individual seed or pod weight [33], something also observed in Year 1 of the current trial where plants were grown on high and low N.

3.5. Flowering Start Date and Duration Do Not Affect Seed Yield

Oilseed rape has been bred to be self-fertile, but it can also be pollinated by wind and insects; the latter was found to enhance seed oil content quality, potentially due to the ability of insects to optimise the timing of fertilisation for the plant [61]. In *Arabidopsis* grown under controlled conditions, initially under short days and then switched to long days to induce flowering, the plants that flowered later had a lower yield [17]. In contrast, in the current field-based study, it was found that the flowering start date and flowering window were related neither to the number of pods per plant, nor to seed yield. Thus, increasing the flowering duration does not appear to increase pollination efficiency, perhaps because the general decline in pollinators within agricultural systems [62] means insufficient exploitation of the additional flowers. The increased early flowering nature of lines with long flowering duration may also have increased the unwanted presence of pollen beetle pests (*Meligethes* spp.), which are attracted by the visual and olfactory properties of oilseed rape flowers [63]. Thus, within the current field trial, the unopened buds on plants of those lines that flowered earlier may have been at increased risk from feeding damage than those on plants not yet flowering, especially if pollen resources were scarce due to high intraspecific competition, and this may partly explain why it did not afford a yield advantage to be flowering earlier or over a longer period.

3.6. The Relationship between Plant Senescence and Seed Yield

In this study, neither chlorophyll concentration nor senescence of the pods were related to seed yield per plant or other yield parameters. This is in contrast to previous work on *Arabidopsis* [64] and wheat [65] where delaying leaf senescence led to an increase in seed yield. The majority of photosynthetic tissue in an oilseed rape plant is raceme material; therefore, a greater biomass would also provide a larger photosynthetic surface area through an increased number of stems. In the present study, canopy leaf chlorophyll b content (a proxy measure for the stage of senescence) had a positive relationship with final yield, suggesting that the leaf resources are mobilised into the developing plant prior to abscission via a route that contributes to seed yield; hence, the parameter was included in the predictive model. Chlorophyll b content is a measure of light-harvesting complex status rather than of reaction centre complexes, and light harvesting complexes (LHC) are second only to Rubisco as sources of recyclable protein [66]. LHC also tend to be in excess, so the chloroplast might be able to lose significant quantities of chlorophyll b and associated proteins without it having a negative effect on photosynthesis. In cereals, photosynthesis of the canopy is largely responsible for the carbon yield of seeds, whereas seed N is mobilized from senescing vegetative tissues [67]. Stems have also been shown to play a pivotal role in *B. napus* seed filling by contributing 31% of the carbon contained within seeds [68]. Experiments in which soy bean pods were removed from mature plants showed that N fixation continued to take place in the roots [69]; elevated levels of N represent a 'metabolic sink' for photosynthate, thus altering carbohydrate metabolism and sugar signalling in source leaves. Leaves therefore accumulated starch products that would normally be mobilised to the pods. De-podding *Arabidopsis* showed that restricting the number of sinks increased the concentration of lipids in the remaining pods [19]; since *B. napus* loses leaves relatively early in its reproductive cycle, it seems reasonable to conclude that canopy photosynthesis is unlikely to contribute to oil/protein content in the seed, but it may make a contribution to carbon accumulation. However, as discussed in Bennett et al. [18], *Arabidopsis* is a weed that treats each pod as an individual unit, and local remobilisation of photosynthates into the seeds makes sense as it is important that each pod shatters as soon as the seed attains viability in order to stand the best chance of reproductive success. In contrast, the artificial selection that occurs within crop species has sought to synchronise seed maturation across the plant; therefore, the photosynthetic window of a pod is less closely related to yield and shatter since the pod sits in 'suspended animation' waiting for the whole plant to be ready to shatter. The extent of developmental coordination is poorer in oilseed rape compared to other crop species that have been domesticated for longer, such as cereal crops [70,71], but artificial selection has started to influence the loss of weedy reproductive traits. Other authors have found that seed maturity, as opposed to pod wall maturity, drives pod shatter in oilseed rape [72], highlighting the uncoupling of pod wall senescence in modern varieties from shattering and seed yield.

An increase in the number of pods per plant does not affect the rate of pod senescence or pod development in oilseed rape, giving credence to the idea that pod reproductive development is not necessarily driving the senescence process, but it is temporally fixed under growing conditions that impart only minimal stress on the plant. Just as leaf senescence has been shown to naturally occur in an age-dependent manner [73,74], so whole plant senescence also appears to be controlled by the age of the plant, an assumption borne out by the fact that there was little variation in the lifespan of different oilseed rape lines despite their genetic diversity. Plant longevity appears to be connected to plant mass, with small plants having a shorter lifespan than larger ones; however, within-species lifespan is very similar to keep the population stable [75]. An as yet unknown biological system in *B. napus* controls the timing of whole plant senescence and accordingly lifespan. Thus, whilst senescence is the visible output of an internal biological clock, the number of pods, plant size and seed yield are the consequences of resource availability.

4. Materials and Methods

In an effort to understand better the physiological basis governing seed yield in oilseed rape, 35 cultivar lines of *B. napus*, showing substantive divergence in seed yield per plant were assessed over two seasons for 133 different traits, covering developmental, architectural and nutritional parameters, generating a powerful dataset for exploring the factors underpinning yield.

4.1. Site Description and Meteorological Conditions

The experiment was carried out over two growing seasons, Year 1: August 2009–July 2010; Year 2: September 2010–August 2011, at Rothamsted Research, Hertfordshire, UK (51°48′32.6376″ N, 0°21′22.5432″ W). The trials in each year were conducted in different fields, but in both cases, the previous crop was spring barley. The soil type for the Year 1 field was flinty silt clay loam (Batcombe–Carstens), and in Year 2, it was clay with flints (Batcombe). Soil samples (30-cm cores) were taken prior to the application of fertilisers. The soil in Year 1 contained a total residual N concentration of 18 kg N ha^{-1}, which was comprised of 9 kg ha^{-1} N-NO$_3^-$ and 9 kg ha^{-1} N-NH$_4^+$, whereas that in Year 2 was 22 Kg N ha^{-1}, comprising 9 kg ha^{-1} N-NO$_3^-$ and 13 kg ha^{-1} N-NH$_4^+$.

Weather parameters (rainfall, maximum and minimum temperature) for the experimental period (2009–2011) and the 30-year average were taken from the Rothamsted Research meteorological station (Figure S1). There was a large variation in the rainfall pattern between the two years; e.g., June rainfall in Year 2 was approximately four-fold higher than the same period in Year 1.

4.2. Crop Management and Field Trial Design

The field trials were arranged in a randomised complete block design and used two replicate plots per treatment. In Year 1 (2009–2010), two soil N supplies were investigated, low (= residual) N (~18 kg ha^{-1}) and high N (148 kg ha^{-1}). The high N treatment was applied in a split dose application; 30 kg ha^{-1} N were applied in the autumn (29 September 2009), and 50 kg ha^{-1} N were subsequently applied in the spring (12 March 2010 and 07 April 2010) to a maximum concentration of 148 kg ha^{-1} total available N. The Year 2 trial used only residual nitrogen (22 kg ha^{-1}) across the whole plot. In Year 1, seeds were drilled on 28 August 2009, and each replicate plot (1 × 1.5 m) was drilled in four rows with 120 seeds per plot. In Year 2, 125 plots were drilled on 6 September 2010, but due to bad weather conditions, the drilling of the remaining 307 plots had to be postponed by one week; ANOVA showed that no significant differences were found in yield parameters as a result of this split drilling time Year 2 plots were 1.8 × 1.5 m, drilled with 160 seeds per plot in 14 rows. Pre-germination herbicides and fungicides were applied to both trials at the manufacturers' recommended rates with pests, diseases and volunteer cereals/weeds controlled according to standard agronomic practice. In Year 1, all of the 61 winter cultivars planted were sampled, and this included 48 winter oilseed rape (WOSR) lines, 8 fodder/forage/salad kales, 4 swedes and 1 synthetic line. In Year 2, of the 84 WOSR varieties planted, a subset of 35 lines was sampled; being selected if they were also present in Year 1 of the trial and had transcriptome sequencing information available.

4.3. Chlorophyll Measurements

Chlorophyll and carotenoid concentration were analysed in early and canopy leaf samples as a proxy measure of senescence, and in Year 2, additional sampling was made from leaf disks from upper and lower leaves and whole pods sampled at set intervals throughout the growing season, March–June and May–July, respectively. The early leaves represented the earliest not fully-expanded leaves, and the canopy leaves were the youngest true leaves on the plant when sampled on 2 February 2011. The lower leaf (sampled 11 March, 6 April and 3 May 2011) was the 5th leaf up from the base of the main raceme, and the upper leaf (sampled 3 May 2011 and 6 June 2011) was that immediately subtending the lowermost pod. All pods were sampled from the main raceme. Chlorophyll was extracted and analysed in the leaf and pod material using the methods described in Bennett et al. [29]. The progression of chlorophyll and carotenoid loss from the leaves and pod walls was calculated by determining the % difference from the previous sampling date. The pod photosynthetic period was calculated as the time from the start of flowering until pod wall chlorophyll fell below 10% of its maximum.

4.4. Leaf Mineral Analysis

Early and canopy leaves from ten plants per plot were dried overnight at 80 °C and stored in a controlled environment room at 15 °C, 15% humidity. Before milling, the leaf samples were re-dried overnight at 60 °C. Plots were analysed in duplicate for their mineral content, and a Kjeldahl digestion was used to analyse total calcium (Ca), potassium (K), magnesium (Mg), manganese (Mn), nitrogen

(N) and phosphorus (P) using the methods described in Broadley et al. [76]. Aqueous extracts were subsequently used to measure the nitrate concentration [76].

4.5. Plant Development

The plots were walked periodically between March–July. Growth stage [77] was visually scored over the whole plot and graded accordingly based on 25% of the plants in a plot reaching a particular stage. Flowering duration was defined by the difference between the start date of flowering (when 25% of the plot had reached GS60) and the date when >95% of the plants within the plot had no more flowers.

Pod development was also visually scored across the whole plot and categorized per plot on the following scale: 0: plot still in flower; 0.5: immature pods; 1: mature green pods fully expanded; 1.5: 25% yellow pods; 2: 50% yellow pods (yellow/green); 2.5: 75% yellow pods; 3: 100% yellow pods; 3.5: 50% brown pods; 4: 100% brown pods just before shatter. A plot was recorded as shattering when 25% of the pods in the whole plot were seen to have dehisced.

4.6. Plant Architecture

Several plant architectural traits were measured at the time of harvest. Five plants per plot were selected at random from each plot and were returned to the laboratory for assessment of the following traits: plant height, length of the raceme containing pods, the total number of branches and number of branches arising from the main raceme only. Pod spacing on the main raceme was calculated by dividing the length of the raceme containing pods by the number of pod sites on the raceme, i.e., including sites where pods had subsequently aborted. Raceme area was calculated using a calliper to measure two raceme diameters 10 cm above the ground at 90° to each other and multiplying these values together. Raceme stiffness was visually scored on a scale of 1–9 where: 1, all plants prostrate; 2, main stems leaning at 20°–30° to the horizontal; 3, main stems leaning at 45° to the horizontal; 4, main stems leaning at 70°–80° to the horizontal; 5, main stems leaning slightly, but branched canopy bent over at 10° to the horizontal; 6, main stems leaning slightly, but branched canopy leaning at 20°–30° to the horizontal; 7, main stems leaning slightly, but branched canopy leaning at 45° to the horizontal; 8, main raceme erect with branched canopy leaning at 70°–80° to the horizontal; and 9, all plants completely erect. The above-ground biomass was calculated from the dry weight of 5 plants per plot at harvest; the vegetative mass was calculated from the above-ground biomass minus the seed yield per plant.

4.7. Pod Physiology

Several measures of pod physiology were taken at harvest from pods on the main raceme at Stage 4 (brown and ready to dehisce) just before the plots were harvested. In each case, ten replicate pod samples were taken per plot, and there were four plots per line. Pod length was determined by measuring from the start of the bivalve to the end of the beak, and the seed cavity length was also determined (the length of the pod minus the beak). The seeds were removed from the pod and aborted seeds discarded before the remainder was counted. The seed packing density was determined by dividing the pod cavity length by the number of seeds per pod. The weight of seeds per pod was determined by placing the seeds in a controlled environment room at 10 °C, 15% humidity until they had reached a constant weight. Seed area was calculated using a MARVIN seed scanner (Gta Sensors Gmbh-Germany).

4.8. Yield

At harvest, ten plants per plot were bagged and dried at 20 °C. The plants were then threshed and winnowed before the seeds were placed in a controlled environment room at 15 °C, 15% humidity until they had reached a constant weight. Seed yield was subsequently determined as total seed weight per plant. To calculate the MARVIN Thousand Grain Weight (TGW), 100 harvested seeds per plot (post threshing, winnowing and storage) were placed into the MARVIN seed scanner with a ± 10 seed tolerance; any seed that was not representative of the batch (i.e., damaged or discoloured) was removed by hand prior to the analysis, and the seeds were subsequently weighed. The following calculations were used to determine the MARVIN TGW, TGW, number of seeds per plant, number of pods per plant and harvest index.

$$\text{MARVIN TGW}: \left[\frac{\text{Weight of seeds (g)}}{\text{Number of seeds}}\right] \times 1000 \text{ (g)} \tag{1}$$

$$\text{TGW}: \left[\frac{\text{Seed weight per pod (g)}}{\text{Number of seeds per pod}}\right] \times 1000 \text{ (g)} \tag{2}$$

$$\text{Number of seeds per plant}: \text{Seed yield per plant (g)} \div \left[\frac{\text{TGW (g)}}{1000}\right] \tag{3}$$

$$\text{Number of pods per plant}: \left[\frac{\text{Seed yield per plant (g)}}{\text{Seed weight per pod (g)}}\right] \tag{4}$$

$$\text{Harvest index}: \left[\frac{\text{Seed yield per plant (g)}}{\text{Total above ground biomass (g)}}\right] \tag{5}$$

4.9. Seed Composition

Seed oil, fatty acid, protein, moisture and glucosinolate content were determined by Near Infrared Spectroscopy (NIRS) using the methods described in Kelly et al. [78]. In the current study, seed samples from five plants per plot were analysed.

4.10. Statistics

Statistical analyses were performed using the statistical software package R (R Core Team, Florham Park, New Jersey, USA, 2013), except for the yield mean per plant and standard error of the mean, which were calculated using IBM SPSS version 19 (IBM Corp., Armonk, NY, USA). For the Pearson's correlation analysis of all traits, with 9180 pairwise tests, a Bonferroni multiple comparison correction was applied making the cut-off value for significant correlations ± 0.496. To determine the main traits influencing seed yield per plant and above-ground biomass, linear regression models were developed using forward selection based on the 133 traits listed in Supplementary Table S1. Starting with the null model (no traits included), forward selection identifies the most significant trait to include in the model based on an F-test. The process was repeated multiple times, at each stage selecting the most significant model with the $n + 1$ variable compared to the current model with n variables based on the F-test. The process was stopped when none of the remaining traits significantly improved the estimation of the response variable at the 5% significance level. Some traits, including yield, underwent a log transformation prior to inclusion in the model to improve the linear relationship between the response variable and the other traits. After selecting the traits identified in Table 3 and standardising, the traits and lines underwent hierarchical clustering using Ward linkage, to compile the dendrogram, heat map and scatter diagram (Figures 4 and 6).

5. Conclusions

We have developed a forward stepwise multiple linear regression model to determine the traits most important in influencing seed yield per plant in *B. napus*. From the 133 traits recorded, that could have potentially influenced seed yield, above-ground biomass and protein were found to be the most important factors, together accounting for 94.36% of the variation in seed yield. Increasing our understanding of resource allocation strategies could lead to improved plant breeding via marker-assisted selection, and this would have important consequences for crop breeding targets and meeting future food security demands. It would appear that year-on-year environmental weather conditions have a bigger impact than soil conditions, since the Year 2 model fitted to Year 1 data was less good than the same model fitted to different soil nitrogen regimes in Year 1. However, the refined model developed in Year 2 included more informative parameters that provided more robust indicators of yield. We therefore propose that this model would require adjustment for climate and farm-specific soil mineral content, but that it could provide the basis of a model for oilseed rape yield across the country.

Overall, our findings demonstrate that there is a diverse array of resource allocation strategies within the *B. napus* gene pool, and different approaches will need to be applied depending on which yield components are to be targeted. Whilst biomass and protein are the main contributors towards the variation observed in seed yield, the traits that underpin these variables, such as raceme width for biomass and early leaf mineral concentration, could provide useful early indicators of seed yield, as well as potential indirect targets for improving yields.

Acknowledgments: This work was supported by funding from AL Tozer Ltd., Pyports, Downside Bridge Road, Cobham, Surrey, KT11 3EH, U.K., and The University of Reading Endowment Trust Fund to enable Emma Bennett to undertake her PhD. The OREGIN Diversity Trials at Rothamsted Research were supported by the U.K. Defra Project IF0144. Rothamsted Research is a national institute of bioscience strategically funded by the U.K. Biotechnology and Biological Sciences Research Council (BBSRC). The authors are grateful to Helen Ougham (IBERS, Aberystwyth) for critical reading of the manuscript. The collection of data described in this paper would not have been possible without the assistance of several dedicated undergraduate students from the University of Reading: Jason Fitzgerald, Manon Morin and Leo Shum, supported by Ian Bennett; and from Rothamsted Research, extensive assistance was provided by Scott Pilkington-Bennett, Stephanie Cooper, Carl Halford, Matthew Skellern, Nigel Watts and the farm staff who grew and maintained the trials. The OREGIN project was managed by Jacqueline Barker, and Sue Welham provided statistical rigour in the experimental design of the field trials.

Author Contributions: E.J.B. performed the majority of trait screening and preliminary analyses and drafted the text. C.J.B. performed the statistical modelling. P.W.C.C. curated all OREGIN data. S.M.C., P.E., J.H., C.L. contributed trait data and to the writing process. G.T. and G.K. obtained funding for the OREGIN trials and managed the project. J.A.R. and C.W. supervised E.J.B., conceived of the experiments, obtained funding for the work and wrote the finished article.

Conflicts of Interest: The authors declare no conflict of interest. The founding sponsors had no role in the design of the study; in the collection, analyses or interpretation of data; in the writing of the manuscript, nor in the decision to publish the results.

References

1.   Wang, E.; Engel, T. Simulation of phenological development of wheat crops. *Agric. Syst.* **1998**, *58*, 1–24. [CrossRef]

2.   Yang, H.S.; Dobermann, A.; Lindquist, J.L.; Walters, D.T.; Arkebauer, T.J.; Cassman, K.G. Hybrid-maize—A maize simulation model that combines two crop modeling approaches. *Field Crops Res.* **2004**, *87*, 131–154. [CrossRef]

3.   Son, N.T.; Chen, C.F.; Chen, C.R.; Minh, V.Q.; Trung, N.H. A comparative analysis of multitemporal MODIS EVI and NDVI data for large-scale rice yield estimation. *Agric. For. Meteorol.* **2014**, *197*, 52–64. [CrossRef]

4.   Habekotte, B. Evaluation of seed yield determining factors of winter oilseed rape (*Brassica napus* L.) by means of crop growth modelling. *Field Crops Res.* **1997**, *54*, 137–151. [CrossRef]

5.   Habekotte, B. A model of the phenological development of winter oilseed rape (*Brassica napus* L.). *Field Crops Res.* **1997**, *54*, 127–136. [CrossRef]

6.   Habekotte, B. Options for increasing seed yield of winter oilseed rape (*Brassica napus* L.): A simulation study. *Field Crops Res.* **1997**, *54*, 109–126. [CrossRef]

7.   Deligios, P.A.; Farci, R.; Sulas, L.; Hoogenboom, G.; Ledda, L. Predicting growth and yield of winter rapeseed in a Mediterranean environment: Model adaptation at a field scale. *Field Crops Res.* **2013**, *144*, 100–112. [CrossRef]

8. Husson, F.; Wallach, D.; Vandeputte, B. Evaluation of CECOL, a model of winter rape (*Brassica napus* L.). *Eur. J. Agron.* **1998**, *8*, 205–214. [CrossRef]

9. Peltonen-Sainio, P.; Jauhiainen, L.; Hyovela, M.; Nissila, E. Trade-off between oil and protein in rapeseed at high latitudes: Means to consolidate protein crop status? *Field Crops Res.* **2011**, *121*, 248–255. [CrossRef]

10. Nesi, N.; Delourme, R.; Bregeon, M.; Falentin, C.; Renard, M. Genetic and molecular approaches to improve nutritional value of *Brassica napus* L. seed. *Comptes Rendus Biol.* **2008**, *331*, 763–771. [CrossRef] [PubMed]

11. Smooker, A.M.; Wells, R.; Morgan, C.; Beaudoin, F.; Cho, K.; Fraser, F.; et al. The identification and mapping of candidate genes and QTL involved in the fatty acid desaturation pathway in *Brassica napus*. *Theor. Appl. Genet.* **2011**, *122*, 1075–1090. [CrossRef] [PubMed]

12. Chalhoub, B.; Denoeud, F.; Liu, S.; Parkin, I.A.P.; Tang, H.; Wang, X.; Chiquet, J.; Belcram, H.; Tong, C.; Samans, B.; et al. Early allopolyploid evolution in the post-Neolithic *Brassica napus* oilseed genome. *Science* **2014**, *345*, 950–953. [CrossRef] [PubMed]

13. Gressel, J. *Genetic Glass Ceilings: Transgenics for Crop Biodiversity*, 1st ed.; Johns Hopkins University Press: Baltimore, MD, USA, 2008; p. 488.

14. FAOSTAT. Available online: http://faostat3.fao.org/faostat-gateway/go/to/download/Q/QC/E (accessed on 31 October 2015).

15. Mawson, R.; Heaney, R.K.; Zdunczyk, Z.; Kozlowska, H. Rapeseed meal-glucosinolates and their antinutritional effects. Part II. Flavor and palatability. *Mol. Nutr. Food Res.* **1993**, *37*, 336–344.

16. Quijada, P.A.; Udall, J.A.; Lambert, B.; Osborn, T.C. Quantitative trait analysis of seed yield and other complex traits in hybrid spring rapeseed (*Brassica napus* L.): 1. Identification of genomic regions from winter germplasm. *Theor. Appl. Genet.* **2006**, *113*, 549–561. [CrossRef] [PubMed]

17. Masclaux-Daubresse, C.; Chardon, F. Exploring nitrogen remobilization for seed filling using natural variation in *Arabidopsis thaliana*. *J. Exp. Bot.* **2011**, *62*, 2131–2142. [CrossRef] [PubMed]

18. Bennett, E.J.; Roberts, J.A.; Wagstaff, C. The role of the pod in seed development: Strategies for manipulating yield. *New Phytol.* **2011**, *190*, 838–853. [CrossRef] [PubMed]

19. Bennett, E.J.; Roberts, J.A.; Wagstaff, C. Manipulating resource allocation in plants. *J. Exp. Bot.* **2012**, *63*, 3391–3400. [CrossRef] [PubMed]

20. Byzova, M.; Verduyn, C.; Brouwer, D.; Block, M. Transforming petals into sepaloid organs in Arabidopsis and oilseed rape: Implementation of the hairpin RNA-mediated gene silencing technology in an organ-specific manner. *Planta* **2004**, *218*, 379–387. [CrossRef] [PubMed]

21. Buchanan-Wollaston, V.; Earl, S.; Harrison, E.; Mathas, E.; Navabpour, S.; Page, T.; Pink, D. The molecular analysis of leaf senescence—A genomics approach. *Plant Biotechnol. J.* **2003**, *1*, 3–22. [CrossRef] [PubMed]

22. Balazadeh, S.; Parlitz, S.; Mueller-Roeber, B.; Meyer, R.C. Natural developmental variations in leaf and plant senescence in *Arabidopsis thaliana*. *Plant Biol.* **2008**, *10*, 136–147. [CrossRef] [PubMed]

23. Levey, S.; Wingler, A. Natural variation in the regulation of leaf senescence and relation to other traits in Arabidopsis. *Plant Cell Environ.* **2005**, *28*, 223–231. [CrossRef]

24. Flood, P.J.; Harbinson, J.; Aarts, M.G.M. Natural genetic variation in plant photosynthesis. *Trends Plant Sci.* **2011**, *16*, 327–335. [CrossRef] [PubMed]

25. Wagstaff, C.; Yang, T.J.W.; Stead, A.D.; Buchanan-Wollaston, V.; Roberts, J.A. A molecular and structural characterization of senescing Arabidopsis siliques and comparison of transcriptional profiles with senescing petals and leaves. *Plant J.* **2009**, *57*, 690–705. [CrossRef] [PubMed]

26. Lewis, G.J.; Thurling, N. Growth, development, and yield of three oilseed *Brassica* species in a water-limited environment. *Aust. J. Exp. Agric.* **1994**, *34*, 93–103. [CrossRef]

27. Robinson, C.K.; Hill, S.A. Altered resource allocation during seed development in Arabidopsis caused by the *abi3* mutation. *Plant Cell Environ.* **1999**, *22*, 117–123. [CrossRef]

28. Diaz, C.; Lemaitre, T.; Christ, A.; Azzopardi, M.; Kato, Y.; Sato, F.; Morot-Gaudry, J.-F.; Le Dily, F.; Masclaux-Daubresse, C. Nitrogen recycling and remobilization are differentially controlled by leaf senescence and development stage in Arabidopsis under low nitrogen nutrition. *Plant Physiol.* **2008**, *147*, 1437–1449. [CrossRef] [PubMed]

29. Bennett, E.J.; Roberts, J.A.; Wagstaff, C. Use of mutants to dissect the role of ethylene signalling in organ senescence and the regulation of yield in *Arabidopsis thaliana*. *J. Plant Growth Regul.* **2014**, *33*, 56–65. [CrossRef]

30. Hopkins, C.J.; Welham, S.J.; Teakle, G.R.; Peplow, K.-S.; Pink, D.; Carion, P.W.C.; King, G.J.; Barker, J. *Oilseed Rape Genetic Improvement Network Diversity Demonstration Trial—Year 1*; Rothamsted Research: Harpenden, UK, 2009–2010.

31. Hopkins, C.J.; Welham, S.J.; Teakle, G.R.; Peplow, K.-S.; Pink, D.; Carion, P.W.C.; King, G.J.; Barker, J. *Oilseed Rape Genetic Improvement Network Diversity Demonstration Trial—Year 2*; Rothamsted Research: Harpenden, UK, 2010–2011.

32. Ali, N.; Javidfar, F.; Elmira, J.Y.; Mirza, A.M.Y. Relationship among yield components and selection criteria for yield improvement in winter rapeseed (*Brassica napus* L.). *Pak. J. Bot.* **2003**, *35*, 167–174.

33. Allen, E.J.; Morgan, D.G. A quantitative analysis of the effects of nitrogen on the growth, development and yield of oilseed rape. *J. Agric. Sci.* **1972**, *78*, 315–324. [CrossRef]

34. Lu, G.-Y.; Zhang, F.; Zheng, P.-Y.; Cheng, Y.; Liu, F.-I.; Fu, G.-P.; Zhang, X.-K. Relationship among yield components and selection criteria for yield improvement in early rapeseed (*Brassica napus* L.). *Agric. Sci. China* **2011**, *10*, 997–1003. [CrossRef]

35. Berry, P.M.; Spink, J.H. A physiological analysis of oilseed rape yields: Past and future. *J. Agric. Sci.* **2006**, *144*, 381–392. [CrossRef]

36. Gabrielle, B.; Denoroy, P.; Gosse, G.; Justes, E.; Andersen, M.N. A model of leaf area development and senescence for winter oilseed rape. *Field Crops Res.* **1998**, *57*, 209–222. [CrossRef]

37. Diepenbrock, W. Yield analysis of winter oilseed rape (*Brassica napus* L.): A review. *Field Crops Res.* **2000**, *67*, 35–49. [CrossRef]

38. Bennett, E.J.; Gawthrop, F.; Yao, C.; Boniface, C.; Ishihara, H.; Roberts, J.A.; Wagstaff, C. Effects of planting density and nitrogen application on seed yield and other morphological traits of the leafy vegetable kale (*Brassica oleracea*). *Ann. Appl. Biol.* **2013**, *119*, 201–216.

39. Navabi, A.; Iqbal, M.; Strenzke, K.; Spaner, D. The relationship between lodging and plant height in a diverse wheat population. *Can. J. Plant Sci.* **2006**, *86*, 723–726. [CrossRef]

40. Islam, N.; Evans, E.J. Influence of lodging and nitrogen rate on the yield and yield attributes of oilseed rape (*Brassica napus* L.). *Theor. Appl. Genet.* **1994**, *88*, 530–534. [CrossRef] [PubMed]

41. Al-Ahmad, H.; Dwyer, J.; Moloney, M.; Gressel, J. Mitigation of establishment of *Brassica napus* transgenes in volunteers using a tandem construct containing a selectively unfit gene. *Plant Biotechnol. J.* **2006**, *4*, 7–21. [CrossRef] [PubMed]

42. Leyser, O. Auxin, self-organisation, and the colonial nature of plants. *Curr. Biol.* **2011**, *21*, R331–R337. [CrossRef] [PubMed]

43. Sachs, T.; Novoplansky, A.; Cohen, D. Plants as competing populations of redundant organs. *Plant Cell Environ.* **1993**, *16*, 765–770. [CrossRef]

44. Grami, B.; Stefansson, B.R.; Baker, R.J. Genetics of protein and oil content in summer rape: Heritability, number of effective factors, and correlations. *Can. J. Plant Sci.* **1977**, *57*, 937–943. [CrossRef]

45. Chardon, F.; Jasinski, S.; Durandet, M.; Lecureuil, A.; Soulay, F.; Bedu, M.; Guerche, P.; Masclaux-Daubresse, C. QTL meta-analysis in Arabidopsis reveals an interaction between leaf senescence and resource allocation to seeds. *J. Exp. Bot.* **2014**, *65*, 3949–3962. [CrossRef] [PubMed]

46. Specht, J.E.; Hume, D.J.; Kumudini, S.V. Soybean yield potential—A genetic and physiological perspective. *Crop Sci.* **1999**, *39*, 1560–1570. [CrossRef]

47. Petersson, A.; Thomsen, M.H.; Hauggaard-Nielsen, H.; Thomsen, A.-B. Potential bioethanol and biogas production using lignocellulosic biomass from winter rye, oilseed rape and faba bean. *Biomass Bioenergy* **2007**, *31*, 812–819. [CrossRef]

48. Pandey, A.; Soccol, C.R.; Nigam, P.; Soccol, V.T. Biotechnological potential of agro-industrial residues. I: Sugarcane bagasse. *Bioresour. Technol.* **2000**, *74*, 69–80. [CrossRef]

49. Orson, J.; Booth, E.; Merritt, C.; Lea, C. *Growing High Oleic Low Linolenic (HOLL) Oilseed Rape for Specialised Markets*; HGCA Project Report 442; AHDB Cereals & Oilseeds: Kenilworth, UK, 2008.

50. Venegas-Caleron, M.; Sayanova, O.; Napier, J.A. An alternative to fish oils: Metabolic engineering of oil-seed crops to produce omega-3 long chain polyunsaturated fatty acids. *Prog. Lipid Res.* **2010**, *49*, 108–119. [CrossRef] [PubMed]

51. Betancor, M.B.; Sprague, M.; Usher, S.; Sayanova, O.; Campbell, P.J.; Napier, J.A.; Tocher, D.R. A nutritionally-enhanced oil from transgenic *Camelina sativa* effectively replaces fish oil as a source of eicosapentaenoic acid for fish. *Sci. Rep.* **2015**, *5*, 8104. [CrossRef] [PubMed]

52. Vigeolas, H.; Waldeck, P.; Zank, T.; Geigenberger, P. Increasing seed oil content in oil-seed rape (*Brassica napus* L.) by over-expression of a yeast glycerol-3-phosphate dehydrogenase under the control of a seed-specific promoter. *Plant Biotechnol. J.* **2007**, *5*, 431–441. [CrossRef] [PubMed]

53. Brennan, R.F.; Bolland, M.D.A. Effect of fertiliser phosphorus and nitrogen on the concentrations of oil and protein in grain and the grain yield of canola (*Brassica napus* L.) grown in south-western Australia. *Aust. J. Exp. Agric.* **2007**, *47*, 984–991. [CrossRef]

54. Brennan, R.F.; Bolland, M.D.A. Comparing the nitrogen and phosphorus requirements of canola and wheat for grain yield and quality. *Crop Pasture Sci.* **2009**, *60*, 566–577. [CrossRef]

55. Krueger, K.; Goggi, A.S.; Mallarino, A.P.; Mullen, R.E. Phosphorus and potassium fertilization effects on soybean seed quality and composition. *Crop Sci.* **2013**, *53*, 602–610. [CrossRef]

56. Lickfett, T.; Matthäus, B.; Velasco, L.; Möllers, C. Seed yield, oil and phytate concentration in the seeds of two oilseed rape cultivars as affected by different phosphorus supply. *Eur. J. Agron.* **1999**, *11*, 293–299. [CrossRef]

57. Mourtzinis, S.; Arriaga, F.J.; Balkcom, K.S.; Ortiz, B.V. Corn grain and stover yield prediction at R1 growth stage. *Agron. J.* **2013**, *105*, 1045–1050. [CrossRef]

58. Zhang, L.; Liu, P.; Hong, D.; Huang, A.; Li, S.; He, Q.; Yang, G. Inheritance of seeds per silique in *Brassica napus* L. using joint segregation analysis. *Field Crops Res.* **2010**, *116*, 58–67. [CrossRef]

59. Yang, P.; Shu, C.; Chen, L.; Xu, J.; Wu, J.; Liu, K. Identification of a major QTL for silique length and seed weight in oilseed rape (*Brassica napus* L.). *Theor. Appl. Genet.* **2012**, *125*, 285–296. [CrossRef] [PubMed]

60. Chay, P.; Thurling, N. Identification of genes controlling pod length in spring rapeseed, *Brassica napus* L.; and their utilization for yield improvement. *Plant Breed.* **1989**, *103*, 54–62. [CrossRef]

61. Bommarco, R.; Marini, L.; Vaissière, B. Insect pollination enhances seed yield, quality, and market value in oilseed rape. *Oecologia* **2012**, *169*, 1025–1032. [CrossRef] [PubMed]

62. Biesmeijer, J.C.; Roberts, S.P.M.; Reemer, M.; Ohlemüller, R.; Edwards, M.; Peeters, T.; Schaffers, A.P.; Potts, S.G.; Kleukers, R.; Thomas, C.D.; et al. Parallel declines in pollinators and insect-pollinated plants in Britain and the Netherlands. *Science* **2006**, *313*, 351–354. [CrossRef] [PubMed]

63. Cook, S.M.; Rasmussen, H.B.; Birkett, M.A.; Murray, D.A.; Pye, B.J.; Watts, N.P. Behavioural and chemical ecology underlying the success of turnip rape (*Brassica rapa*) trap crops in protecting oilseed rape (*Brassica napus*) from the pollen beetle (*Meligethes aeneus*). *Arthropod-Plant Interact.* **2007**, *1*, 57–67. [CrossRef]

64. Gan, S.S.; Amasino, R.M. Inhibition of leaf senescence by autoregulated production of cytokinin. *Science* **1995**, *270*, 1986–1988. [CrossRef] [PubMed]

65. Verma, V.; Foulkes, M.J.; Worland, A.J.; Sylvester-Bradley, R.; Caligari, P.D.S.; Snape, J.W. Mapping quantitative trait loci for flag leaf senescence as a yield determinant in winter wheat under optimal and drought-stressed environments. *Euphytica* **2004**, *135*, 255–263. [CrossRef]

66. Makino, A.; Osmond, B. Effects of nitrogen nutrition on nitrogen partitioning between chloroplasts and mitochondria in pea and wheat. *Plant Physiol.* **1991**, *96*, 355–362. [CrossRef] [PubMed]

67. Gifford, R.M.; Evans, L.T. Photosynthesis, carbon partitioning, and yield. *Ann. Rev. Plant Physiol.* **1981**, *32*, 485–509. [CrossRef]

68. Brar, G.; Thies, W. Contribution of leaves, stem, siliques and seeds to dry matter accumulation in ripening seeds of rapeseed, *Brassica napus* L. *Z. Pflanzenphysiol.* **1977**, *82*, 1–13. [CrossRef]

69. Nakano, H.; Muramatsu, S.; Makino, A.; Mae, T. Relationship between the suppression of photosynthesis and starch accumulation in the pod-removed bean. *Aust. J. Plant Physiol.* **2000**, *27*, 167–173.

70. Hay, R.; Kirby, E. Convergence and synchrony-a review of the coordination of development in wheat. *Aust. J. Agric. Res.* **1991**, *42*, 661–700. [CrossRef]

71. Price, J.S.; Hobson, R.N.; Neale, M.A.; Bruce, D.M. Seed losses in commercial harvesting of oilseed rape. *J. Agric. Eng. Res.* **1996**, *65*, 183–191. [CrossRef]

72. Meakin, P.J.; Roberts, J.A. Anatomical and biochemical changes associated with the induction of oilseed rape (*Brassica napus*) pod dehiscence by *Dasineura brassicae* (Winn.). *Ann. Bot.* **1991**, *67*, 193–197. [CrossRef]

73. Hensel, L.L.; Grbic, V.; Baumgarten, D.A.; Bleecker, A.B. Developmental and age-related processes that influence the longevity and senescence of photosynthetic tissues in Arabidopsis. *Plant Cell* **1993**, *5*, 553–564. [CrossRef] [PubMed]

74. Nooden, L.D.; Penney, J.P. Correlative controls of senescence and plant death in *Arabidopsis thaliana* (Brassicaceae). *J. Exp. Bot.* **2001**, *52*, 2151–2159. [CrossRef] [PubMed]

75. Marbà, N.; Duarte, C.M.; Agustí, S. Allometric scaling of plant life history. *Proc. Natl. Acad. Sci. USA* **2007**, *104*, 15777–15780. [CrossRef] [PubMed]

76. Broadley, M.R.; Hammond, J.P.; King, G.J.; Astley, D.; Bowen, H.C.; Meacham, M.C.; Mead, A.; Pink, D.A.; Teakle, G.R.; Hayden, R.M.; et al. Shoot calcium and magnesium concentrations differ between subtaxa, are highly heritable, and associate with potentially pleiotropic loci in *Brassica oleracea*. *Plant Physiol.* **2008**, *146*, 1707–1720. [CrossRef] [PubMed]

77. Lancashire, P.; Bleiholder, H.; Boom, T.; van den Bout-van den Beukel, C.J.P.; Langelüddeke, P.; Stauss, R.; Weber, E.; Velde, T.; Lancashire, P.D.; Boom, T.V.D.; et al. A uniform decimal code for growth stages of crops and weeds. *Ann. Appl. Biol.* **1991**, *119*, 561–601. [CrossRef]

78. Kelly, A.A.; Shaw, E.; Powers, S.J.; Kurup, S.; Eastmond, P.J. Suppression of the SUGAR-DEPENDENT1 triacylglycerol lipase family during seed development enhances oil yield in oilseed rape (*Brassica napus* L.). *Plant Biotechnol. J.* **2013**, *11*, 355–361. [CrossRef] [PubMed]

10

CO$_2$-Induced Changes in Wheat Grain Composition: Meta-Analysis and Response Functions

Malin C. Broberg [1],*, Petra Högy [2] and Håkan Pleijel [1]

[1] Department of Biological and Environmental Sciences, University of Gothenburg, P.O. Box 461, SE-40530 Göteborg, Sweden; hakan.pleijel@bioenv.gu.se

[2] Institute of Landscape and Plant Ecology, University of Hohenheim, Ökologiezentrum 2, August-von-Hartmann Str. 3, D-70599 Stuttgart, Germany; petra.hoegy@uni-hohenheim.de

* Correspondence: malin.broberg@bioenv.gu.se

Academic Editor: Hans-Joachim Weigel

Abstract: Elevated carbon dioxide (eCO$_2$) stimulates wheat grain yield, but simultaneously reduces protein/nitrogen (N) concentration. Also, other essential nutrients are subject to change. This study is a synthesis of wheat experiments with eCO$_2$, estimating the effects on N, minerals (B, Ca, Cd, Fe, K, Mg, Mn, Na, P, S, Zn), and starch. The analysis was performed by (i) deriving response functions to assess the gradual change in element concentration with increasing CO$_2$ concentration, (ii) meta-analysis to test the average magnitude and significance of observed effects, and (iii) relating CO$_2$ effects on minerals to effects on N and grain yield. Responses ranged from zero to strong negative effects of eCO$_2$ on mineral concentration, with the largest reductions for the nutritionally important elements of N, Fe, S, Zn, and Mg. Together with the positive but small and non-significant effect on starch concentration, the large variation in effects suggests that CO$_2$-induced responses cannot be explained only by a simple dilution model. To explain the observed pattern, uptake and transport mechanisms may have to be considered, along with the link of different elements to N uptake. Our study shows that eCO$_2$ has a significant effect on wheat grain stoichiometry, with implications for human nutrition in a world of rising CO$_2$.

Keywords: *Triticum aestivum*; carbon dioxide; minerals; protein; starch; baking properties; crop quality; food security

1. Introduction

The atmospheric concentration of carbon dioxide (CO$_2$) has steadily increased since the 19th century, from the pre-industrial level of 280 ppm to the current level of 400 ppm [1]. Latest projections by the Intergovernmental Panel on Climate Change [1] suggest that concentrations are likely to reach levels in the range of 420 ppm (RCP2.6) to 1300 ppm (RCP8.5) by the year 2100.

The effects of elevated CO$_2$ (eCO$_2$) on plants are well studied, in particular on food crops due to the strong concern for future food security. Photosynthesis and growth in C3 plants are often enhanced by eCO$_2$ resulting in a higher yield, which has been observed for many crops [2]. The magnitude of yield response has been shown to vary between different crops [3] and crop varieties [4,5], but also to depend on differences in experimental systems [6]. It has been argued that yield stimulation is overestimated due to unrealistic growing conditions in enclosure systems, including open-top chambers (OTCs) [7,8]. In contrast, Ziska and Bunce [9] found that there were no significant differences in yield response for rice, soybean, and wheat when comparing experiments using enclosure methodologies with Free-Air-CO$_2$-Enrichment (FACE) technology in a single experiment. According to Körner [10], carbon is rarely the limiting factor for plant growth but soil resources, e.g., nutrients and water, are more likely to determine plant performance and the observed positive effects of eCO$_2$ are according to this argument consequently a result of improved water use efficiency. Comparing the eCO_2 effects on plants grown in different experimental systems could possibly reveal if these statements are valid also for effects on wheat crop quality.

Wheat is a major food crop globally, being the second most important energy source for the human population with an annual global production of approximately 700 million tons [11]. The main source of food energy within the wheat grain is starch, accounting for 50%–70% of total grain mass. It has been proposed that eCO_2 could enhance concentration of carbohydrates, starch being the major component, and thus reduce the concentrations of other constituents, often referred to as the "dilution hypothesis" [12]. Photosynthetic nitrogen (N) use efficiency can potentially increase under eCO_2 [13], and consequently more carbon can be assimilated with the same amount of N, resulting in a relative decrease in N content in the leaf. Since most of the grain N is translocated from non-reproductive parts of the plant during grain filling [14], grain N content could also be affected under eCO_2 by this mechanism.

Changes in crop quality, like nutritional aspects, are of great importance for assessments of climate change and eCO_2 effects on future food production [15]. Loladze [16] pointed out that eCO_2 is likely to induce a shift in the stoichiometry, i.e., the elemental balance of plants, promoting higher concentrations of C and lower concentrations of e.g., N, Fe, and Zn, with important implications for human nutrition. The average effect on protein (hereafter referred to as N) content, estimated in a meta-analysis by Taub et al. [17], showed a significant decrease for several crops, including wheat, barley, rice, and soybean. Along with the "dilution hypothesis" a few more hypotheses have been proposed to explain the observed pattern of decreasing N concentration in plants exposed to eCO_2, such as a reduction in transpiration driven mass flow [18] and impaired N acquisition [19], processes that both can result in a reduced N uptake under eCO_2 even without yield stimulation. According to the mechanism put forward by Bloom [19], the decrease in photorespiration under eCO_2 leads to a reduced malate export from the chloroplasts, and the nicotinamide adenine dinucleotide hydride (NADH) generated from this malate in the cytoplasm powers the reduction of nitrate (NO_3^-) to nitrite (NO_2^-), which is the first step of plant NO_3^- assimilation. In line with this, Pleijel and Uddling [20] found that the dilution hypothesis is likely to exist, but cannot fully explain the reduction in N concentration in wheat under eCO_2, since N concentration is reduced also where grain yield is unaffected. This suggests a role for the mechanism proposed by Bloom [17]. Another important and related question is if there is a level of CO_2 where the effect of eCO_2 on grain N concentration saturates, analogous to the saturation seen in the response of photosynthesis under eCO_2 of C3 plants [21].

The effects on N content in wheat grains have been observed in a rather large number of studies with wheat grown under eCO_2, while observations of effects on other elements are limited. As suggested by Loladze [16], the decrease in concentrations of some essential mineral nutrients (Fe and Zn) have been documented [22,23], while it is still uncertain to what extent other elements are affected by eCO_2 and the mechanism behind the observed changes. Reduction in concentrations of N and nutrient elements are of great concern for future food security and the issue of so called 'hidden hunger', where the amount of calories might be sufficient but with undernourishment with respect to essential nutrients. A modelling study by Myers et al. [24] estimated that the CO_2-induced reduction in Zn concentration in staple crops could substantially increase the number of people at risk of Zn deficiency by 138 million by 2050. Cereals, including wheat, are also an important source of dietary Cadmium (Cd) exposure [25], which could cause injury to kidney and bones [26], hence the CO_2 effect on Cd content is also of importance.

N is often considered to be one of the most limiting elements for crop growth, and thus have the potential to be primarily affected by eCO_2. If there are common mechanisms behind eCO_2 effects on N and other elements it should be possible to detect the correlation of effects. Assuming that dilution is the main process that acts to reduce mineral concentration, the eCO_2 effect on grain yield would be closely related to effects on minerals, where a negative effect on mineral concentration will only occur in association with yield stimulation.

Since wheat is used for baking to a large extent, it is also relevant to study how different baking properties are affected by eCO_2, where alteration in quality may affect the market value and quality of products (e.g., review by Högy et al. [27]). Many measures of baking properties are related to the content and quality of protein, such as gluten concentration and composition, dough elasticity/resistance, and bread loaf volume, and consequently these variables are likely to be impaired by eCO_2 following the pattern of grain N concentration. Negative effects on various baking properties have been observed in individual experiments [28–32], but to our knowledge no meta-analysis has been made on this aspect.

This study intends to provide an up-to–date review of observed effects of eCO_2 on wheat grain quality, based on all available ecologically realistic experiments. Meta-analysis is used to test the overall magnitude and statistical significance of the effects. There is, however, also a need to understand the gradual change in the dietary value of wheat crops as CO_2 concentration increases. To meet this need, a novel aspect of our study is that we provide response functions for the effects of eCO_2 on the concentration of N and other minerals in wheat and test to what extent the data suggest responses to be linear or non-linear. We also assess the eCO_2 impact on the total production of starch, N, and other nutritionally important elements, by estimating the eCO_2 effect on the content (mass per unit area) of those constituents. As a further novel contribution, we relate the effect of eCO_2 on the concentration of a range of minerals to the eCO_2 effect on N concentration and grain yield. This is done in order to understand to what extent eCO_2 effects are consistent among different minerals and the degree to which they are related to the effects on N concentration and grain yield stimulation. By these three approaches our study aims to examine the following research questions:

1. Are the negative effects of eCO_2 on N concentration and N content independent of the experimental setup, such as exposure system, rooting environment, and concentration level of CO_2 treatment?
2. Is the negative effect of eCO_2 on N concentration saturating at high CO_2?
3. To what extent are the nutritional and baking quality of wheat grain negatively affected by eCO_2?
4. Can starch dilution explain the reduction in concentration of N and minerals under eCO_2?
5. Are effects of eCO_2 on mineral concentration linked to the effect on N concentration and grain yield stimulation?

2. Results

2.1. Nitrogen and Starch

Grain N concentration was significantly reduced by eCO_2 with an overall effect of -8.4% (confidence interval (CI) -9.8 -7.4; Figure 1a). The magnitude of effect was shown to be dependent on the experimental setup where significant differences were observed between exposure systems (FACE < OTC) and the rooting environment (pots > field soil). There was, however, no significant difference between OTC and FACE when excluding eCO_2 treatments >600 ppm (only OTC experiments). A comparison of concentration levels (above or below 600 ppm) in OTC experiments did not show any significant difference, but indicated a larger effect with higher CO_2. Even though N concentration was reduced by eCO_2 there was a significant increase in N content, with an overall effect of 12% (CI 7.93 15.90; Figure 1a), associated with a strong grain yield stimulation. Subgroup analysis revealed that experiments performed in field tunnels (FT) and pots did not show a significant CO_2 effect on N content; however, it should be noted that those groups have few observations and thus larger CIs. There were no significant differences with regard to the effect on N content when comparing OTC with FACE or different CO_2 concentrations.

The response function for the relationship between N concentration and CO_2 (Figure 1b) showed a strong non-linear relationship ($r^2 = 0.57$), with an initial reduction in N concentration with increasing CO_2, but reaching a minimum at ~600 ppm. N content ($g \cdot m^{-2}$) was positively affected by eCO_2, but showed a rather weak relationship with CO_2 ($r^2 = 0.19$). Details of the regression models are presented in Table 1.

(a)

(b)

$$y = 49.9 - 0.20x + 0.00017x^2$$
$$R^2 = 0.57$$

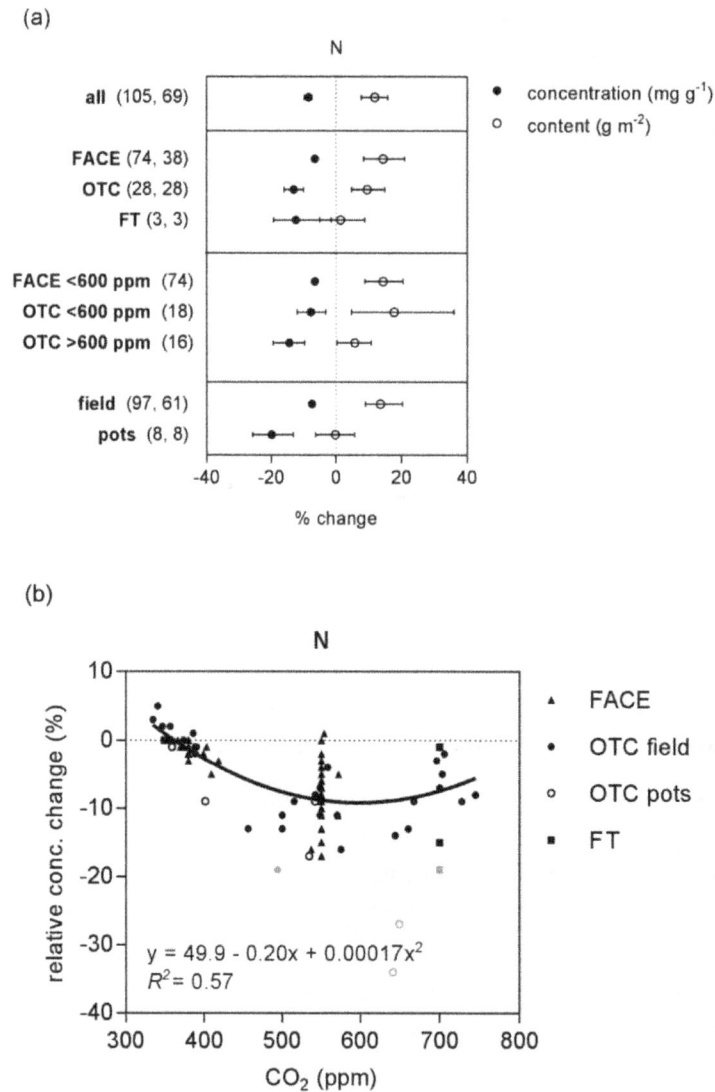

Figure 1. (a) Meta-analysis of eCO_2 effects on N concentration and N content using ambient CO_2 as the reference, with subgroup-analysis of exposure systems, rooting environment, and concentration level for eCO_2 treatment. Number of comparisons for concentration and content, respectively, are given within brackets. (b) Response function for N concentration (relative to 350 ppm) with CO_2 concentration, grey markers show data points identified as outliers not included in the curve fitting.

Table 1. Response functions for regression of concentration ($mg \cdot g^{-1}$) and content ($g \cdot m^{-2}$) of N, starch, and minerals with CO_2.

	Variable	Observations	Regression Model	B0	B1	B2	r^2	Sign.	Preferred Model
N	concentration	132 (4)	linear	9.9	−0.031		0.43	*	
			quadratic	49.9	−0.198	1.66×10^{-4}	0.57		preferred
	content	96 (11)	linear	−5.7	0.025		0.12	*	
			quadratic	−51.9	0.216	-1.85×10^{-4}	0.18		preferred
starch	concentration	30 (3)	linear	0.4	0.001		0.00083	ns	preferred
			quadratic	10.3	−0.039	3.72×10^{-5}	0.028		
	content	30 (2)	linear	−14.3	0.052		0.35	*	preferred
			quadratic	−7.7	0.026	2.46×10^{-5}	0.35		

Table 1. *Cont.*

	Variable	Observations	Regression Model	B0	B1	B2	r^2	Sign.	Preferred Model
B	concentration	68 (2)	linear	0.9	−0.002		0.00046	ns	preferred
			quadratic						n.a.
	content	32 (4)	linear	−66.4	0.196		0.40	*	preferred
			quadratic						n.a.
Ca	concentration	83 (4)	linear	12.9	−0.037		0.32	*	preferred
			quadratic						n.a.
	content	47 (7)	linear	−16.1	0.056		0.16	*	preferred
			quadratic						n.a.
Cd	concentration	13	linear	12.4	−0.039		0.31	*	preferred
			quadratic	64.0	−0.253	2.07×10^{-4}	0.39		
	content	13	linear	−1.4	0.003		0.0025	ns	preferred
			quadratic	10.3	−0.045	4.71×10^{-5}	0.0068		
Cu	concentration	80 (2)	linear	7.3	−0.020		0.14	*	preferred
			quadratic						n.a.
	content	44 (5)	linear	−33.3	0.104		0.27	*	preferred
			quadratic						n.a.
Fe	concentration	86 (7)	linear	13.7	−0.039		0.51	*	preferred
			quadratic						n.a.
	content	50 (4)	linear	−12.3	0.047		0.07	ns	
			quadratic	−211.7	0.911	-9.00×10^{-4}	0.17		preferred
K	concentration	83 (7)	linear	3.2	−0.008		0.07	*	preferred
			quadratic						n.a.
	content	47 (7)	linear	−37.6	0.116		0.51	*	preferred
			quadratic						n.a.
Mg	concentration	83 (8)	linear	11.7	−0.033		0.61	*	preferred
			quadratic						n.a.
	content	47 (3)	linear	8.2	−0.024		0.39	*	preferred
			quadratic						n.a.
Mn	concentration	84 (3)	linear	6.4	−0.019		0.13	*	
			quadratic	58.9	−0.247	2.39×10^{-4}	0.20		preferred
	content	48 (8)	linear	−20.1	0.067		0.36	*	preferred
			quadratic	−52.3	0.205	-1.41×10^{-4}	0.36		
P	concentration	83 (4)	linear	7.3	−0.022		0.20	*	preferred
			quadratic						n.a.
	content	47 (7)	linear	−27.9	0.018		0.38	*	preferred
			quadratic						n.a.
S	concentration	83 (3)	linear	9.9	−0.028		0.32	*	preferred
			quadratic						n.a.
	content	47 (7)	linear	−19.9	0.065		0.20	*	preferred
			quadratic						n.a.
Zn	concentration	90 (5)	linear	11.3	−0.033		0.18	*	preferred
			quadratic	51.1	−0.205	1.78×10^{-4}	0.21		
	content	54 (6)	linear	−18.0	0.062		0.28	*	
			quadratic	−143.7	0.596	-5.42×10^{-4}	0.43		preferred

Model parameters are presented for both linear ($y = B1x + B0$) and quadratic ($y = B2x^2 + B1x + B0$) curve fits, x being the CO_2 concentration and y the response variable. Values within brackets are the number of data points identified as outliers that were excluded from regressions. Sign: * denotes that the slope of linear model (B1) is significantly ($p \leq 0.05$) different from zero. Model fit is compared for each variable and the simpler (linear) model is preferred unless the p-value of the quadratic term is less than 0.05. ns denotes non-significant; n.a. denotes not applicable.

Figure 2 shows the eCO_2 effect on various baking properties, where a significant negative effect is observed for the Hagberg falling number (−5.8%, CI −9.9 −1.7), Zeleny value (−21.2%, CI −25.5 −16.9), dry gluten content (−16.5%, CI −22.0 −11.2), wet gluten content (−17.0%, CI −23.0 −11.5), peak resistance (−11.4%, CI −17.3 −3.0), and bread loaf volume (−11.9%, CI −21.3 −2.3). Mixing time significantly increased (11.2%, CI 0.6 21.5) under eCO_2, while resistance breakdown remained unaffected (−2.6%, CI −12.5 9.0).

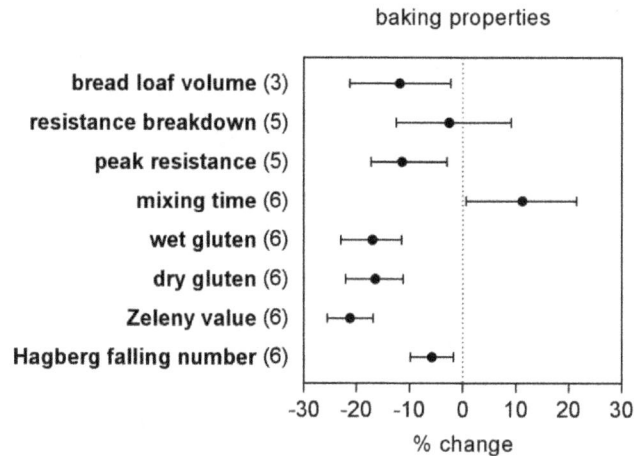

Figure 2. Meta-analysis showing the effect of eCO_2 on various baking properties using ambient CO_2 as the reference. Number of comparisons are given within brackets.

Meta-analysis for the eCO_2 effect on grain starch concentration (Figure 3a) showed a non-significant positive effect of 2.2% (CI −0.6 6.2). In line with this result, the response function for starch concentration with CO_2 did not reveal any relationship (Figure 3b). Starch content was significantly positively affected by 20.8% (CI 12.4 30.9). Due to limited amount of data (19 observations), subgroup analysis was not performed for starch concentration and starch content.

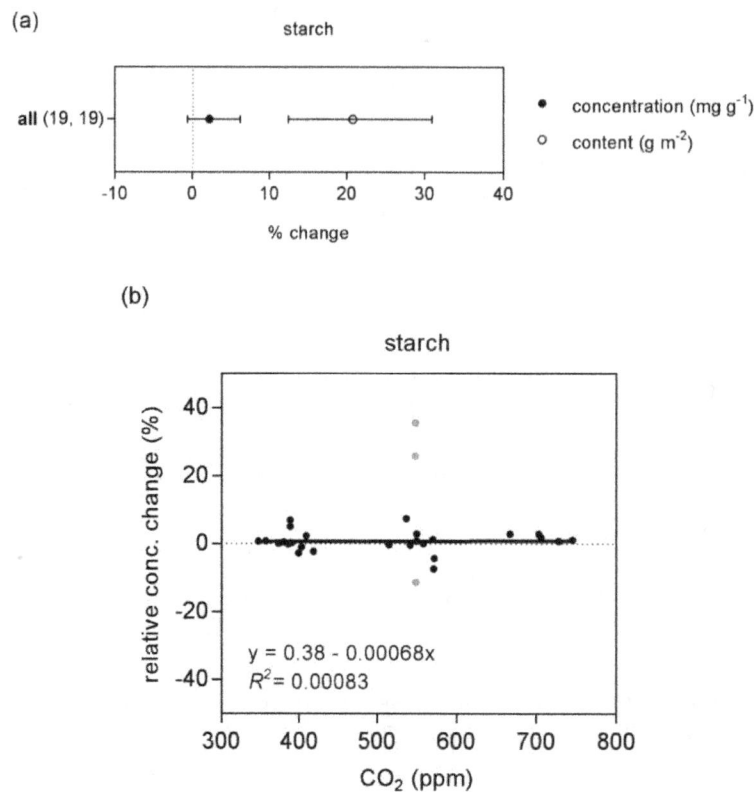

Figure 3. (a) Meta-analysis of eCO_2 effects on starch concentration and content using ambient CO_2 as a reference. Number of comparisons for concentration and content, respectively, are given within brackets. (b) Response function for starch concentration (relative to 350 ppm) with CO_2. Grey markers show data points identified as outliers and that were not included in the curve fitting.

2.2. Minerals

Meta-analysis (Figure 4) showed that eCO_2 significantly reduced the concentration of various minerals (Ca, Cd, Cu, Fe, Mg, Mn, P, S, and Zn) in wheat grains, while others were unaffected (B and Na) or significantly increased by a small amount (K). A significant increase in content was observed for all minerals except for Cd. It should be noted that there was a considerable variation in the magnitude of response (concentration and content) among the different elements.

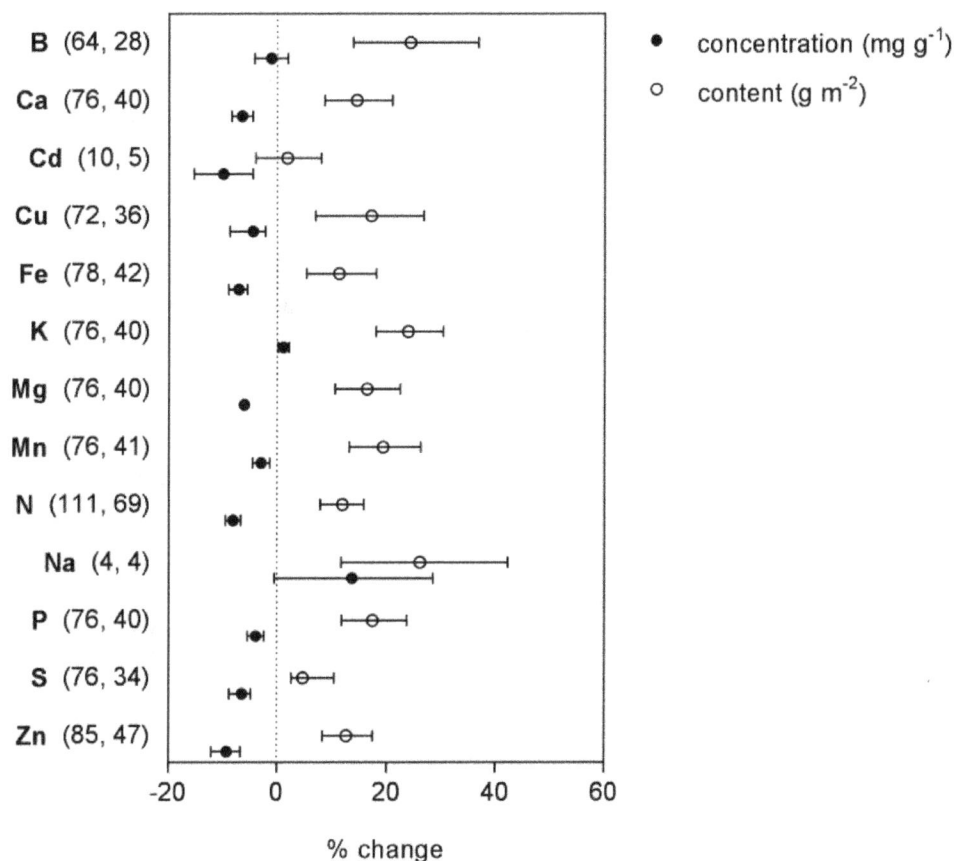

Figure 4.   Meta-analysis output for mineral concentration and content using ambient CO_2 as the Numbers within brackets gives the number of comparisons for the concentration and content, respectively. content, respectively.

Response functions in Figure 5 show that concentrations of several mineral nutrients had a strong linear relationship with increasing CO_2, with a significant negative slope for all elements (Fe, Mg, P, S, and Zn) except K. Regression models for the remaining elements are presented in Table 1. Concentrations of Ca, Cd, and Cu also showed a significant linear decrease with higher CO_2, however, a quadratic model had a better fit for Mn, while B did not show any relationship with CO_2. Na was excluded from this analysis due to the small number of observations. The slope of the linear regression line suggests a reduction in mineral concentration of about 2%–4% per 100 ppm for all minerals except for B and K, which had a non-significant slope close to zero. Mineral content showed a positive relationship with CO_2 and a significant slope for all elements except for Cd and Fe (Table 1). The strongest relationships were found for B, K, Mg, and P with an r^2 between 0.40 and 0.68.

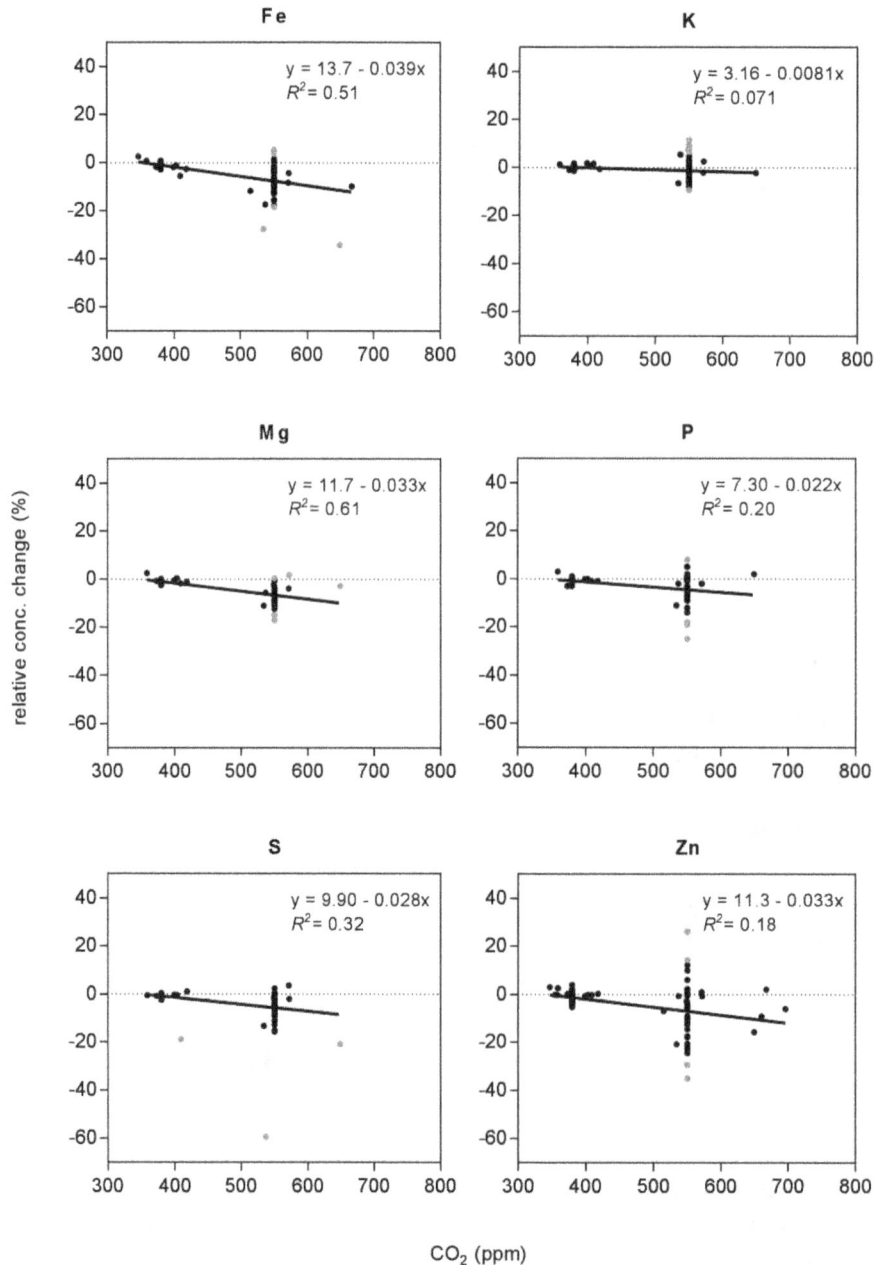

Figure 5. Response-functions for mineral concentrations of P, Mg, Fe, K, Zn, and S (relative to 350 ppm) with CO_2 concentration. Grey markers show data points identified as outliers and not included in the curve fitting.

2.3. Effects on Minerals in Relation to the Effects on N Concentration and Grain Yield

Figure 6 shows the relationship between eCO_2 effects on the concentration of various minerals and the eCO_2 effect on the N concentration. The correlation coefficient provides an estimate of the association of effects, and a strong association ($r > 0.75$) is found for S and Fe ($r = 0.87$ and $r = 0.79$, respectively). Ca, Cd, Mg, P, and Zn show a moderate association ($0.5 < r < 0.75$), while it was rather weak for the remaining elements (B, Cu, K, and Mn).

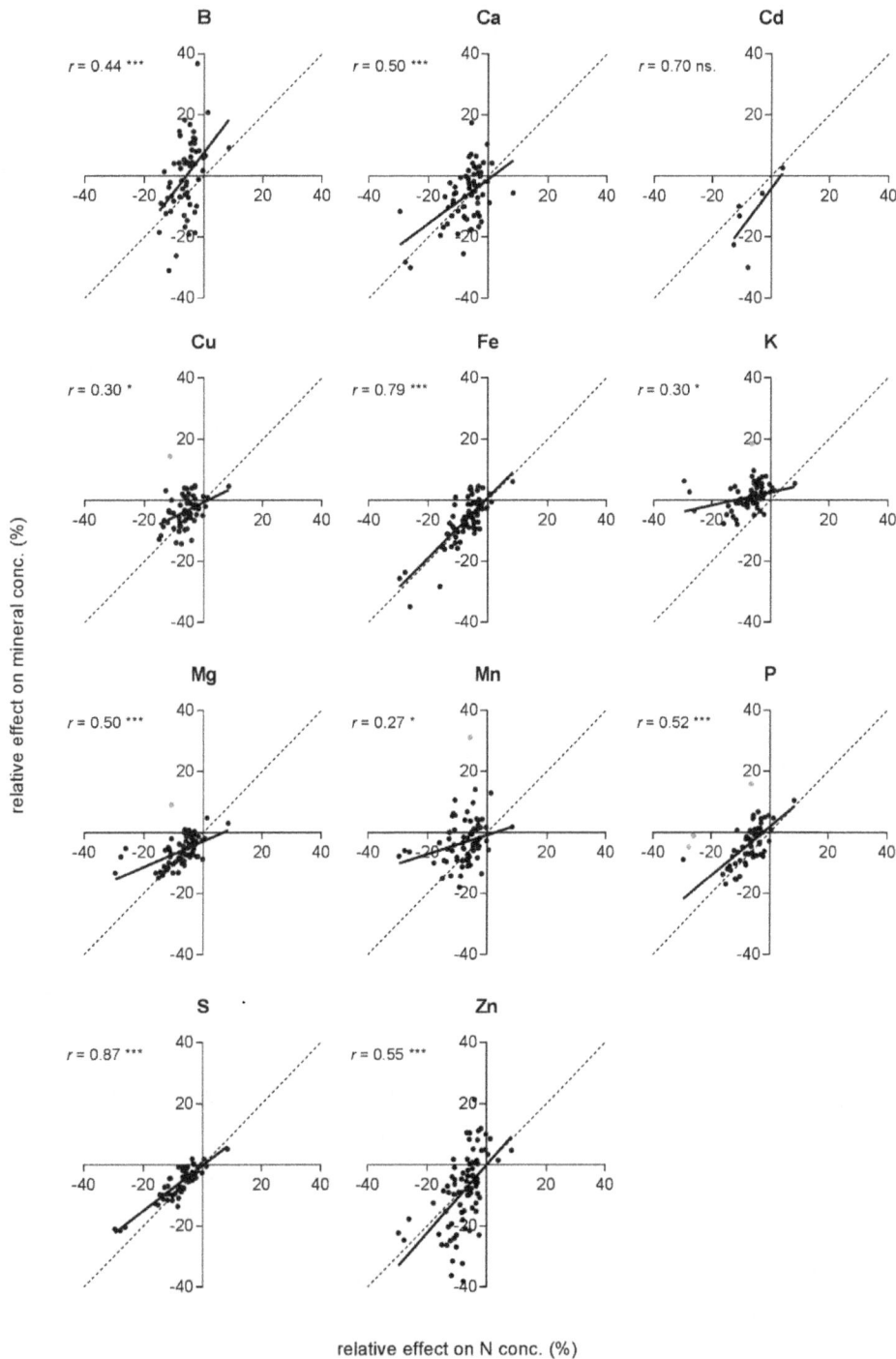

Figure 6. Relative effect of eCO_2 on mineral concentration (B, Ca, Cd, Cu, Fe, K, Mg, Mn, P, S, and Zn) related to the relative effect on N concentration. Correlation coefficient (r) and its significance is presented in each plot. Black solid lines represent the linear regression model, for which parameters and model performance are presented in Table 2. Grey markers show data points identified as outliers not included in the curve fitting. Dashed lines represent the hypothetical situation where the effect of eCO_2 on mineral concentration is equal to the effect on N concentration.

Regression analysis of the eCO_2 effect on mineral concentrations with the CO_2 effect on the N concentration (Table 2) showed a strong relationship for S ($R^2 = 0.75$) and Fe ($R^2 = 0.63$), while the relationships were rather weak for B, Cu, K, and Mn ($R^2 < 0.25$). Remaining elements were found in the intermediate range ($0.25 < R^2 < 0.50$). A deviation of the fitted line from the 1:1 line indicates that the element:N ratio was affected by eCO_2, hence the grain stoichiometry was altered. Cu, K, Mg, Mn, and S had a slope smaller and significantly different from 1 (Table 2), while the remaining elements did not having regression slopes significantly different from the 1:1 line. Relating the eCO_2 effect on minerals to the effect on grain yield showed a weak and non-significant relationship for most elements (Table 2), except for the concentrations of K, N, P, and Zn that had significantly negative relationships with the effects on grain yield (Figure 7).

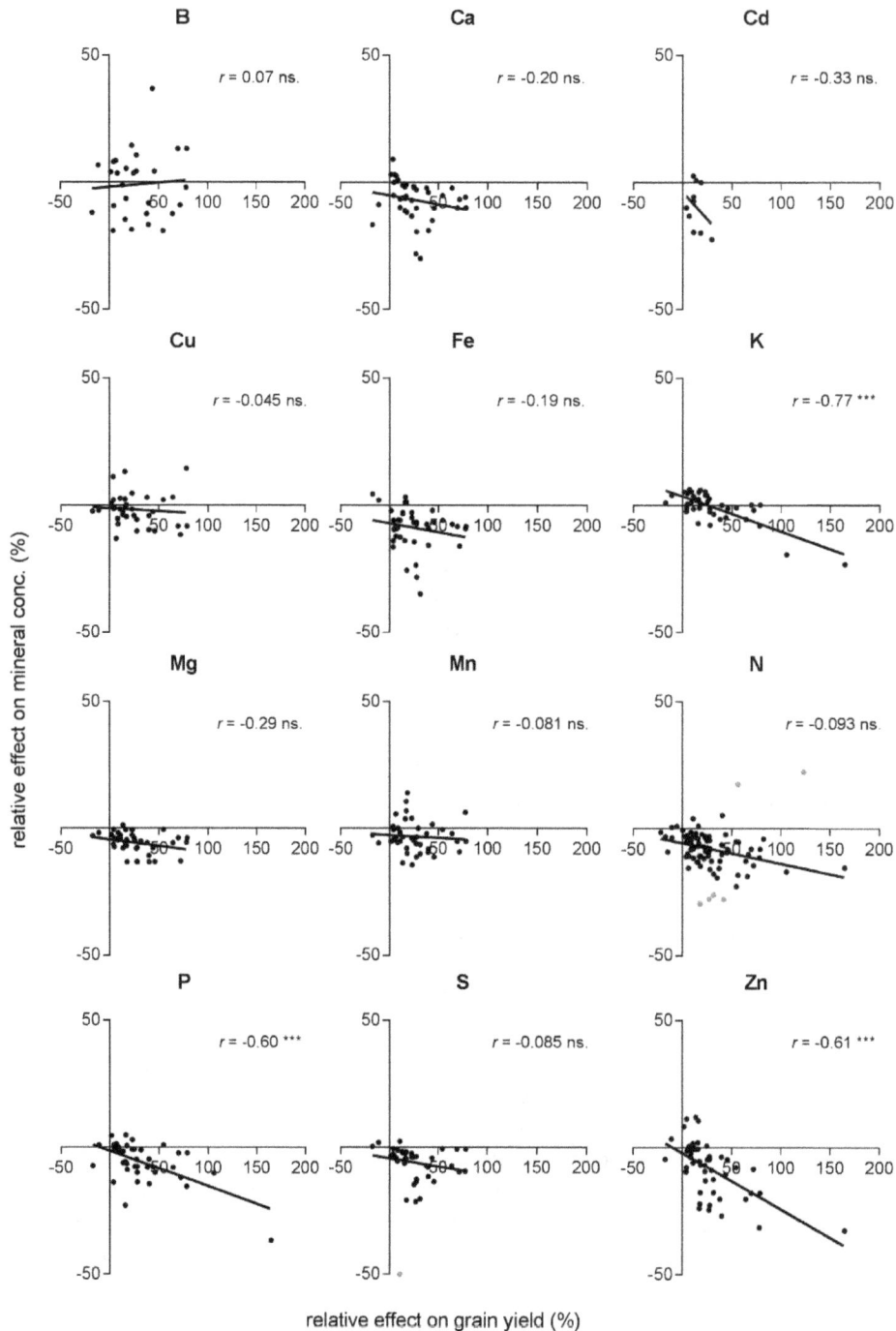

Figure 7. Relative effect of eCO_2 on the mineral concentration of B, Ca, Cd, Cu, Fe, K, Mg, Mn, N, P, S, and Zn vs. the relative effect on grain yield ($g \cdot m^{-2}$). Black solid lines represent the linear regression model, for which parameters and model performance are presented in Table 2.

Table 2. Response functions for the linear regression between the relative eCO_2 effect on the concentration of various minerals with the eCO_2 effect on N concentration and grain yield.

x	Element	Observations	r^2	B0	B1	Sign.
N	B	64	0.20	7.65	1.28	ns.
	Ca	69	0.46	1.80	0.80	ns.
	Cd	6	0.49	−4.17	1.29	ns.
	Cu	65 (1)	0.17	−0.66	0.48	*
	Fe	70	0.63	0.80	0.99	ns.
	K	69 (1)	0.11	2.57	0.21	*
	Mg	76	0.32	−2.86	0.42	*
	Mn	74 (1)	0.084	−0.84	0.31	*
	P	69 (3)	0.46	1.80	0.80	ns.
	S	68	0.75	−0.18	0.74	*
	Zn	83	0.30	−0.084	1.11	ns.
Grain yield	B	28	0.0049	−1.87	0.034	ns.
	Ca	40	0.042	−5.10	−0.070	ns.
	Cd	10	0.11	−3.90	−0.42	ns.
	Cu	36 (1)	0.0021	−1.18	−0.24	ns.
	Fe	42	0.038	−7.08	−0.069	ns.
	K	42	0.59	3.61	−0.14	*
	Mg	40	0.087	−4.42	−0.51	ns.
	Mn	43	0.0065	−2.63	−0.022	ns.
	P	42	0.36	−1.38	−0.14	*
	S	40 (1)	0.063	−4.26	−0.065	ns.
	Zn	50	0.37	−21.1	−0.22	*
	N	87 (6)	0.18	−5.70	−0.081	*

Values within brackets are the number of data points identified as outliers that have been excluded from the regressions. B0 gives the intercept and B1 gives the slope of regression line. Sign.: * denotes that slope (B1) is significantly ($p \leq 0.05$) different from 1, when x = effect on N, and significantly different from zero when x = effect on grain yield. ns, denotes non-significant.

3. Discussion

The overall results from this study suggest that eCO_2 can cause an overall significant shift in wheat grain stoichiometry, with concentration reductions for N and several nutritionally important minerals, in line with the conclusion of Loladze [12,16], together with a decreased baking quality and thus lower commodity value. This is the most comprehensive synthesis of eCO_2 effects on mineral elements in wheat, with meta-analyses including more than 60 pair of observations for most mineral elements and 105 for N.

Our results showed a significant negative effect of eCO_2 on N concentration regardless of experimental setup. The negative effect of eCO_2 on N concentration observed in some recent studies was estimated to be between 6.3% [23] and 9.8% [17], which is in line with the overall results in this study (8.4%). The large amount of data gives robust results (small CIs) and allows for subgroup analysis to unravel the sources of variation within the data.

The response function for the relationship between N concentration and CO_2 indicates that there is a gradual reduction in N concentration that saturates around 600 ppm. This has not been highlighted before and is of importance e.g., for scenarios of how the nutritional value in crops will gradually change over the present century in response to rising CO_2. The meta-analysis, however, points to a stronger response in experiments using an eCO_2 level above 600 ppm compared to that below 600 ppm, although this difference was not statistically significant. The significant difference detected when comparing OTC and FACE for all data was indicated to be a consequence of the different levels of eCO_2 used, and not the exposure system itself, since the difference was not found when comparing FACE with the subset of OTC experiment data with eCO_2 concentrations below 600 ppm (average eCO_2 550 ppm and 528 ppm, respectively).

The comparison of the rooting environments showed that there was a much stronger negative effect on N concentration in potted plants compared to those grown in field soil. This is in line with the results from Taub et al. [17] where wheat grown in OTCs showed a similar difference in response between rooting environments. Assuming that experiments in field soil are more realistic, this suggests that potted experiments may strongly overestimate the negative effect of eCO_2 on N concentration. Potted plants are more likely to suffer from nutrient limitation due to their restricted rooting space, thus nutrient uptake cannot increase with the same rate as photosynthesis under eCO_2. It should, however, be noted that only eight pairs of observations from potted plants were included in this study, compared to 97 observations for field soil, and the large CIs for potted plants indicate that conclusions about them are uncertain.

As a consequence of the decrease in N concentration, eCO_2 had a significant negative effect on most baking properties (Figure 2), even though the number of observations is rather small. A reduction in gluten proteins results in lower elasticity and resistance of the dough and smaller bread loaf volume, but also longer mixing time [33]. In addition, the falling number was reduced under eCO_2 reflecting an increase in α-amylase activity, which is associated with poor baking properties, such as sticky dough and poorly structured loaves [34], but also shortens the storage time of flour and grains [35].

No significant effect of eCO_2 on starch concentration could be demonstrated and consequently the negative effect on N could not be explained by starch dilution, thus the dilution hypothesis was not supported. On the other hand, the number of observations is rather small, resulting in large CIs and low statistical power. Since starch is a major component of the wheat grain (50%–70%) even a small change in its concentration could alter the grain stoichiometry considerably. To detect an effect with small magnitude a large sample size is required and the non-significant results found here could be a consequence of power failure. Even with a small effect size, a dilution effect by starch is likely to be of importance for all elements although other factors, such as transpiration-driven mass flow, N acquisition, and variation in plant demand-to-availability, can modify and even overshadow the dilution effect [12]. Further investigations would be needed to outline the relative importance of different mechanisms for different elements and growing conditions.

The analysis of the eCO_2 effect on mineral concentrations (Figures 4, 6 and 7) showed that there was a variation in the magnitude of effects, ranging from effects close to zero to reductions of about 10%. Together with the non-significant effect on starch concentration, this indicates that CO_2-induced responses cannot be explained only by a simple growth dilution model. In addition, almost all elements (except K) showed a weak relationship when comparing eCO_2 effects on mineral concentration with grain yield stimulation. If dilution was the only mechanism operating, the reduction in mineral concentration would closely follow the increase in biomass and would be the same for all elements.

The eCO_2 effects on Fe and S were strongly correlated to the effects on N (Figure 5) and those elements were also among the ones most strongly negatively affected by eCO_2 in the meta-analysis (Figure 4). In contrast, the effect on minerals (B, Cu, K, and Mn) that showed a weak relationship with effects on N, were observed to be little (B, Cu, Mn) or not significantly (K) affected by eCO_2. This suggests that eCO_2 effects on N may play a role also for other minerals such as Fe and S. The regression of effects between B and N gives a slope >1, however, this should not be interpreted as a stronger effect on B than N since it is mainly a result of the large response range in B (with both positive and negative effects) compared to N. As shown in the meta-analysis (Figure 4) the large variation of eCO_2 effects on B cancel each other out, resulting in a net zero effect.

The different response patterns of mineral elements could possibly be attributed to their different functions in the plant. In a study by Ågren and Weih [36] stoichiometric clusters of mineral elements were identified in the leaves of six *Salix* genotypes grown under altered water and nutrient supply. Changes in concentration for one group of elements (N, P, S, and Mn) were associated with growth, the second group (K, Ca, and Mg) followed changes in biomass, while the third group (Fe, B, Zn, and Al) were believed to be limited by soil availability. It was also suggested that these groups could be associated with different biochemical functions, where elements of the first group are linked to nucleic acids/proteins, the second group is related to structure/photosynthesis, and the third group is associated with enzymes. The significant relationship between K and grain yield stimulation (Figure 7) confirms that K concentration is associated with changes in biomass, while the corresponding relationships were rather weak (non-significant) for Ca and Mg. With the current data it is not possible to test if the elements most strongly affected by eCO_2 in our study, Fe and Zn, are reduced due to soil limitation or if they are functionally linked to N. It is also important to note that effects on element concentration in leaves do not necessarily translate to the same response in seeds.

The mineral concentration in wheat grains is generally a result of total plant uptake, biomass accumulation, and the rate of translocation from vegetative tissues during grain filling. Waters et al. [37] showed that the translocation of Fe, Zn, and N from vegetative tissues to grain is partly regulated by the same proteins in wheat plants. eCO_2 could possibly affect translocation rates indirectly through higher leaf temperatures due to lower transpiration rates [38]. Increase in leaf temperature can lead to heat stress, which is known to promote senescence [39], and thus shorten the grain filling period [40]. This is, however, likely to increase the concentrations of minerals since starch accumulation is often more strongly reduced than N and minerals [39]. If the rate and efficiency of translocation were strongly affected by eCO_2, Fe, Zn, and N could be expected to follow the same response pattern. Our results (Figure 6) show a strong correlation between eCO_2 effects on Fe and N ($r = 0.79$), while the relationship is moderately strong for Zn and N ($r = 0.55$), suggesting that additional mechanisms are of importance in terms of wheat grain concentrations for Zn.

In line with other minerals, Cd concentration was significantly reduced under eCO_2, which could be considered a positive effect due to the toxicity of Cd. A reduction in Cd concentration was also observed for wheat grown under CO_2 enrichment [20] and ozone exposure [41]. Cd is a non-essential element for the plant and the uptake is known to be dependent on transpiration driven mass-flow [42], therefore lower concentrations could be expected since transpiration rates are often reduced under both eCO_2 and high ozone [43].

The content of N (Figure 1) and all minerals, except for Cd (Figure 4), were significantly increased under eCO_2, which indicates that there is an increase in total soil uptake of these elements. As a potential mitigation strategy, more fertilizers could be added to the agricultural system, however, with the risk of also increasing the leaching of nutrients and enhanced emissions of nitrous oxide (N_2O), ammonia (NH_3), and nitrogen oxide (NO).

In order to fully understand mechanisms behind the shift in wheat grain composition, further research is needed. The response of nutrient concentration under eCO_2 has to be tested under different levels of fertilizers and water supply to identify possible interaction of these factors, which has been done for N in a few experiments (e.g., Li et al. [44]) but not for other nutrients. It would also be possible to follow translocation rates of elements from straw and leaves, by measuring element composition of all plant parts during growth. Simultaneous measurements of transpiration could test the strength of the link between eCO_2 effects on different minerals in crop yield and transpiration-driven mass flow.

eCO_2-induced reductions in the concentration of N, as a proxy for protein, and essential minerals can have significant impacts on human nutrition. Fe and Zn deficiency is already an urgent issue in many parts of the world. An estimated two billion people suffer from these deficiencies [45], especially in regions where people depend on C3 grains such as wheat as their primary dietary source of Zn and Fe. Consequently, these factors are also important to take into account when assessing the effects of CO_2 and climate change on global food security.

4. Materials and Methods

4.1. Database

Web of Science, Scopus, and Google Scholar were used to survey all peer-reviewed literature published between 1980 and 2016 (May) related to the response of wheat grain quality to eCO_2. Experimental data were included in the database if at least one of the following variables were reported: grain protein concentration (or N concentration), grain starch concentration, grain mineral concentrations (B, Ca, Cd, Cu, Fe, K, Mg, Mn, Na, P, S, and Zn), grain yield, and baking properties (Hagberg falling number, Zeleny value, gluten content, mixing time, peak resistance, resistance breakdown, bread loaf volume). In order to only include ecologically realistic data, experiments performed in greenhouse or closed growth chambers were excluded. For factorial design, experiments with elevated ozone only treatments without ozone fumigation were included, since ozone is known to have significant effects on both yield and grain quality [41]. Data sources for the included experiments are presented in the Supplementary Information (Tables S1 and S2).

Data from figures were extracted using the software GetData Graphic Digitizer [46]. For experiments where the ambient CO_2 were not reported, it was assumed to be equal to the global mean for the year the study was conducted, with the Mauna Loa record used as reference (retrieved from the National Oceanic & Atmospheric Administration (NOAA) [47].

4.2. Meta-Analysis

Meta-analysis was performed using a meta-analytical software package MetaWin [48]. The experimental treatment with ambient CO_2 was used as control, and parameter values were considered independent if they were made on different cultivars, different (CO_2), or different years, in line with previous meta-analysis [43,49]. The effect size used was the natural log of the response ratio (r, the ratio of the means of two groups, experimental and control) reported as percentage change from the control [43,48,50]. All variables were analyzed using an un-weighted approach due to lack of data for the computation of sample variance (standard deviation or standard error with degree of replication). In line with previous meta-analyses [41,48], variance of the effect size was calculated using a resampling method with 9999 iterations, and confidence intervals (CI) were calculated using the bootstrap method. If the 95% CI did not overlap zero, the average effect size for each variable was considered to be significant, and for subgroup analysis the different groups were considered significantly different if the 95% CI did not overlap [49].

Experiments with additional treatments were included, such as different application levels of N, water supply, temperature, and time of sowing. However, only the effect of eCO_2 was tested in the meta-analysis, and interactions of eCO_2 and additional treatments were not further examined. Subgroup analysis was performed for the N concentration and N content, for which a substantial amount of data was available, where data was categorized by (1) exposure system, Free-Air-CO_2-Enrichment (FACE), Open-Top-Chamber (OTC), and Field Tunnel (FT), (2) rooting environment, pots or field soil, and (3) the concentration level of the eCO_2 treatment, above or below 600 ppm (only applicable for OTC experiments).

4.3. Response Functions

Response functions were derived through regression between the relative effect of each variable and the corresponding CO_2 concentration for the treatment. The response was related to the effect estimated at 350 ppm by linear regression for each individual experiment. At 350 ppm the variables were set to take the value of 0 on a relative scale. Both a linear (first order polynomial) and quadratic (second order polynomial) model was fitted to the data, and the simpler model was preferred if the second parameter (in quadratic model) did not significantly improve the model fit. All additional treatments, such as low N, drought, and high temperature, were excluded from the response functions since they were observed to cause large scatter not related to the effect of eCO_2. All response functions were derived using automatic outlier removal [51].

4.4. Comparison of CO_2 Effects on Different Response Variables

The eCO_2 effect was related to the control treatment (ambient CO_2) when relating the effects on minerals to the effects on N or grain yield. The correlation coefficient was calculated to estimate the association of effects, while regression was used to test if effects on minerals are dependent on effects on N or grain yield. Only linear regression was used to explore the relationship with N, since the slope of a linear trend line could be compared to a 1:1 line that represents the theoretical situation where the mineral and N concentrations are equally affected. The deviation from the 1:1 line was tested for each regression model. For regressions between the eCO_2 effect on minerals and the effect on grain yield, it was tested if the slope deviated from zero, where a slope close to zero indicates a poor relationship between the effects on mineral concentrations with grain yield stimulation.

5. Conclusions

Our study, based on an extensive database, shows that eCO_2 has significant negative effects on the concentration of several minerals and N (as a proxy for protein) in wheat grain, and that the effects on N translates into reduced baking quality. Subgroup analysis of experimental systems reveals that N concentration was more strongly affected in potted plants than plants grown in field soil. Also, the significant difference found between FACE and OTC studies could be attributed to the different concentration levels used and not the enclosure system itself. The pattern of effects by eCO_2 on different minerals was complex, showing that a single mechanism cannot account for the diversity of responses. Although the positive effect on starch concentration was not statistically significant, a dilution effect by starch may be of importance for element concentration. However, for most of the minerals the eCO_2 effect was not strongly related to the effect on grain yield, suggesting that dilution was not of large importance. The association with N was strong for eCO_2 effects on S and Fe, elements that are important components of proteins, and fairly strong also for P. The response functions and relationships between different elements and N presented in this study show a gradual change in nutritional quality and can be used in risk assessments of the effects on nutrition in a future high CO_2 world.

Acknowledgments: The work by M.B. and H.P. was supported by the strategic research area, Biodiversity and Ecosystem Services in a Changing Climate (BECC, http://www.becc.lu.se/).

Author Contributions: M.C.B. and H.P. conceived and designed the study; data collection was performed by M.C.B. in close collaboration with P.H.; all authors participated in the analysis of the data; M.C.B. wrote the paper with substantial input from P.H. and H.P.

Conflicts of Interest: The authors declare no conflict of interest.

References

1. IPCC. *Climate Change 2013: The Physical Science Basis*; World Meteorological Organization: Geneva, Switzerland, 2013.

2. Long, S.P.; Ainsworth, E.A.; Rogers, A.; Ort, D.R. Rising atmospheric carbon dioxide: Plants face the future. *Annu. Rev. Plant Biol.* **2004**, *55*, 591–628. [CrossRef] [PubMed]

3. Parry, M.L.; Rosenzweig, C.; Iglesias, A.; Livermore, M.; Fischer, G. Effects of climate change on global food production under sres emissions and socio-economic scenarios. *Glob. Environ. Chang.* **2004**, *14*, 53–67. [CrossRef]

4. Manderscheid, R.; Weigel, H.J. Photosynthetic and growth responses of old and modern spring wheat cultivars to atmospheric CO_2 enrichment. *Agric. Ecosyst. Environ.* **1997**, *64*, 65–73. [CrossRef]

5.   Schmid, I.; Franzaring, J.; Muller, M.; Brohon, N.; Calvo, O.C.; Hogy, P.; Fangmeier, A. Effects of CO_2 enrichment and drought on photosynthesis, growth and yield of an old and a modern barley cultivar. *J. Agron. Crop Sci.* **2016**, *202*, 81–95. [CrossRef]

6.   Amthor, J.S. Effects of atmospheric CO_2 concentration on wheat yield: Review of results from experiments using various approaches to control CO_2 concentration. *Field Crop. Res.* **2001**, *73*, 1–34. [CrossRef]

7.   Long, S.P.; Ainsworth, E.A.; Leakey, A.D.B.; Morgan, P.B. Global food insecurity. Treatment of major food crops with elevated carbon dioxide or ozone under large-scale fully open-air conditions suggests recent models may have overestimated future yields. *Phil. Trans. R. Soc. B* **2005**, *360*, 2011–2020. [CrossRef] [PubMed]

8.   Schimel, D. Climate change and crop yields: Beyond cassandra. *Science* **2006**, *312*, 1889–1890. [CrossRef] [PubMed]

9.   Ziska, L.H.; Bunce, J.A. Predicting the impact of changing CO_2 on crop yields: Some thoughts on food. *New Phytol.* **2007**, *175*, 607–617. [CrossRef] [PubMed]

10. Korner, C. Plant CO_2 responses: An issue of definition, time and resource supply. *New Phytol.* **2006**, *172*, 393–411. [CrossRef] [PubMed]

11. Food and Agriculture Organization of the United Nations. FAO Cereal Supply and Demand Brief. Available online: http://www.fao.org/worldfoodsituation/csdb/en/ (aceessed on 12 December 2016).

12. Loladze, I. Hidden shift of the ionome of plants exposed to elevated co_2 depletes minerals at the base of human nutrition. *Elife* **2014**, *3*, e02245. [CrossRef] [PubMed]

13. Leakey, A.D.B.; Ainsworth, E.A.; Bernacchi, C.J.; Rogers, A.; Long, S.P.; Ort, D.R. Elevated CO_2 effects on plant carbon, nitrogen, and water relations: Six important lessons from face. *J. Exp. Bot.* **2009**, *60*, 2859–2876. [CrossRef] [PubMed]

14. Simpson, R.J.; Lambers, H.; Dalling, M.J. Nitrogen redistribution during grain-growth in wheat (*Triticum aestivum* L.): 4. Development of a quantitative model of the translocation of nitrogen to the grain. *Plant Physiol.* **1983**, *71*, 7–14. [CrossRef] [PubMed]

15. IPCC. *Climate Change 2014: Impacts, Adaptation and Vulnerability*; World Meteorological Organization: Geneva, Switzerland, 2014.

16. Loladze, I. Rising atmospheric CO_2 and human nutrition: Toward globally imbalanced plant stoichiometry? *Trends Ecol. Evol.* **2002**, *17*, 457–461. [CrossRef]

17. Taub, D.R.; Miller, B.; Allen, H. Effects of elevated CO_2 on the protein concentration of food crops: A meta-analysis. *Glob. Chang. Biol.* **2008**, *14*, 565–575. [CrossRef]

18. McGrath, J.M.; Lobell, D.B. Reduction of transpiration and altered nutrient allocation contribute to nutrient decline of crops grown in elevated CO_2 concentrations. *Plant Cell Environ.* **2013**, *36*, 697–705. [CrossRef] [PubMed]

19. Bloom, A.J. Photorespiration and nitrate assimilation: A major intersection between plant carbon and nitrogen. *Photosynth. Res.* **2015**, *123*, 117–128. [CrossRef] [PubMed]

20. Pleijel, H.; Uddling, J. Yield vs. Quality trade-offs for wheat in response to carbon dioxide and ozone. *Glob. Chang. Biol.* **2012**, *18*, 596–605. [CrossRef]

21. Ainsworth, E.A.; Rogers, A. The response of photosynthesis and stomatal conductance to rising (CO_2): Mechanisms and environmental interactions. *Plant Cell Environ.* **2007**, *30*, 258–270. [CrossRef] [PubMed]

22. Högy, P.; Wieser, H.; Kohler, P.; Schwadorf, K.; Breuer, J.; Franzaring, J.; Muntifering, R.; Fangmeier, A. Effects of elevated CO_2 on grain yield and quality of wheat: Results from a 3-year free-air CO_2 enrichment experiment. *Plant Biol.* **2009**, *11*, 60–69. [CrossRef] [PubMed]

23. Myers, S.S.; Zanobetti, A.; Kloog, I.; Huybers, P.; Leakey, A.D.B.; Bloom, A.J.; Carlisle, E.; Dietterich, L.H.; Fitzgerald, G.; Hasegawa, T.; et al. Increasing CO_2 threatens human nutrition. *Nature* **2014**, *510*, 139–142. [CrossRef] [PubMed]

24. Myers, S.S.; Wessells, K.R.; Kloog, I.; Zanobetti, A.; Schwartz, J. Effect of increased concentrations of atmospheric carbon dioxide on the global threat of zinc deficiency: A modelling study. *Lancet Glob. Health* **2015**, *3*, E639–E645. [CrossRef]

25. European Food Safety Authority (EFSA). Cadmium in Food. *EFSA J.* **2009**, *980*, 1–139.

26. Satarug, S.; Garrett, S.H.; Sens, M.A.; Sens, D.A. Cadmium, environmental exposure, and health outcomes. *Environ. Health Perspect.* **2010**, *118*, 182–190. [CrossRef] [PubMed]

27. Högy, P.; Fangmeier, A. Effects of elevated atmospheric CO_2 on grain quality of wheat. *J. Cereal Sci.* **2008**, *48*, 580–591. [CrossRef]

28. Blumenthal, C.; Rawson, H.M.; McKenzie, E.; Gras, P.W.; Barlow, E.W.R.; Wrigley, C.W. Changes in wheat grain quality due to doubling the level of atmospheric CO_2. *Cereal Chem.* **1996**, *73*, 762–766.

29. Högy, P.; Wieser, H.; Kohler, P.; Schwadorf, K.; Breuer, J.; Erbs, M.; Weber, S.; Fangmeier, A. Does elevated atmospheric CO_2 allow for sufficient wheat grain quality in the future? *J. Appl. Bot. Food Qual.* **2009**, *82*, 114–121.

30. Kimball, B.A.; Morris, C.F.; Pinter, P.J.; Wall, G.W.; Hunsaker, D.J.; Adamsen, F.J.; LaMorte, R.L.; Leavitt, S.W.; Thompson, T.L.; Matthias, A.D.; et al. Elevated CO_2, drought and soil nitrogen effects on wheat grain quality. *New Phytol.* **2001**, *150*, 295–303. [CrossRef]

31. Piikki, K.; De Temmerman, L.; Ojanpera, K.; Danielsson, H.; Pleijel, H. The grain quality of spring wheat (*Triticum aestivum* L.) in relation to elevated ozone uptake and carbon dioxide exposure. *Eur. J. Agron.* **2008**, *28*, 245–254. [CrossRef]

32. Fernando, N.; Panozzo, J.; Tausz, M.; Norton, R.M.; Neumann, N.; Fitzgerald, G.J.; Seneweera, S. Elevated CO_2 alters grain quality of two bread wheat cultivars grown under different environmental conditions. *Agric. Ecosyst. Environ.* **2014**, *185*, 24–33. [CrossRef]

33. Wrigley, C.W.; Békés, F.; Bushuk, W. Chapter 1 gluten: A balance of gliadin and glutenin. In *Gliadin and Glutenin: The Unique Balance of Wheat Quality*; AACC International, Inc.: St. Paul, MN, USA, 2006; pp. 3–32.

34. Kindred, D.R.; Gooding, M.J.; Ellis, R.H. Nitrogen fertilizer and seed rate effects on hagberg failing number of hybrid wheats and their parents are associated with alpha-amylase activity, grain cavity size and dormancy. *J. Sci. Food Agric.* **2005**, *85*, 727–742. [CrossRef]

35. Hruskova, M.; Skodova, V.; Blazek, J. Wheat sedimentation values and falling number. *Czech J. Food Sci.* **2004**, *22*, 51–57.

36. Agren, G.I.; Weih, M. Plant stoichiometry at different scales: Element concentration patterns reflect environment more than genotype. *New Phytol.* **2012**, *194*, 944–952. [CrossRef] [PubMed]

37. Waters, B.M.; Uauy, C.; Dubcovsky, J.; Grusak, M.A. Wheat (*Triticum aestivum*) nam proteins regulate the translocation of iron, zinc, and nitrogen compounds from vegetative tissues to grain. *J. Exp. Bot.* **2009**, *60*, 4263–4274. [CrossRef] [PubMed]

38. Ainsworth, E.A.; Long, S.P. What have we learned from 15 years of free-air CO_2 enrichment (face)? A meta-analytic review of the responses of photosynthesis, canopy. *New Phytol.* **2005**, *165*, 351–371. [CrossRef] [PubMed]

39. Wang, Y.X.; Frei, M. Stressed food—The impact of abiotic environmental stresses on crop quality. *Agric. Ecosyst. Environ.* **2011**, *141*, 271–286. [CrossRef]

40. Gelang, J.; Pleijel, H.; Sild, E.; Danielsson, H.; Younis, S.; Sellden, G. Rate and duration of grain filling in relation to flag leaf senescence and grain yield in spring wheat (*Triticum aestivum*) exposed to different concentrations of ozone. *Physiol. Plant.* **2000**, *110*, 366–375. [CrossRef]

41. Broberg, M.C.; Feng, Z.Z.; Xin, Y.; Pleijel, H. Ozone effects on wheat grain quality—A summary. *Environ. Pollut.* **2015**, *197*, 203–213. [CrossRef] [PubMed]

42. Salt, D.E.; Prince, R.C.; Pickering, I.J.; Raskin, I. Mechanisms of cadmium mobility and accumulation in indian mustard. *Plant Physiol.* **1995**, *109*, 1427–1433. [CrossRef] [PubMed]

43. Feng, Z.Z.; Kobayashi, K.; Ainsworth, E.A. Impact of elevated ozone concentration on growth, physiology, and yield of wheat (*Triticum aestivum* L.): A meta-analysis. *Glob. Chang. Biol.* **2008**, *14*, 2696–2708.

44. Li, W.L.; Han, X.Z.; Zhang, Y.Y.; Li, Z.Z. Effects of elevated CO_2 concentration, irrigation and nitrogenous fertilizer application on the growth and yield of spring wheat in semi-arid areas. *Agric. Water Manag.* **2007**, *87*, 106–114. [CrossRef]

45. Tulchinsky, T.H. Micronutrient deficiency conditions: Global health issues. *Publ. Health Rev.* **2010**, *32*, 243–255.

46. Federov, S. Getdata Graph Digitizer 2.26.0.20. Available online: http://www.getdata-graph-digitizer.com/ (accessed on 20 May 2015).

47. NOAA. Available online: http://www.noaa.gov/ (accessed on 19 September 2014).

48. Rosenberg, M.S.; Adams, D.C.; Gurevitch, J. *Metawin: Statistical Software for Meta-Analysis*; Version 2.0; Sinauer Associates, Inc.: Sunderland, MA, USA, 2000.

49. Curtis, P.S.; Wang, X.Z. A meta-analysis of elevated CO_2 effects on woody plant mass, form, and physiology. *Oecologia* **1998**, *113*, 299–313. [CrossRef] [PubMed]

50. Koricheva, J.; Gurevitch, J.; Mengersen, K. *Handbook of Meta-Analysis in Ecology and Evolution*; Princeton University Press: Princeton, NJ, USA, 2013.

51. Motulsky, H.J.; Brown, R.E. Detecting outliers when fitting data with nonlinear regression—A new method based on robust nonlinear regression and the false discovery rate. *BMC Bioinf.* **2006**, *7*, 1–20. [CrossRef] [PubMed]

11

Spore Density of Arbuscular Mycorrhizal Fungi is Fostered by Six Years of a No-Till System and is Correlated with Environmental Parameters in a Silty Loam Soil

Julien Verzeaux [1,†], Elodie Nivelle [1,†], David Roger [1], Bertrand Hirel [2,*], Frédéric Dubois [1] and Thierry Tetu [1]

[1] Ecologie et Dynamique des Systèmes Anthropisés (EDYSAN, FRE 3498 CNRS UPJV), Laboratoire d'Agroécologie, Ecophysiologie et Biologie intégrative, Université de Picardie Jules Verne, 33 rue St Leu, Amiens CEDEX 80039, France; julienverzeaux@gmail.com (J.V.); elodienivelle@gmail.com (E.N.); david.roger@u-picardie.fr (D.R.); frederic.dubois@u-picardie.fr (F.D.); thierry.tetu@u-picardie.fr (T.T.)

[2] Adaptation des Plantes à leur Environnement, Unité Mixte de Recherche 1318, Institut Jean-Pierre Bourgin, Institut National de la Recherche Agronomique, Centre de Versailles-Grignon, R.D. 10, Versailles CEDEX F-78026, France

* Correspondence: bertrand.hirel@inra.fr

† These authors contributed equally to this work.

Academic Editor: Ilan Stavi

Abstract: Arbuscular mycorrhizal fungi (AMF) play major roles in nutrient acquisition by crops and are key actors of agroecosystems productivity. However, agricultural practices can have deleterious effects on plant–fungi symbiosis establishment in soils, thus inhibiting its potential benefits on plant growth and development. Therefore, we have studied the impact of different soil management techniques, including conventional moldboard ploughing and no-till under an optimal nitrogen (N) fertilization regime and in the absence of N fertilization, on AMF spore density and soil chemical, physical, and biological indicators in the top 20 cm of the soil horizon. A field experiment conducted over six years revealed that AMF spore density was significantly lower under conventional tillage (CT) combined with intensive synthetic N fertilization. Under no-till (NT) conditions, the density of AMF spore was at least two-fold higher, even under intensive N fertilization conditions. We also observed that there were positive correlations between spore density, soil dehydrogenase enzyme activity, and soil penetration resistance and negative correlations with soil phosphorus and mineral N contents. Therefore, soil dehydrogenase activity and soil penetration resistance can be considered as good indicators of soil quality in agrosystems. Furthermore, the high nitrate content of ploughed soils appears to be detrimental both for the dehydrogenase enzyme activity and the production of AMF spores. It can be concluded that no-till, by preventing soil from structural and chemical disturbances, is a farming system that preserves the entire fungal life cycle and as such the production of viable spores of AMF, even under intensive N fertilization.

Keywords: tillage; nitrogen fertilization; arbuscular mycorrhizal fungi; spore density; microbial activity

---

1. Introduction

Symbiosis between arbuscular mycorrhizal fungi (AMF) and plants arose on earth more than 400 million years ago [1]. The roots of the majority of land plant species are colonized by these fungi, extending the prospecting capacity of plants into the surrounding soil through mycorrhizosphere (a network formed by root-like extensions of the fungi known as hyphae), where the fungal spores are also formed [2]. It is commonly accepted that AMF develop obligate symbioses with at least 65% of vascular plants [3], including a large number of cereals grain crops and legumes, thus increasing their ability to acquire nutrients. However, AMF are particularly sensitive to physical, chemical, and biological disturbances caused by human activities that limit their establishment in agrosystems. Although it has been shown that tillage (through aggregate disruption [4]) and N fertilization can reduce the colonization of crops by arbuscular mycorrhizal fungi [5–7], our knowledge of the factors that determine the successful establishment of an AMF symbiosis remains limited. Spore density of AMF together with the level of hyphal growth and branching are critical for successful root colonization [6,8]. The density of spores has been defined as an early and useful indicator of AMF colonization potential [9]. Spore density determines the capability of AMF to resist ecological and physical disturbances, such as periods during which suitable host plants are not present or following intensive tillage, as they both limit the ability of AMF to colonize the subsequent cultivated crops [10]. In addition, among AMF species, there are different levels of tolerance to the disruption of hyphae resulting from tillage [11]. Inoculation with AMF spores has recently been shown to increase plant growth [12] and the expression of nitrate and phosphate transporter genes in wheat roots [13]. Moreover, inoculation can be used to offset tillage effects on the number of indigenous spores. However, there is still a paucity of knowledge concerning the potential ecological impact resulting from propagule inoculation [6], as such methods could increase the competition between inoculated fungi and previously established indigenous AMF communities [14]. Therefore, the implementation of alternative farming practices that promote natural production of viable AMF spores, and thus the occurrence of an efficient symbiosis with crop plants, needs to be studied further. The aim of this study was to evaluate the impact of alternative farming practices on AMF spore density in order to propose if such practices could be a way to increase root mycorrhizal colonization and thus crop mineral nutrient use efficiency in the context of agricultural sustainability.

2. Results and Discussion

Monitoring the effects of contrasting agrosystems, both in terms of mineral fertilization and tillage practices on AMF development could lead to a better understanding of such an important biological process and improve soil fertility. In the present work, the effect of tillage under zero or optimal N fertilization on AMF spore density was investigated over a six-year period. Moreover, we have examined if there was any relationship between the AMF spore density and soil chemical as well as biological parameters.

When multiple comparisons were performed using the Conover post-hoc tests, soil DH activity was significantly higher (35%) in the absence of tillage and N fertilization (Table 1). In ploughed soils, the concentration of PO_4^{3-} was also slightly higher, but only in the absence of N fertilization. We also observed that both plant NO_3^- and soil soluble NO_3^- contents were approximately 3- and 2-fold higher, respectively, in ploughed soils as compared to those managed under no-till (NT) conditions, irrespective of the N fertilization regime. The higher NO_3^- content found in the water collected from lysimeters suggests that inorganic N availability for microorganisms was higher in ploughed soils during the inter-cropping period. The lower soil NO_3^- content in the NT plots could result from a greater uptake by soil microorganisms. Moreover, it has been shown that substantial amounts of soil inorganic N can be taken up by microorganisms [15] and that NT enhances microbial biomass as compared to conventional tillage (CT) [16]. Another hypothesis is that increased mineralization of soil organic matter pools, caused by the annual moldboard ploughing [17,18], leads to an accumulation of NO_3^- in CT plots. However, the DH enzyme activity, which reflects the metabolic state of microorganisms in soils [19], was higher in NT and N0 plots as compared to those that were ploughed or fertilized. The finding that NT-N0 soils, characterized by the lowest concentration of nutrients (i.e., available phosphorus and nitrate), are also the most microbiologically active is supported by the results obtained by Nivelle et al. [20], and is consistent with those obtained by Das et al. [21], who also observed that the DH enzyme activity was approximately twice as high under NT than under CT conditions.

The highest value of penetration resistance (SPR) was obtained under NT conditions in N-fertilized soils (1.265 MPa), while the lowest value was found in ploughed soils without N fertilization (0.57 MPa) (Table 1). Such a result is not surprising as the upper soil layer under NT conditions is more compact [22], while annual moldboard ploughing destroys soil structure by mechanical aggregate disruption [4]. In contrast, annual moldboard ploughing could cause a strong subsoil compaction through the formation of a plough pan at a depth of 25–30 cm [23,24]. Such a soil compaction is a major agronomic concern that is detrimental for root system growth, soil aeration, and water infiltration. However, in our study, values for SPR at the 0–20 cm depth soil horizon under NT are not a problem from an agronomic point of view, as they do not reach levels that could limit crop production potential [25]. One can also hypothesize that the increased SPR induced by N fertilization in NT soils could be due either to chemically induced changes in soil physical properties or to changes in root development under low or high N fertilization inputs.

Table 1. Impact of tillage and nitrogen fertilization on the biological, chemical, and physical parameters of soil at a depth of 0–20 cm.

Soil Parameter	H (p)	NT-N0			NT-NX			CT-N0			CT-NX		
DH (μg TPF g^{-1} 24 h^{-1})	7.35 (0.042)	12.58	$\pm$	1.22 a	8.35	$\pm$	1.6 b	8.22	$\pm$	0.36 b	7.63	$\pm$	0.55 b
NO$_3$$^-$s (mg kg^{-1})	12.93 (0.005)	2.91	$\pm$	0.06 b	3.35	$\pm$	0.18 b	11.25	$\pm$	1.12 a	8.49	$\pm$	1.21 a
NO$_3$$^-$w (mg L^{-1})	13.26 (0.004)	18.13	$\pm$	2.04 b	27.62	$\pm$	3.67 b	44.67	$\pm$	2.65 a	61.76	$\pm$	3.81 a
PO$_4$$^{3-}$ (mg kg^{-1})	8.28 (0.040)	38.08	$\pm$	2.21 b	41.58	$\pm$	2.11 ab	51.37	$\pm$	2.38 a	42.55	$\pm$	3.45 ab
TN (g kg^{-1})	NS	1.18	$\pm$	0.04	1.21	$\pm$	0.06	1.18	$\pm$	0.02	1.25	$\pm$	0.07
TOC (g kg^{-1})	NS	11.54	$\pm$	0.48	11.75	$\pm$	0.51	12.04	$\pm$	0.19	11.59	$\pm$	0.89
C:N ratio	7.57 (0.048)	9.74	$\pm$	0.12 b	9.74	$\pm$	0.16 b	10.17	$\pm$	0.08 a	9.29	$\pm$	0.31 b
SPR (MPa)	12.11 (0.007)	0.88	$\pm$	0.06 b	1.265	$\pm$	0.08 a	0.57	$\pm$	0.08 c	0.83	$\pm$	0.04 bc

In H: Values of the Kruskal–Wallis test with its probability in brackets. Different letters a, b, and c indicate significant differences between treatments according to the Conover post-hoc test ($p < 0.05$), following a significant Kruskal–Wallis test. NS = not significant. CT: conventional tillage; NT: no-till; NX: with mineral N fertilization; N0: without N fertilization. NS: not significant. DH: dehydrogenase activity; NO$_3$$^-$s: nitrates extracted from soil; NO$_3$$^-$w: nitrates analyzed in soil solution; PO$_4$$^{3-}$: available phosphorus; TN: total nitrogen; TOC: total organic carbon; C:N ratio: carbon to nitrogen ratio; SPR: soil penetration resistance. Results are presented as mean values for soil parameters in the four treatments with standard errors.

Spore density was approximately two-fold higher ($p < 0.01$) after six years of experimentation under NT conditions irrespective of the N fertilization regime (Figure 1A). In contrast, under conventional ploughing less AMF spores were present in the soil, notably when mineral N fertilization occurred. It has already been shown that both spore density and AMF colonization are representative markers of the effect of tillage on AMF biological activity. For example, in comparison to CT, NT had a positive impact on both AMF colonization of wheat roots [7] and AMF spore density [26]. The lower density of AMF spores in CT treatments could be explained by hyphal networks disturbance resulting from ploughing [26]. However, Hu et al. [27] did not find any impact of tillage on AMF spore density in a sandy-loam soil in a temperate monsoon climate, which could be due to specific physical characteristics of such type of soil. In agreement with our results, Curaqueo et al. [28] showed that there was a larger density of AMF spores after six years of experimentation under NT conditions as compared to CT. However, after four additional years of NT, Curaqueo et al. [28] observed a reduction in the density of spores, likely due to instability of the favorable soil characteristics when NT was extended. Moreover, the results of Curaqueo et al. [28] were obtained under a continuous durum wheat–maize rotation, without any cover crop between the two main crops and with a cultivation system based on the use of intensive urea and triple super phosphate fertilization. It is therefore difficult to compare Curaqueo et al. results [28] with those obtained in the present investigation since we used cover crops as a healthier agronomic practice instead of fertilizer application. Indeed, it has been observed that the addition of mineral P fertilizers usually decreases the densities of AMF spores and hyphae under different soil and climatic conditions [29,30]. In the present study, cover crops containing leguminous species were cultivated between the main crops during winter periods. Therefore, cultivating mixtures of cover crops including leguminous species, which are known to be effective AMF hosts within the tripartite symbiosis with rhizobia [31,32], appears to be beneficial for a more efficient fungal colonization.

We observed a significant decrease of spore density in CT under high N fertilization conditions. In the present study, N fertilization consisted in the application of urea, ammonium, and nitrate. Cornejo et al. [33] emphasized the importance of the type of applied fertilizer on spore density in an andosol, since the authors observed that, compared to an ammonium-based fertilization, there were more spores of *Glomus etunicatum* in soils fertilized with NO_3^-. In our study, we can only speculate that the frequent application of urea and ammonium in CT-NX plots may induce deleterious effects on the AMF life cycle, thus limiting the density of viable spores. Such a result is consistent with the findings of Bhadalung et al. [34], Egerton-Warburton and Allen [35], and Mbuthia et al. [36], who showed that the use of ammonium sulfate plus triple superphosphate or ammonium nitrate caused a decrease in the total density of AMF spores and in the level of expression of AMF biomarkers. This decrease is probably due to a lower root colonization of potential hosts throughout the crop rotation. However, in our study, the lack of difference between N0 and NX plots in terms of spore density under NT conditions (Figure 1A) indicates that the detrimental effect of N fertilization on AMF spore density is somehow buffered by the better soil quality over the six-year period.

Further analysis of the data showed that the AMF spore density was positively correlated with soil penetration resistance ($\rho = 0.69$) and DH activity ($\rho = 0.62$) and was negatively correlated with the nitrate content of the soil ($\rho = -0.62$) or the nitrate content present in the soil solution ($\rho = -0.76$) (Figure 1B).

The results of the correlation studies are further supported by the principal component analysis (PCA) on which the first axis (36.5% of explained variance) clearly separated ploughed soils from those managed under NT (Figure 1C). Interestingly, in NT-managed soils, there was a high DH, SPR, and spore density, irrespective of the N fertilization regime. In contrast, annually ploughed soils were characterized by a high phosphate and nitrate content (Figure 1D). Although a positive correlation between AMF spore density and alkaline phosphatase activity has already been reported in the literature [27], such a positive correlation has never been described for soil DH activity. Nevertheless, the finding that a decrease in AMF spore density is associated with soils exhibiting the lowest value of DH activity (i.e., CT-NX) is not surprising because both parameters are known to be disrupted by intensive tillage and high levels of inorganic N [17]. In contrast, the negative correlation found between spore density and both soil and water NO_3^- content has already been observed by Egerton-Warburton and Allen [35]. These authors observed that an anthropogenic N deposition gradient, characterized by increased concentrations of soil NO_3^-, led to a significant reduction in AMF spore density. Additionally, Egerton-Warburton et al. [37] reported that the decrease in AMF productivity in N-fertilized soils was associated with an enrichment in inorganic P and a lower inorganic N/P ratio. However, in the present study, performed over six years without any addition of inorganic P, we showed that neither nitrate concentration nor the inorganic N/P ratio were related to the fertilizer management regime (Table 1). We therefore suggest that, among the tested soil parameters, CT, by strongly and permanently modifying both the soil physicochemical properties and the microbial response, is the main driver causing the decrease in AMF spore density. Such a decrease is even more important under optimal N fertilizer input.

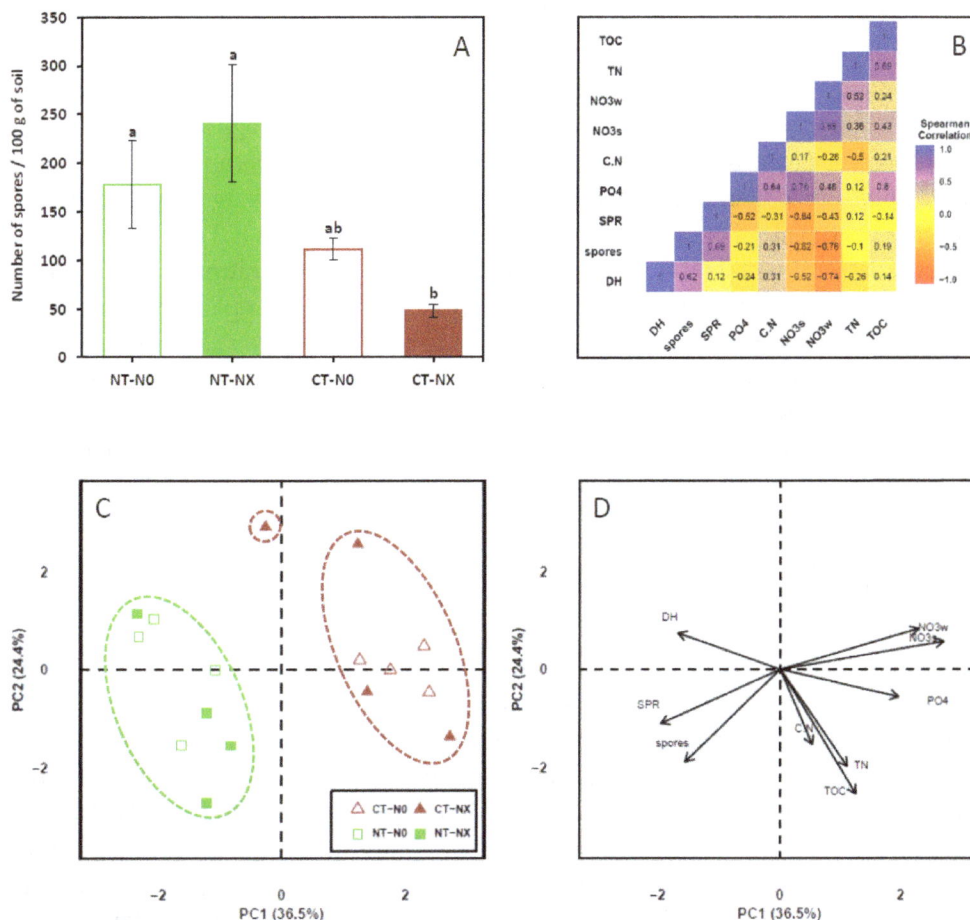

Figure 1. (**A**) Average number of spores per 100 g of soil ± standard error in each of the four treatments. Letters indicate differences among treatments according to a Conover post-hoc test ($p < 0.05$) following a significant Kruskal–Wallis test ($p < 0.01$). The filled bars correspond to the presence of nitrogen fertilization. The empty bars correspond to the absence of nitrogen fertilization. No-till is indicated in green while conventional tillage is indicated in brown. (**B**) Spearman correlations between measured parameters. Values of Spearman correlations are colored by blue as positive, red as negative, and yellow as neutral. Numbers in the squares are correlation coefficients. Factor map of the individuals (**C**) and variables (**D**) as obtained by principal component analysis based on spore density, dehydrogenase activity, soil penetration resistance and nutrient content. The green circle (- - - - -) groups samples collected in no-tilled soils while brown circles (- - - - -) group samples collected in ploughed soils. CT-N0: conventional tillage without nitrogen fertilization; CT-NX: conventional tillage with nitrogen fertilization; NT-N0: no-till without nitrogen fertilization; NT-NX: no-till with nitrogen fertilization; spores: spore density; DH: dehydrogenase activity; SPR: soil penetration resistance; NO_3w: NO_3^--N content measured from soil solution collected in lysimeters; NO_3s: NO_3^--N content extracted from soil samples; PO4: soil available phosphorus; TN: soil total nitrogen; TOC: soil total organic carbon; C:N: soil carbon/nitrogen ratio.

3. Materials and Methods

3.1. Site Description and Experimental Design

The field experiment was conducted at the *La Woestyne* experimental site, in Northern France (50°44' N, 2°22' E, 40 m above sea level). The average annual air temperature and total rainfall were 10.5 °C and 675 mm, respectively, with amounts of rainfall relatively homogeneous across the four seasons, these values being considered as normal for the region. The silty loam soil particle size composition was as follows: 66.7% silt, 21.2% clay, and 12% sand. The soil pH of 6.7 was homogeneous in the entire field.

Prior to the establishment of the field experiment in 2010, the field was managed with chisel ploughing and a rotary power system. In order to study the effect of tillage and N fertilization on the density of arbuscular mycorrhizal fungi (AMF) spores in soils, the experimental field was split into four replicated plots placed randomly for each of the four treatments making 16 plots in total. The four different treatments consisted of conventional tillage (CT) with nitrogen (CT-NX) or without nitrogen (CT-N0) fertilization and no-till (NT) with nitrogen (NT-NX) or without nitrogen (NT-N0) fertilization. Each experiment plot measured 2 m $\times$ 2 m and was separated from other plots by 3-m-wide corridors. The crop rotation before the sampling date consisted of wheat (*Triticum aestivum* L.) in 2011, bean (*Phaseolus vulgaris* L.) in 2012, wheat in 2013, green peas (*Pisum sativum* L.) in 2014, and maize (*Zea mays* L.) in 2015 for the entire field. As maize was grown for silage, all the aboveground material was removed from the field. Bean and pea haulm as well as wheat straw, were mechanically ground and returned to the soil. In the whole field, winter cover crops were cultivated each year between the growing seasons of the major crops except in 2013 before wheat. The cover crop mixture was composed of 60 seeds m^{-2} of oats (*Avena sativa* L.), 200 seeds m^{-2} of phacelia (*Phacelia tanacetifolia* Benth.), 80 seeds m^{-2} of flax (*Linum usitatissimum* L.), 50 seeds m^{-2} of vetch (*Vicia sativa* L.), 30 seeds m^{-2} of faba bean (*Vicia faba* L.), and 400 seeds m^{-2} of Egyptian clover (*Trifolium alexandrinum* L.). All cover crop seeds were mixed and simultaneously sown in line using a conventional seeder. Each year, cover crops were sown immediately after the harvest of the previous crop and were terminated by grinding following a frost period. Before the main crops were sown, cover crop residues were buried by annual conventional moldboard ploughing to a depth of 30 cm in CT plots and left on the soil surface in NT plots. In the NX plots, wheat received 160 kg of N ha^{-1} in 2011 and 2013, bean received 80 kg of N ha^{-1} in 2012 and maize received 108 kg of N ha^{-1} in 2015 (50% urea, 25% ammonium, 25% nitrate). In 2014, green peas did not receive any N fertilization in NX plots in accordance with European policies. The amount of N fertilizer applied under NX conditions was determined according to the N budget method [38]. The N0 plots were not fertilized with N for the whole 6 years of the experiment. Apart from N, no additional fertilizer was added to the experimental fields.

3.2. Sample Collection and Analyzes

In April 2016 (6 years after the beginning of the different soil management practices) during the intercropping period (before sowing of the subsequent main crop), three 0–20-cm-deep soil cores were randomly collected using an auger 5 cm in diameter from each of the 16 plots. In each of the four replicated plots for each of the four treatments, the three soil cores were pooled to form a single sample. The pooled soil samples were then homogenized and sieved using a 2 mm mesh.

Plant-available nitrate (NO_3^-) was extracted from 20 g of fresh soil with 100 mL of 2 M KCl. After shaking for 1 h, the extracts were centrifuged for 10 min at 4000 rpm, and the supernatants were analyzed by using an Alpkem Flow Solution IV continuous flow analytical system (Alpkem, Wilsonville, VT, USA). One week before soil sampling, 3 lysimeters (Sdec France, Reignac-sur-Indre, France) were inserted at a 20 cm depth in each of the 16 plots to extract the soil solution. Samples used to measure NO_3^- content in the soil solution were collected simultaneously with the soil samples. The 3 samples of soil solution collected from the 3 lysimeters were pooled to form a single sample in each of the 16 plots. The NO_3^- content in the pooled soil samples solution was directly analyzed with the Alpkem Flow Solution IV continuous flow analytical system. Soil phosphorus (PO_4^{3-}) was extracted with 0.5 M $NaHCO_3$, pH 8.5, and quantified using the colorimetric method described by Olsen et al. [39]. Total soil carbon (TOC) and total soil N (TN) contents were determined using an elemental analyzer Flash EA 1112 series (Thermo Fisher Scientific, Waltham, MA, USA) after drying at 35 °C for 48 h and ball-milled using a grinder MM 400 (Retsch, Haan, Germany). The soil C:N ratio corresponds to the ratio TOC on TN.

Soil dehydrogenase activity (DH) was measured as described by Casida et al. [40]. Soil sub-samples were adjusted with $CaCO_3$ to a final mass ratio of 100:1 (soil:$CaCO_3$) using 5.94 g of fresh soil. One mL of 3% 2,3,5-triphenyltetrazolium chloride solution and 2.5 mL of ultrapure water were added to the soil sub-samples. After mixing, the tubes were incubated at 37 °C for 24 h. The resulting triphenylformazan (TPF) was extracted with 30 mL of pure methanol by stirring for 1 min. The solution was then filtered in a dark room and the intensity of TPF was measured at 485 nm using an Eon spectrophotometer (BioTek Instruments Inc., Winooski, VT, USA). The AMF spore density was determined using the sucrose extraction method described by McKenney and Lindsey [41]. Only bright, apparently viable spores were counted on a gridded Petri dish under a binocular stereomicroscope.

Near the soil sampling areas, penetration resistance (SPR) was measured with a penetrologger (Eijkelkamp, Giesbeek, The Netherlands) fitted with a 60 deg and 1 cm^2 base area cone. Measuring the resistance to penetration of the soil was executed according to the manufacturer's instructions by applying the electronic penetrometer together with a datalogger, allowing for immediate storage and processing of the data in the datalogger.

3.3. Statistical Analysis

All statistical analyzes were performed using R software, v. 3.1.2 (R Development Core Team, 2014 [42]). Mean values are given with their standard error. Because of the low number of replicates, spore density was compared among treatments by using a non-parametric Kruskal–Wallis one-way analysis of variance followed by a Conover post-hoc test whenever significant (*PMCMR* package, [43]). A principal component analysis (PCA) including spore density, DH, SPR, and soil chemical parameters was performed using the *vegan* package [44].

4. Conclusions

To summarize, this work demonstrates that NT under continuous cover cropping system is an agricultural practice that maintains a chemical and a biological soil environment that is favorable for AMF spore production. Such an environment appears to be a key factor for increasing the potential of root mycorrhizal colonization and thus for enhancing mineral nutrient use efficiency in crop plants.

Acknowledgments: Peter J. Lea is thanked for its critical reading of the manuscript. This research was funded by Bonduelle and Syngenta companies within the framework of the collaborative project *VEGESOL* with the University of Picardy Jules Verne.

Author Contributions: J.V., E.N., and D.R. performed all the experiments. J.V. and E.N. computed data and wrote the manuscript. T.T., F.D., and B.H. have supervised the work and have participated in the interpretation and critical discussion of the results.

Conflicts of Interest: The authors declare no conflict of interest.

References

1. Selosse, M.A.; Strullu-Derrien, C.; Martin, F.M.; Kamoun, S.; Kenrick, P. Plants, fungi and oomycetes: A 400-million year affair that shapes the biosphere. *New Phytol.* **2015**, *206*, 501–506. [CrossRef] [PubMed]
2. Jansa, J.; Wiemken, A.; Frossard, E. The effects of agricultural practices on arbuscular mycorrhizal fungi. *Geol. Soc.* **2006**, *266*, 89–115. [CrossRef]
3. Brundrett, M.C. Mycorrhizal associations and other means of nutrition of vascular plants: Understanding the global diversity of host plants by resolving conflicting information and developing reliable means of diagnosis. *Plant Soil* **2009**, *320*, 37–77. [CrossRef]
4. Six, J.; Elliott, E.T.; Paustian, K. Soil macroaggregate turnover and microaggregate formation: A mechanism for C sequestration under no-tillage agriculture. *Soil Biol. Biochem.* **2000**, *32*, 2099–2103. [CrossRef]
5. Corkidi, L.; Rowland, D.L.; Johnson, N.C.; Allen, E.B. Nitrogen fertilization alters the functioning of arbuscular mycorrhizas at two semiarid grasslands. *Plant Soil* **2002**, *240*, 299–310. [CrossRef]
6. Brito, I.; Goss, M.J.; de Carvalho, M. Effect of tillage and crop on arbuscular mycorrhiza colonization of winter wheat and triticale under Mediterranean conditions. *Soil Use Manag.* **2012**, *28*, 202–208. [CrossRef]

7. Verzeaux, J.; Roger, D.; Lacoux, J.; Nivelle, E.; Adam, C.; Habbib, H.; Hirel, B.; Dubois, F.; Tetu, T. In winter wheat, no-till increases mycorrhizal colonization thus reducing the need for nitrogen fertilization. *Agronomy* **2016**, *6*, 38. [CrossRef]

8. Schalamuk, S.; Cabello, M. Arbuscular mycorrhizal fungal propagules from tillage and no-tillage systems: Possible effects on Glomeromycota diversity. *Mycologia* **2010**, *102*, 261–268. [CrossRef] [PubMed]

9. Castillo, C.G.; Rubio, R.; Rouanet, J.L.; Borie, F. Early effects of tillage and crop rotation on arbuscular mycorrhizal fungal propagules in an Ultisol. *Biol. Fertil. Soils* **2006**, *43*, 83–92. [CrossRef]

10. Oruru, M.B.; Njeru, E.M. Upscaling Arbuscular Mycorrhizal Symbiosis and Related Agroecosystems Services in Smallholder Farming Systems. *BioMed Res. Int.* **2016**, *2016*, 4376240. [CrossRef] [PubMed]

11. Jansa, J.; Mozafar, A.; Kuhn, G.; Anken, T.; Ruh, R.; Sanders, I.R.; Frossard, E. Soil tillage affects the community structure of mycorrhizal fungi in maize roots. *Ecol. Appl.* **2003**, *13*, 1164–1176. [CrossRef]

12. Berruti, A.; Lumini, E.; Balestrini, R.; Bianciotto, V. Arbuscular mycorrhizal fungi as natural biofertilizers: Let's benefit from past successes. *Front. Microbiol.* **2016**, *6*, 1559. [CrossRef] [PubMed]

13. Saia, S.; Rappa, V.; Ruisi, P.; Abenavoli, M.R.; Sunseri, F.; Giambalvo, D.; Frenda, A.S.; Martinelli, F. Soil inoculation with symbiotic microorganisms promotes plant growth and nutrient transporter genes expression in durum wheat. *Front. Plant. Sci.* **2015**, *6*, 815. [CrossRef] [PubMed]

14. Janoušková, M.; Krak, K.; Wagg, C.; Štorchová, H.; Caklová, P.; Vosátka, M. Effects of inoculum additions in the presence of a preestablished arbuscular mycorrhizal fungal community. *Appl. Environ. Microbiol.* **2013**, *79*, 6507–6515. [CrossRef] [PubMed]

15. Liu, Q.; Qiao, N.; Xu, X.; Xin, X.; Han, J.Y.; Tian, Y.; Ouyang, H.; Kuzyakov, Y. Nitrogen acquisition by plants and microorganisms in a temperate grassland. *Sci. Rep.* **2016**, *6*, 22642. [CrossRef] [PubMed]

16. Zuber, S.M.; Villamil, M.B. Meta-analysis approach to assess effect of tillage on microbial biomass and enzyme activities. *Soil Biol. Biochem.* **2016**, *97*, 176–187. [CrossRef]

17. Van den Bossche, A.; de Bolle, S.; de Neve, S.; Hofman, G. Effect of tillage intensity on N mineralization of different crop residues in a temperate climate. *Soil Tillage Res.* **2009**, *103*, 316–324. [CrossRef]

18. Pandey, C.B.; Chaudhari, S.K.; Dagar, J.C.; Singh, G.B.; Singh, R.K. Soil N mineralization and microbial biomass carbon affected by different tillage levels in a hot humid tropic. *Soil Tillage Res.* **2010**, *110*, 33–41. [CrossRef]

19. Watts, D.B.; Torbert, H.A.; Feng, Y.; Prior, S.A. Soil microbial community dynamics as influenced by composted dairy manure, soil properties, and landscape position. *Soil Sci.* **2010**, *175*, 474–486. [CrossRef]

20. Nivelle, E.; Verzeaux, J.; Habbib, H.; Kuzyakov, Y.; Decocq, G.; Roger, D.; Lacoux, J.; Duclercq, J.; Spicher, F.; Nava-Saucedo, J.E.; et al. Functional response of soil microbial communities to tillage, cover crops and nitrogen fertilization. *Appl. Soil Ecol.* **2016**, *108*, 147–155. [CrossRef]

21. Das, A.; Lal, R.; Patel, D.P.; Idapuganti, R.G.; Layek, J.; Ngachan, S.V.; Ghosh, P.K.; Bordoloi, J.; Kumar, M. Effects of tillage and biomass on soil quality and productivity of lowland rice cultivation by small scale farmers in North Eastern India. *Soil Tillage Res.* **2014**, *143*, 50–58. [CrossRef]

22. Cid, P.; Carmona, I.; Murillo, J.M. No-tillage permanent bed planting and controlled traffic in a maize-cotton irrigated system under Mediterranean conditions: Effects on soil compaction, crop performance and carbon sequestration. *Eur. J. Agron.* **2014**, *61*, 24–34. [CrossRef]

23. Lipiec, J.; Hatano, R. Quantification of compaction effects on soil physical properties and crop growth. *Geoderma* **2003**, *116*, 107–136. [CrossRef]

24. Bertolino, A.V.F.A.; Fernandes, N.F.; Miranda, J.P.L.; Souza, A.P.; Lopes, M.R.S.; Palmieri, F. Effects of plough pan development on surface hydrology and on soil physical properties in Southeastern Brazilian plateau. *J. Hydrol.* **2010**, *393*, 94–104. [CrossRef]

25. Nawaz, M.F.; Bourrié, G.; Trolard, F. Soil compaction impact and modelling. A review. *Agron. Sustain. Dev.* **2013**, *33*, 291–309. [CrossRef]

26. Kabir, Z. Tillage or no-tillage: Impact on mycorrhizae. *Can. J. Plant Sci.* **2005**, *85*, 23–29. [CrossRef]

27. Hu, J.; Yang, A.; Wang, J.; Zhu, A.; Dai, J.; Wong, M.H.; Lin, X. Arbuscular mycorrhizal fungal species composition, propagule density, and soil alkaline phosphatase activity in response to continuous and alternate no-tillage in Northern China. *Catena* **2015**, *133*, 215–220. [CrossRef]

28. Curaqueo, G.; Barea, J.M.; Acevedo, E.; Rubio, R.; Cornejo, P.; Borie, F. Effects of different tillage system on arbuscular mycorrhizal fungal propagules and physical properties in a Mediterranean agroecosystem in central Chile. *Soil Tillage Res.* **2011**, *113*, 11–18. [CrossRef]

29. Douds, D.D.; Millner, P. Biodiversity of arbuscular mycorrhizal fungi in agroecosystems. *Agric. Ecosyst. Environ.* **1999**, *74*, 77–93. [CrossRef]

30. Kahiluoto, H.; Ketoja, E.; Vestberg, M.; Saarela, I. Promotion of AM utilization through reduced P fertilization 2. Field studies. *Plant Soil* **2001**, *231*, 65–79. [CrossRef]

31. De Varennes, A.; Goss, M.J. The tripartite symbiosis between legumes, rhizobia and indigenous mycorrhizal fungi is more efficient in undisturbed soil. *Soil Biol. Biochem.* **2007**, *39*, 2603–2607.

32. Mortimer, P.E.; Pérez-Fernandez, M.A.; Valentine, A.J. The role of arbuscular mycorrhizal colonization in the carbon and nutrient economy of the tripartite symbiosis with nodulated *Phaseolus vulgaris. Soil Biol. Biochem.* **2008**, *40*, 1019–1027. [CrossRef]

33. Cornejo, P.; Borie, F.; Rubio, R.; Azcon, R. Influence of nitrogen source on the viability, functionality and persistence of *Glomus etunicatum* fungal propagules in an Andisol. *Appl. Soil Ecol.* **2006**, *35*, 423–431. [CrossRef]

34. Bhadalung, N.N.; Suwanarit, A.; Dell, B.; Nopamornbodi, O.; Thamchaipenet, A.; Rungchuang, J. Effects of long-term NP-fertilization on abundance and diversity of arbuscular mycorrhizal fungi under a maize cropping system. *Plant Soil* **2005**, *270*, 371–382. [CrossRef]

35. Egerton-Warburton, L.M.; Allen, E.B. Shifts in arbuscular mycorrhizal communities along an anthropogenic nitrogen deposition gradient. *Ecol. Appl.* **2000**, *10*, 484–496. [CrossRef]

36. Mbuthia, L.W.; Acosta-Martínez, V.; De Bruyn, J.; Schaeffer, S.; Tyler, D.; Odoi, E.; Mpheshea, M.; Walker, F.; Eash, N. Long term tillage, cover crop, and fertilization effects on microbial community structure, activity: Implications for soil quality. *Soil Biol. Biochem.* **2015**, *89*, 24–34. [CrossRef]

37. Egerton-Warburton, L.M.; Johnson, N.C.; Allen, E.B. Mycorrhizal community dynamics following nitrogen fertilization: A cross-site test in five grasslands. *Ecol. Monogr.* **2007**, *77*, 527–544. [CrossRef]

38. Machet, J.M.; Dubrulle, P.; Louis, P. Azobil1: A computer program for fertiliser N recommendations based on a predictive balance sheet method. In Proceedings of the 1st Congress of the European Society of Agronomy, Paris, France, 5–7 December 1990.

39. Olsen, S.R.; Cole, C.V.; Watanabe, F.S.; Dean, L.A. *Estimation of Available Phosphorus in Soils by Extraction with Sodium Bicarbonate*; U.S. Department of Agriculture: Washington, DC, USA, 1954.

40. Casida, L.E.; Klein, D.A.; Santoro, T. Soil dehydrogenase activity. *Soil Sci.* **1964**, *98*, 371–376. [CrossRef]

41. McKenney, M.C.; Lindsey, D.L. Improved method for quantifying endomycorrhizal fungi spores from soil. *Mycologia* **1987**, *79*, 779–782. [CrossRef]

42. The R Project for Statistical Computing. Available online: http://www.R-project.org (accessed on 23 January 2017).

43. Pohlert, T. PMCMR: Calculate Pairwise Multiple Comparisons of Mean Rank Sums Version 4.1. 2016. Available online: http://cran.r-project.org/ (accessed on 23 January 2017).

44. Oksanen, J.; Blanchet, F.G.; Kindt, R.; Legendre, P.; Minchin, P.R.; O'Hara, R.B.; Simpson, G.L.; Solymos, P.; Stevens, M.H.H.; Wagner, H. Vegan: Community Ecology Package Version 2.3-2. 2015. Available online: http://cran.r-project.org/ (accessed on 23 January 2017).

12

N$_2$ Fixation of Common and Hairy Vetches when Intercropped into Switchgrass

Amanda J. Ashworth [1,*], **Fred L. Allen** [2], **Kara S. Warwick** [2], **Patrick D. Keyser** [3], **Gary E. Bates** [2], **Don D. Tyler** [4], **Paris L. Lambdin** [5] and **Dan H. Pote** [6]

[1] United States Department of Agriculture, Agricultural Research Service, Poultry Production and Product Safety Research Unit, Fayetteville, AR 72927, USA

[2] Department of Plant Sciences, University of Tennessee, Knoxville, TN 37996, USA; fallen@utk.edu (F.L.A.); kspivey@utk.edu (K.S.W.); gbates@utk.edu (G.E.B.)

[3] Department of Forestry, Wildlife & Fisheries, University of Tennessee, Knoxville, TN 37996, USA; pkeyser@utk.edu

[4] Department of Biosystems Engineering and Soil Science, University of Tennessee, Knoxville, TN 37996, USA; dyler@utk.edu

[5] Department of Entomology and Plant Pathology, University of Tennessee, Knoxville, TN 37996, USA; plambdin@utk.edu

[6] United States Department of Agriculture, Agricultural Research Service, Dale Bumpers Small Farms Research Center, 6883 S. Hwy 23, Booneville, AR 72927, USA; Dan.Pote@ars.usda.gov

* Correspondence: Amanda.Ashworth@ars.usda.gov;

Academic Editor: Bertrand Hirel

Abstract: Interest in sustainable alternatives to synthetic nitrogen (N) for switchgrass (*Panicum virgatum* L.) forage and bioenergy production, such as biological N$_2$ fixation (BNF) via legume-intercropping, continues to increase. The objectives were to: (i) test physical and chemical scarification techniques (10 total) for common vetch (*Vicia sativa* L.); (ii) assess whether switchgrass yield is increased by BNF under optimum seed dormancy suppression methods; and (iii) determine BNF rates of common and hairy vetch (*Vicia villosa* L.) via the N-difference method. Results indicate that chemical scarification (sulfuric acid) and mechanical pretreatment (0.7 kg of pressure for one minute) improve common vetch germination by 60% and 50%, respectively, relative to controls. Under optimum scarification methods, BNF was 59.3 and 43.3 kg·N·ha^{-1} when seeded at 7 kg pure live seed ha^{-1} for common and hairy vetch, respectively. However, at this seeding rate, switchgrass yields were not affected by BNF ($p > 0.05$). Based on BNF rates and plant density estimates, seeding rates of 8 and 10 kg pure live seed (PLS) ha^{-1} for common and hairy vetch, respectively, would be required to obtain plant densities sufficient for BNF at the current recommended rate of 67 kg·N·ha^{-1} for switchgrass biomass production in the Southeastern U.S.

Keywords: biological nitrogen fixation; legume intercropping; biomass sustainability; N-difference method

---

1. Introduction

Legumes are agronomically beneficial because they fix atmospheric nitrogen (N$_2$) through a symbiotic relationship with *Rhizobia* bacteria, which form nodules in leguminous roots. These beneficial bacteria enhance soil fertility by increasing N through rhizodeposition, which reduces the amount of synthetic N fertilizer needed for switchgrass growth [1]. However, biological N$_2$ fixation (BNF) can be affected by weather, inorganic-N present in soils, as well as legume vigor [2,3]. Furthermore, the decay of legumes may not be in synchrony with peak N demand by the main crop [4], and matching legumes with companion crops can be challenging [5]. Annual switchgrass yields average 15.9 Mg·ha^{-1} in the upper Southeast [6], with only modest responses to greater N fertilization [7]. Consequently, switchgrass N fertilization is recommended at an annual rate of 67 kg·ha^{-1} [8], or approximately half the rate for corn (*Zea mays* L.) [9].

Legumes interseeded into switchgrass may fix N required for biomass production [10]. Experiments with legume-switchgrass mixtures (e.g., red clover (*Trifolium pretense* L.)) reported yields that exceed those of N-only, even at inorganic-N rates of 240 kg·ha^{-1} [11]. Similarly, common and hairy vetch are reportedly effective at increasing soil N and can fix N_2 required for a single biomass-cut system [12,13]. Specifically, common and hairy vetches have been reported to fix between 50 and 350 kg·N·ha^{-1} and 25 and 190 kg·N·ha^{-1}, respectively, in aboveground growth [14–17]. Common and hairy vetches are cool-season legumes, and as such, peak photosynthesis and subsequent fixation occur from winter until switchgrass' spring green-up. There are several potential advantages of using common vetch in lieu of hairy vetch. Common vetch is frequently found growing throughout the Southeastern U.S. [18] and typically has fewer hard seeds than most varieties of hairy vetch [19,20]. Hairy vetch hard seeds can range between 5 and 30%, last 5+ years in the soil and be a noxious weed [20–22].

Myriad methods are used to determine BNF, including acetylene reduction and hydrogen evolution [2,3]. These techniques must be performed in a controlled environment and are therefore unsuitable for quantifying N_2 fixation of field-grown legumes [23,24]. On the other hand, ^{15}N isotope dilution, ^{15}N natural abundance, N-balance and N-difference methods all are suitable for in situ experiments; however, each technique has inherent advantages and disadvantages. The N-difference method estimates amounts of N supplied from symbiosis by comparing N_2-fixing legumes to neighboring non-fixing reference plants. This method is simple and inexpensive and works best under low soil-N conditions [25,26]. The disadvantages are that the N-difference method assumes that legumes and non-fixing plants exploit equal amounts of soil N [2,27] and that plant sizes and/or root morphologies do not differ [28,29]. However, estimates obtained by the N-difference method are comparable to those from more expensive techniques [30,31].

Proper seeding rates for interseeding legumes into lowland switchgrass stands are not well defined. Rates used for previous studies with upland switchgrass have been for frost-seeding into grass pastures, and a reduction of rates has been recommended [32]. Therefore, legume seeding rates need to be developed to establish persistent legume stands that increase N availability without inducing spatial and resource competition with switchgrass. Consequently, legume symbiotic relationships and their interaction with the soil environment were assessed via a comparison of switchgrass dry matter yields to help determine the effectiveness of N_2-fixation by legume hosts. The specific objectives of this study were to: (i) determine the efficacy of physical and chemical seed scarification for common vetch germination; (ii) determine whether or not switchgrass yields are increased by vetch intercrops; and (iii) determine N-fixation rates of common and hairy vetch via the N-difference method in switchgrass production systems.

2. Materials and Methods

2.1. Switchgrass Stands and Site Descriptions

Switchgrass cv. Alamo was planted in spring 2007 at 9 kg·ha^{-1} pure live seed (PLS) at three field sites, two at the East Tennessee Research and Education Center (ETREC): the Plant Sciences Unit [(ETREC-PS (35°8′ N, 83°9′ W)] and the Holston Unit [(ETREC-H (35.53° N 83.57° W)], as well as at the Plateau Research and Education Center (PREC), Grasslands Unit in Crossville, TN (36.1° N 85.8° W). Soils at the Plant Sciences Unit are classified as a Huntington silt loam (fine-silty, mixed, active, mesic Fluventic Hapludolls), and soils at the Holston Unit are classified as a Huntington silt loam (fine-silty, mixed, active, mesic Fluventic Hapludolls). The Plant Sciences Unit has a 30-year mean annual temperature of 14.4 °C, with average precipitation of 1240 mm. Soil PREC is classified as a Lily silt loam (fine-loamy, siliceous, semi-active, mesic Typic Hapludults), with 30-year average annual precipitation of 1400 mm and an average temperature of 12.6 °C. Switchgrass plots had no soil amendments applied during this study.

2.2. Nitrogen Fixation of Legume Intercrops

Nitrogen content of common and hairy vetch plants at ETREC-Plant Sciences Unit were compared to monocots [wheat (*Triticum* spp.) and switchgrass]. The authors previously found that switchgrass and wheat assimilate soil-N similar to other commonly-used non-N_2 fixing reference plants and therefore are adequate reference plants for the N-difference method [33]. N_2 fixation of vetches was determined by using the N-difference method. Sample shoots of common vetch, hairy vetch and non-N_2-fixing reference plants wheat and switchgrass were gathered by cutting plants flush to the soil with pruning shears in late spring 2010. Sample tissue-N (grass separated from legumes) was then analyzed with near-infrared reflectance spectroscopy (NIR) using a LabSpec® Pro Spectrometer (Analytical Spectral Devices, Boulder, CO, USA) by Land O'Lakes/Sure-Tech (Indianapolis, IN, USA). Equations were standardized and checked for accuracy using grass hay and alfalfa (*Medicago sativa* L.) equations (for each legume of interest), which were developed by the Near Infrared Spectroscopy (NIRS) Forage and Feed Consortium (NIRSC, Hillsboro, WI, USA).

Plant aboveground-N was determined by multiplying plant dry matter (DM) by its percent N content [Equation (1)]. Reference plant N yield (non-nodulating species) was then subtracted from legume plant N yield to obtain the amount of legume fixed N on a per ha basis (Equation (2); [25]). The N-difference between vetches and reference plants was multiplied by average plant weights of legume plants sampled in late spring to obtain the aboveground-N per vetch plant. Total aboveground legume-N mass was determined by legume-N (mass per plant) × plant density (plants·m^{-2}) and expressed as kg·ha^{-1} to determine fixed N accumulated in legume and potentially available to switchgrass (value assuming complete bioavailability).

$$\text{Plant N yield (kg·ha}^{-1}) = \text{Plant DM} \times \%N/100 \qquad (1)$$

$$\text{N-difference (}N_2 \text{ fixed)} = [\text{legume N mass (g·kg}^{-1})] - \text{reference plant N mass (g·kg}^{-1}) \qquad (2)$$

Estimated seeding rates required for common and hairy vetch to fix the recommended rate of 67 kg ha^{-1} N fertilizer were obtained from N_2-fixation rates determined by the N-difference method in this study. Specifically, total vetch aboveground plant N·m^{-2} was calculated by multiplying the average vetch density (planted at a seeding rate of 7 kg·PLS·ha^{-1}) by the aboveground, per plant vetch N and divided by 50% (assuming half is bioavailable, [34,35]). To calculate the seeding rate required for vetch to supply 67 kg·N·ha^{-1} to the companion crop, target N level was divided by bioavailable vetch N, thus developing a ratio to multiply the current seeding rate that would give the suggested seeding rates of common and hairy vetch (Equation (3)).

$$\text{Seeding rate for target N} = [(\text{Target N·kg·ha}^{-1}) * (\text{kg PLS ha}^{-1})]/(\text{legume N·kg·ha}^{-1}) \qquad (3)$$

2.3. Legume Seed Treatment and Establishment Techniques

Common and hairy vetches were seeded in fall 2009 into established (3-year-old) Alamo switchgrass stands at the two locations. Legumes were seeded into approximately 20-cm-tall switchgrass stubble on 22 and 29 October 2009 at PREC and ETREC, respectively, with a Hege™ plot drill (Colwich, KS) at a planting depth ranging from 0.6 to 1.3 cm. At ETREC and PREC, plot sizes were 7.6 × 1.5 m and 7.6 and 1.8 m, respectively, with 18 cm-wide row spacing. Seeding rates for both common and hairy vetch were 7 kg·PLS·ha^{-1}, and the control was represented by a 0 kg·N·ha^{-1} rate. Seeding rates of common and hairy vetch were lowered from the pure stand rates of 34 kg·ha^{-1} used for forage [36], thus reducing competition with switchgrass early in the season.

Legume seed treatments were tested for germination efficacy in a one-factor (scarification treatment method) completely randomized design. Seeds used for common vetch plantings were collected from volunteer populations at ETREC Holston and Plant Science Units in early summer 2009 and treated by stratification and scarification to break dormancy. Seeds collected from the Plant Science Unit were divided into two lots. Lot 1 was dried at room temperature (approximately 25 °C), and Lot 2 was dried at 49 °C in a batch oven (Wisconsin Oven Corporation, East Troy, WI, USA). Seeds collected from the Holston unit were dried at room temperature (Lot 3). The three seed lots were treated for dormancy by dry cold stratification in a cooler at an average of 8 °C for 1 to 6 weeks, plus a control treatment (7 treatments), resulting in 21 treatments (including controls). The stratified vetch was then seeded into sand trays in a greenhouse for germination assessment.

Common vetch seeds from the Holston Unit (Lot 3) were treated for dormancy by physical and chemical scarification with 10 different treatments. Treatments included a control; physical scarification with 100-grit sandpaper (0.5 kg for 30 s, 0.5 kg for 1 min, 0.7 kg for 30 s, 0.7 kg for 1 min, 0.9 kg for 30 s, 0.9 kg for 1 min); treatment with 3% bleach (sodium hypochlorite) for 10 min, treatment with 98% sulfuric acid (H_2SO_4) for 1 min; and treatment with 1% hydrogen peroxide (H_2O_2) for 24 h. Physical scarification was applied with sandpaper attached to wood boards while using a weigh scale to ensure that target pressure was applied.

Scarified common vetch seeds were seeded into sand trays in a greenhouse for germination testing. Because the sulfuric acid seed treatment resulted in the greatest germination rate among all chemical seed scarification methods (Table 1), remaining common vetch seeds (Lots 1, 2, and 3) were treated with sulfuric acid (98% for 1 min), rinsed for 15 min, force-air-dried for 10 min and direct-seeded into switchgrass plots.

In early-June, a frequency grid [37] was used to measure legume stand densities on switchgrass plots interseeded with vetches. Four density counts were taken in each legume treatment plot. Plant densities were averaged from three replications at each location to determine legume density (m^2). The count was multiplied by 0.4 according to [37] based on the likelihood of one plant per cell to estimate plant density per m^2 and averaged over three blocks at each location. Switchgrass height was measured per frequency grid observation, with an average height calculated for each plot.

Table 1. Physical and chemical seed scarification methods, treatments and average germination rates (number and %) of common vetch seed.

Scarification Method	Treatment [b]	Mean [a]	%
Control	Air dried seed (Holston)	8 cde	16
Sandpaper [c]	0.5 kg for 30 s	10 bcd	20
Sandpaper	0.5 kg for 1 min	9 cde	17
Sandpaper	0.7 kg for 30 s	7 cde	14
Sandpaper	0.7 kg for 1 min	16 ab	31
Sandpaper	0.9 kg for 30 s	12 bc	23
Sandpaper	0.9 kg for 1 min	10 bcd	20
Chlorine Bleach	3% sodium hypochlorite/10 min	4 de	7
Sulfuric Acid	98%H_2SO_4/1 min	20 a	40
Hydrogen Peroxide	1% H_2O_2/24 h	2 e	4

[a] Mean separations based on Tukey's test followed by the same letter are not significantly different at $p < 0.05$ level; [b] treatments had three replications consisting of 17 air-dried seeds, each from Holston Unit seed collection; [c] 100 grit sandpaper.

2.4. Switchgrass Yield Measurements

Two harvest systems were tested in a two-factor (harvest system and N source) randomized complete block design to determine how canopy removal affects legume intercrop vigor and included a single, post-dormancy harvest at ETREC on 8 November 2010 and a two-cut harvest system at PREC on 9 June 2010 (early-boot stage) and 21 October 2010 (post-dormancy). Switchgrass plots were harvested using a Carter™ plot harvester (Brookston, IN, USA). The harvested plot area was 0.9 × 7.6 m, and the cutting height was 20 cm. Grab samples (1 to 2 kg) of switchgrass were collected from all plots at harvest and were weighed, dried in a batch oven at 49 °C and re-weighed to determine moisture content.

2.5. Soil Tests

Preliminary (prior to experimentation) soil nutrient levels were quantified on a per-plot basis for both locations to a 0 to 15 cm depth to determine nutrient concentrations of P, K, Mg and Ca. Samples were ground to pass through a 1-mm sieve on a Wiley soil crusher (Thomas Scientific, Swedesboro, NJ, USA), and Mehlich-1 extractable nutrients were measured by inductively-coupled plasma (ICP) using a 7300 ICP-OES DV (Perkin-Elmer, Waltham, MA, USA), respectively.

2.6. Data Analysis

Switchgrass yields and common vetch seed germination following chemical and physical scarification treatments were analyzed using PROC Mixed with SAS v. 9.1.3 [38]. Tukey's honestly significant difference test was used to determine differences in switchgrass yields and seed germination rates at an alpha level of 0.05. Fixed effects were legume and seed treatments, and locations and replications were assigned as random effects.

3. Results and Discussion

3.1. Seed Treatment

Cold stratification and scarification were assessed for their abilities to break the dormancy of common vetch. Stratification by chilling seeds for one to six weeks did not greatly induce vetch germination (Table 2). Germination rates of oven-dried seed were $\leq 4\%$, while germination of air-dried seed averaged 7% (Holston Unit collection, Lot 3) and 6% (Plant Science Unit, Lots 1 and 2), indicating adverse effects from high temperatures and subsequent seed desiccation. Cold stratifying (8 °C and at 41% relative humidity) did not increase common vetch germination. Other studies involving seed chilling at constant temperatures have not shown accelerated legume seed germination, but increased germination in spring has been achieved by moderate winter temperatures [39]. Similarly, hairy vetch germination has improved when subjected to warmer temperatures [40].

Table 2. Average [a] seed germination rates (number and %) of dry and cold stratification treatments of common vetch seed.

| Treatments [b] | Holston Unit | | Plant Science Unit | | | |
| | Air-Dried | | Air-Dried | | Oven-Dried (49 °C) | |
	Means[a]	%	Means	%	Means	%
1 Week	1.5	6	2.5	10	0	0
2 Weeks	3.5	14	1.5	6	1	4
3 Weeks	2.5	10	2.0	8	1	4
4 Weeks	2.0	8	1.0	4	0.5	2
5 Weeks	1.0	4	1.0	4	0	0
6 Weeks	0.5	2	1.5	6	0.5	2
Control [c]	2.5	10	1.5	6	1	4

[a] Means across treatments and replications; [b] treatments used 25 seeds for each of two replications from three different seed collections [Holston Unit and Plant Science Unit (oven- and air-dried)] and were chilled at 8 °C with 41% humidity; [c] air-dried and control seed were stored at ambient temperature that averaged 24 °C with 58% humidity.

Conversely, mechanical and chemical seed scarification treatments did result in differences in germination (Table 1). Sulfuric acid (40% germination) and sandpaper treatments (0.7 kg of pressure for one minute (31% germination)) resulted in germination that exceeded that of the control (8%). Consequently, stronger solutions or longer soaking times may be required for the hydrogen peroxide and bleach treatments to become effective. Hydrogen peroxide, bleach and physical scarification with sandpaper are safer alternatives to sulfuric acid and should be considered further. Similarly,

Larson, J.A. et al. [4] found that sulfuric acid treatments of 15 and 30 min produced 100% germination rates of vetch seed.

3.2. Nitrogen Fixation

Nitrogen fixation rates of common and hairy vetches were similar at 59.3 and 55.2 $kg \cdot N \cdot ha^{-1}$ based on a plant density of 8.5 $plants \cdot m^2$ and 43.3 and 37.6 $kg \cdot N \cdot ha^{-1}$ based on 7.0 $plants \cdot m^2$ (using wheat and switchgrass as non-fixing reference plants, respectively; Table 3); which was less than the current recommended N rate of 67 $kg \cdot ha^{-1}$ for switchgrass [7]. In previous studies, hairy vetch supplied 90 to 150 $kg \cdot N \cdot ha^{-1}$ to subsequent crops [41–44]. Common vetch also has the potential to supply 106 to 146 $kg \cdot N \cdot ha^{-1}$ to subsequent crops [41,43]. Given the preceding N_2-fixation rates and plant densities, estimated seeding rates of common and hairy vetch should be 8 and 10 $kg \cdot PLS \cdot ha^{-1}$, respectively, to achieve 67 $kg \cdot N \cdot ha^{-1}$ contribution. If achieved, these rates should supply the recommended N rate for switchgrass biomass production [7].

Table 3. Estimated seeding rates for common and hairy vetch to obtain the recommended rate of N fertilizer for switchgrass using the N-difference method to calculate N_2-fixation rates of common and hairy vetch at the East Tennessee Research and Education Center.

Reference Plant	Vetch Aboveground $N \cdot plant^{-1}$	Observed Average Vetch Density [a]	Total Aboveground Vetch N	Bioavailable Vetch N	Target N	Vetch Seeding Rate to Supply 67 $kg \cdot N \cdot ha^{-1}$ to Switchgrass
	g	m^{-2}	$g\,m^{-2}$	$kg \cdot ha^{-1}$		$kg \cdot PLS \cdot ha^{-1}$
			Common Vetch			
Wheat	1.4	8.5	11.9	59.3	67	7.6
Switchgrass	1.3	8.5	11.0	55.2	67	8.2
			Hairy Vetch			
Wheat	1.2	7.0	8.7	43.3	67	10.4
Switchgrass	1.1	7.0	7.5	37.6	67	12.0

[a] Common vetch plant density was averaged from both East Tennessee Research and Education Center (ETREC) and Plateau Research and Education Center (PREC) locations in 2010. Hairy vetch plant density was taken from ETREC due to no seedling emergence at PREC. Both common and hairy vetch densities were obtained with seeding rates of 7 $kg \cdot ha^{-1}$.

3.3. Legume Establishment

Recommended seeding dates for cool-season legumes are early fall or the last two weeks in February through the end of March [36]. In this study, legumes could not be planted into uncut, mature switchgrass stands and, thus, were not seeded until late fall following switchgrass biomass harvest. Late seeding, combined with harsh weather conditions, may have led to late germination, and thus, small seedlings that were not winter-hardy had reduced survival. Establishment of common and hairy vetch for seeding rates of 7 $kg \cdot ha^{-1}$ at ETREC averaged 10 and 7 $plants \cdot m^2$, respectively, as measured on 12 and 13 May. At PREC, legume densities were 7 and 0 $plants \cdot m^2$, respectively, on 25 May 2010 (Table 4). In general, common vetch had greater mass than that of hairy vetch across both locations (Table 4).

Lower densities of vetch seedlings at PREC could have been caused by low soil nutrient levels, considering that P (phosphorus) is an essential element for legume nodulation. Levels of soil P and potassium were considerably lower at PREC than at ETREC (Table 5). Successful incorporation of legumes into switchgrass stands will require a soil test and amending nutrient and pH levels before planting. Therefore, the reduced vetch fixation rates observed in this study could, in part, be due to low soil test P levels and the late planting of legume seed.

Table 4. Average [a] common and hairy vetch legume (LG) plant densities [b] and heights and switchgrass (SG) plant heights at ETREC and PREC in 2010.

Location	Common Vetch				Hairy Vetch			Control	
	Plant Density [b]	Height		Weight	Plant Density	Height		Weight	Height
		LG	SG	LG		LG	SG	LG	SG
	No. m^{-2}	cm		g	No. m^{-2}	cm		g	cm
ETREC	10	59	119	9.9	7	54	84	7.5	110
PREC	7	43	86	9.6	0	4	109	7.8	87

[a] Means across treatments and replications; [b] plant density = [(frequency of occurrence × 0.4) × 100] (Vogel and Masters, 2001).

Table 5. Soil nutrient levels of phosphorus (P), potassium (K), calcium (Ca) and magnesium (Mg) in common and hairy vetch plots at East TN (ETREC) and Plateau (PREC) Research and Education Centers (determined via Mechich-1 extractant).

Nutrient	ETREC			PREC		
	Common Vetch	Hairy Vetch	Control	Common Vetch	Hairy Vetch	Control
	kg·ha^{-1}					
P	86	117	63	6	7	9
K	170	102	218	86	81	127
Ca	3375	3408	3019	2439	2374	1276
Mg	437	482	427	205	182	72

3.4. Switchgrass Yield Impacts

Switchgrass yields did not vary among treatments or number of harvests after only one year when compared to the 0 N control (Table 6); considering that common vetch (13 Mg·ha^{-1}), hairy vetch (12.4 Mg·ha^{-1}) and grass-only yields (10.7 Mg·ha^{-1}) were not different ($p > 0.05$). Neither common nor hairy vetch seeds were inoculated prior to planting, although the presence of legume nodules following establishment indicated that *Rhizobia* were present in soils; however, concentrations prior to seeding were not determined. When seeding common or hairy vetches into established stands of switchgrass, it is advisable to inoculate seeds with appropriate species of *Rhizobia* prior to planting to ensure nodulation and effective legume stands to achieve target levels of N$_2$-fixation. In cases of low *Rhizobia* populations in soils and/or no inoculation when seeding, it may take two to three years to naturally develop proper soil *Rhizobia* levels for vetch to fix the N required to elevate switchgrass yield.

Based on the results herein, interseeding switchgrass with inoculated common or hairy vetch seed at rates of 8 and 10 kg·PLS·ha^{-1}, respectively, is expected to illicit a greater switchgrass yield response (Table 3). These estimated seeding rates are predicted to be necessary to achieve the recommended rate of 67 kg·N·ha^{-1} for switchgrass.

Table 6. Average [a] dry matter yields of switchgrass per common vetch or hairy vetch treatment from a one-cut biomass harvest system at ETREC and a two-cut forage/biomass harvest system at PREC in 2010.

Treatment	ETREC		PREC		All Locations
	Biomass	Forage	Biomass	F + B [b]	Both Harvests
	Mg·ha^{-1}				
Common Vetch	15.9 a [a]	3.8 a	6.2 a	10.0 a	13.0 a
Hairy Vetch	12.7 a	3.5 a	5.1 a	8.6 a	12.4 a
Control	11.6 a	3.3 a	5.4 a	8.7 a	10.7 a

[a] Mean separations based on Tukey's test at $p < 0.05$ applied to individual columns across treatments; [b] summation of forage and biomass yields.

4. Conclusions

Under proper seeding rates and P-management, common and hairy vetches could potentially be viable alternatives for offsetting inorganic-N fertilizer inputs for switchgrass production. Common vetch germination can be increased through a sulfuric acid pretreatment before seeding, but such pretreatment may be unsafe and cost-prohibitive. A cost-effective alternative to breaking seed dormancy and increasing vetch seed germination is mechanical scarification (i.e., 100 grit sandpaper at 0.7 kg of pressure for one minute) or via a mechanical drum for large-scale systems.

Relatively similar aboveground N_2-fixation rates of common and hairy vetch plants (59.3 and 43.3 kg·N·ha^{-1}, respectively) were measured in this study. Both common and hairy vetch can theoretically supply 67 kg·N·ha^{-1}, the recommended rate of N fertilizer for switchgrass, if sufficient plant densities are achieved and adequate *Rhizobia* populations are present. Based on the results reported herein, it is estimated that switchgrass yield will increase beyond the control with common or hairy vetch seeded at rates of 8 and 10 kg·PLS·ha^{-1}, respectively. Proper legume management guidelines that address legume varieties compatible with switchgrass, appropriate bacterium for seed inoculation and seeding dates and rates that minimize the competition of vetch when intercropped with switchgrass need to be further developed to make this legume a viable option for displacing inorganic-N in switchgrass biofuel and forage production systems.

Acknowledgments: Mention of tradenames or commercial products in this publication is solely for the purpose of providing specific information and does not imply recommendation or endorsement by the U.S. Department of Agriculture or the University of Tennessee. The authors would like to thank The University of Tennessee Agricultural Experiment Station and the University of Tennessee Soil, Plant and Pest Center, (Nashville, TN, USA), as well as Stacy Warwick for providing technical assistance.

Author Contributions: Fred L. Allen conceived and designed the experiments; Kara S. Warwick performed the experiments; Patrick D. Keyser, Gary E. Bates, Don D. Tyler, Paris L. Lambdin and Dan H. Pote contributed expertise and helped with carrying out the experiments; Amanda J. Ashworth wrote the paper.

Conflicts of Interest: The authors declare no conflict of interest.

References

1. Ashworth, A.J.; Allen, F.; Keyser, P.; Tyler, D.; Saxton, A.; Taylor, A. Switchgrass yield and stand dynamics from legume intercropping based on seeding rate and harvest management. *J. Soil Water Conserv.* **2015**, *70*, 375–385. [CrossRef]

2. Ledgard, S.F.; Steele, K.W. Biological nitrogen fixation in mixed legume/grass pastures. *Plant Soil* **1992**, *141*, 137–153. [CrossRef]

3. Graham, P.H. Biological Dinitrogen Fixation: Symbiotic. In *Principles and Applications of Soil Microbiology*; Sylvia, D.M., Fuhrmann, J.J., Hartel, P.G., Zuberer, D.A., Eds.; Pearson Education Inc.: Upper Saddle River, NJ, USA, 2005; pp. 405–432.

4. Larson, J.A.; Jaenicke, E.C.; Roberts, R.K.; Tyler, D.D. Risk effects of alternative winter cover crop, tillage, and nitrogen fertilization systems in cotton production. *J. Agron. Appl. Econ.* **2001**, *33*, 445–457.

5. Warwick, K.; Allen, F.; Keyser, P.; Ashworth, A.J.; Tyler, D.; Saxton, A.; Taylor, A. Biomass and forage/biomass yields of switchgrass as affected by intercropped cool and warm-season legumes. *J. Soil Water Conserv.* **2016**, *71*, 21–28. [CrossRef]

6. Lemus, R.; Parrish, D.J.; Wolf, D.D. Nutrient uptake by 'Alamo' switchgrass used as an energy crop. *Bioenerg. Res.* **2009**, *2*, 37–50. [CrossRef]

7. Mooney, D.F.; Roberts, R.K.; English, B.C.; Tyler, D.D.; Larson, J.A. Yield and breakeven price of 'Alamo' switchgrass for biofuels in Tennessee. *Agron. J.* **2009**, *101*, 1234–1242. [CrossRef]

8. Garland, C.D. *SP701-A-Growing and Harvesting Switchgrass for Ethanol Production in Tennessee*; University of Tennessee Extension publication: Knoxville, TN, USA, 2008; Available online: https://extension.tennessee.edu/publications/Documents/SP701-A.pdf (accessed on 7 June 2017).

9. Sanderson, M.A.; Reed, R.L.; McLaughlin, S.B.; Wullschleger, S.D.; Conger, B.V.; Parrish, D.J.; Wolf, D.J.; Taliaferro, D.D.; Hopkins, C.; Ocumpaugh, A.A.; et al. Switchgrass as a sustainable energy crop. *Bioresour. Technol.* **1996**, *56*, 83–93. [CrossRef]

10. Ashworth, A.J.; West, C.P.; Allen, F.L.; Keyser, P.D.; Weiss, S.; Tyler, D.D.; Taylor, A.M.; Warwick, K.L.; Beamer, K.P. Biologically fixed nitrogen in legume intercropped systems: comparison of N-difference and 15N enrichment techniques. *Agron. J.* **2015**, *107*, 2419–2430. [CrossRef]

11. George, J.R.; Blanchet, K.M.; Gettle, R.M.; Buxton, D.R.; Moore, K.J. Yield and botanical composition of legume-interseeded vs. nitrogen-fertilized switchgrass. *Agron. J.* **1995**, *87*, 1147–1153. [CrossRef]

12. Opitz von Boberfeld, W.; Beckmann, E.; Laser, H. Nitrogen transfers from *Vicia sativa* L. and *Trifolium resupinatum* L. to the companion grass and the following crop. *Plant Soil Environ.* **2005**, *51*, 267–275.

13. Tyler, D.D.; Duck, B.N.; Graveel, J.G.; Bowen, J.F. Estimating response curves of legume nitrogen contribution to no-till corn. In *The Role of Legumes in Conservation Tillage Systems*; Power, J.F., Ed.; Soil Conservation Society of America: Ankeny, IA, USA, 1987; pp. 50–51.

14. Clark, A.J.; Decker, A.M.; Meisinger, J.J.; Mulford, F.R.; McIntosh, M.S. Hairy vetch kill date effects on soil water and corn production. *Agron. J.* **1995**, *87*, 579–585. [CrossRef]

15. Holderbaum, J.F.; Decker, A.M.; Meisinger, J.J.; Mulford, F.R.; Vough, L.R. Fall seeded legume cover crops for no-tillage corn in the humid east. *Agron. J.* **1990**, *82*, 117–124. [CrossRef]

16. Ranells, N.N.; Wagger, M.G. Grass-legume bicultures as winter annual cover crops. *Agron. J.* **1997**, *89*, 659–665. [CrossRef]

17. Blanchet, K.M.; George, J.R.; Gettle, R.M.; Buxton, D.R.; Moore, K.J. Establishment and persistence of legumes interseeded into switchgrass. *Agron. J.* **1995**, *87*, 935–941. [CrossRef]

18. UC SAREP. *Common Vetch*; University of California Sustainable Agriculture Research & Education Program: Oakland, CA, USA, 2006; Available online: http://www.sarep.ucdavis.edu/cgi-bin/ccrop.EXE/show_crop_14 (accessed on 12 May 2011).

19. Matic, R.; Nagel, S.; Saunders, R. *Vetch Variety Sowing Guide 2015*; National Vetch Breeding Program and SARDI: Australia, 2015. Available online: http://www.sardi.sa.gov.au/__data/assets/pdf_file/0010/45964/vetch.pdf (accessed on 12 February 2015).

20. Sattell, R.; Luna, D.J.; McGrath, D. *Hairy Vetch (Vicia villosa)*; EM 8704; Oregon Cover Crops, Oregon State University: Corvallis, OR, USA, 2004; Available online: http://forages.oregonstate.edu/php/fact_sheet_print_legume.php?SpecID=41 (accessed on 12 Feberary 2015).

21. Hanaway, D.; Larson, C. *Hairy Vetch (Vicia villosa Roth)*; Oregon State University: Corvallis, OR, USA, 2004; Available online: http://ohioline.osu.edu//agf-fact/0006.html (accessed on 26 May 2012).

22. Myers, D.; Underwood, J. *Hairy Vetch as An Ohio Cover Crop*; OSU Extension publication. AGF-006; Ohio State University: Columbus, OH, USA, 1990. Available online: http://ohioline.osu.edu//agf-fact/0006.html (accessed on 26 May 2012).

23. Myrold, D.D.; Ruess, R.W.; Klug, M.J. Dinitrogen fixation. In *Standard Soil Methods for Long-Term Ecological Research*; Robertson, G.P., Coleman, D.C., Bledsoe, C.S., Sollins, P., Eds.; Oxford University Press: New York, NY, USA, 1999.

24. Minchin, F.R.; Witty, J.F.; Mytton, L.R. Reply to 'Measurement of nitrogenase activity in legume root nodules: In defense of the acetylene reduction assay' by J.K. Vessey. *Plant Soil* **1994**, *158*, 163–167. [CrossRef]

25. Danso, S.K.A. Assessment of biological nitrogen fixation. *Fert. Res.* **1995**, *42*, 33–41. [CrossRef]

26. Ebelhar, S.A.; Frye, W.W.; Blevins, R.L. Nitrogen from legume cover crops for no-till corn. *Agron. J.* **1984**, *76*, 51–55. [CrossRef]

27. Zuberer, D.A. Biological Dinitrogen Fixation: Introduction and Non-Symbiotic. In *Principles and Applications of Soil Microbiology*; Sylvia, D.M., Fuhrmann, J.J., Hartel, P.G., Zuberer, D.A., Eds.; Pearson Education Inc.: Upper Saddle River, NJ, USA, 2005; pp. 373–404.

28. Segundo, S.U.; Boddey, R.M. Theoretical considerations in the comparison of total nitrogen difference and ^{15}N isotope dilution estimates of the contribution of nitrogen fixation to plant nutrition. *Plant Soil* **1987**, *102*, 291–294.

29. Boddey, R.M.; Chalk, P.M.; Victoria, R.L.; Matsui, E. Nitrogen fixation by nodulated soybean under tropical field conditions estimated by the ^{15}N isotope technique. *Soil Biol. Biochem.* **1984**, *16*, 583–588. [CrossRef]

30. Chalk, P.M. Dynamics of biologically fixed N in legume-cereal rotation: A review. *Aust. J. Agric. Res.* **1998**, *49*, 303–316. [CrossRef]

31. Phillips, D.A.; Jones, M.B.; Centre, D.M.; Vaughan, C.E. Estimating symbiotic N fixation by *trifolium subterraneum* L. during regrowth. *Agron. J.* **1983**, *75*, 736–741. [CrossRef]

32. Bell, M.J.; Wright, G.C.; Peoples, M.B. The N_2-fixing capacity of peanut cultivars with differing assimilate partitioning characteristics. *Aust. J. Agric. Res.* **1994**, *45*, 1455–1468. [CrossRef]

33. Gettle, R.M.; George, J.R.; Blanchet, K.M.; Buxton, D.R.; Moore, K.J. Frost seeding legumes into established switchgrass: Forage yield and botanical composition of the stratified canopy. *Agron. J.* **1996**, *88*, 555–560. [CrossRef]

34. Ashworth, A.J.; Keyser, P.D.; Allen, F.L.; Tyler, D.D.; Taylor, A.M.; West, C.P. Displacing inorganic-nitrogen in lignocellulosic feedstock production systems. *Agron. J.* **2016**, *108*, 1–8. [CrossRef]

35. Jorgensen, F.V.; Ledgard, S.T. Contribution from stolons and roots to estimate the total amount of N_2 fixed by white clover (*Trifolium repens* L.). *Ann. Bot. (Lond.)* **1997**, *80*, 641–648. [CrossRef]

36. McNeill, A.M.; Zhu, C.; Fillery, I.R.P. Use of *in situ* ^{15}N-labelling to estimate the total below-ground nitrogen of pasture legumes in intact soil-plant systems. *Aust. J. Agric. Res.* **1997**, *48*, 295–304. [CrossRef]

37. Bates, G.; Harper, C.; Allen, F. *PB 378 Forage and Field crop Seeding Guide for Tennessee*; The University of Tennessee Agricultural Extension Service, UT Extension publication: Knoxville, TN, USA, 2008. Available online: http://forages.tennessee.edu/Page%204-%20Planting/pb378.pdf (accessed on 26 May 2014).

38. Vogel, K.P.; Masters, R.A. Frequency grid—A simple tool for measuring grassland establishment. *J. Range Manage.* **2001**, *54*, 653–655. [CrossRef]

39. SAS Institute. *SAS/STAT 9.3 User's guide*; SAS Institute, Inc.: Cary, NC, USA, 2007.

40. Van Assche, J.A.; Debucquoy, K.L.A.; Rommens, W.A.F. Seasonal cycles in the germination capacity of buried seeds of some Leguminosae (Fabaceae). *New Phytol.* **2003**, *158*, 315–323. [CrossRef]

41. Ortega-Olivencia, A.; Devesa, J.A. Seed set and germination in some wild species of *Vicia* from SW Europe (Spain). *Nord. J. Bot.* **1997**, *17*, 639–648. [CrossRef]

42. Hargrove, W.L. Winter legumes as a nitrogen source in no-till grain sorghum. *Agron. J.* **1986**, *78*, 70–74. [CrossRef]

43. Peoples, M.B.; Ladha, J.K.; Herridge, D.F. Enhancing legume N_2 fixation through plant and soil management. *Plant Soil.* **1995**, *174*, 83–101. [CrossRef]

44. Ranells, N.N.; Wagger, M.G. Nitrogen release from grass and legume cover crop monocultures and bicultures. *Agron. J.* **1996**, *88*, 777–782. [CrossRef]

13

Biochar for Horticultural Rooting Media Improvement: Evaluation of Biochar from Gasification and Slow Pyrolysis

Chris Blok [1],*, **Caroline van der Salm** [1], **Jantineke Hofland-Zijlstra** [1], **Marta Streminska** [1], **Barbara Eveleens** [1], **Inge Regelink** [2], **Lydia Fryda** [3] and **Rianne Visser** [3]

[1] Wageningen Plant Research, Glasshouse Horticulture, Violierenweg 1, 2665 MV Bleiswijk, The Netherlands; Caroline.vanderSalm@wur.nl (C.v.d.S.); jantineke.hofland-zijlstra@wur.nl (J.H.-Z.); marta.streminska@wur.nl (M.S.); barbara.eveleens@wur.nl (B.E.)

[2] Wageningen Environmental Research, Wageningen University & Research, Droevendaalsesteeg 3, 6708 PB Wageningen, The Netherlands; inge.regelink@wur.nl

[3] Energy Research Centre of the Netherlands (ECN), P.O. Box 1, 1755 ZG Petten, The Netherlands; fryda@ecn.nl (L.F.); h.visser@ecn.nl (R.V.)

* Correspondence: chris.blok@wur.nl

Academic Editors: Marcus Hardie and Peter Langridge

Abstract: Peat is used as rooting medium in greenhouse horticulture. Biochar is a sustainable alternative for the use of peat, which will reduce peat derived carbon dioxide emissions. Biochar in potting soil mixtures allegedly increases water storage, nutrient supply, microbial life and disease suppression but this depends on feedstock and the production process. The aim of this paper is to find combinations of feedstock and production circumstances which will deliver biochars with value for the horticultural end user. Low-temperature (600 °C–750 °C) gasification was used for combined energy and biochar generation. Biochars produced were screened in laboratory tests and selected biochars were used in plant experiments. Tests included dry bulk density, total pore space, specific surface area, phytotoxicity, pH, EC, moisture characteristics and microbial stability. We conclude that biochars from nutrient-rich feedstocks are too saline and too alkaline to be applied in horticultural rooting media. Biochars from less nutrient-rich feedstocks can be conveniently neutralized by mixing with acid peat. The influence of production parameters on specific surface area, pH, total pore space and toxicity is discussed. Biochar mildly improved the survival of beneficial micro-organisms in a mix with peat. Overall, wood biochar can replace at least 20% v/v of peat in potting soils without affecting plant growth.

Keywords: alkalinity; biochar; gasification; pH; phytotoxicity; pyrolysis; salinity; stability; degradability

1. Introduction

Biochar is the carbon rich co-product of pyrolysis or gasification of biomass. Biochar application to soils is of public and agricultural interest [1]. The public interest in biochar application to soil is focused on the potential to decrease global net carbon dioxide emission by an increased soil storage of carbon [2]. The agricultural interest is focused on a number of positive properties [3], the most striking being plant growth stimulation by increased water storage [4], increased nutrient supply [5,6], increased beneficial microbial life [7,8] and disease suppression [7,9]. Just as for agriculture, biochar application to horticultural rooting media (soilless substrates) is of public and agricultural interest. (1) The public interest is to use biochars from renewable organic residual streams to substitute part of the peat used in rooting media in greenhouse horticulture [10,11]. Peat bogs are important carbon (C) stocks and regulate the local water quality and water regime [12]. In the light of environmental concerns, peat substitution by biochar will preserve peat bogs and lower global carbon dioxide emissions linked with the use of peat extraction and use [13]; (2) The horticultural interest in biochar apart from peat substitution is the use and manipulation of bacterial communities for the protection of plants against diseases, either by direct protection or by induced plant resilience [14–16]. In certain plant growth media, biochar amendment results in chemical responses in the plant as well as shifts in the rhizosphere microbiome [17]. In greenhouse horticulture, the use of high input fertigation systems makes biochar related increases in water storage and nutrient supply of less economic consequence than for agricultural applications. An advantage of greenhouse testing is the improved control over climate effects including rain related water content and nutrient concentration fluctuations.

Several researchers have looked into the potential of biochar as a rooting medium for horticulture [14,18,19], including the substitution of peat. The study presented in [17] demonstrates the effect of oak wood pyrolysis biochar on strawberry grown in white peat and lettuce grown in field soil. In the strawberry bioassay, addition of 3% w/w biochar to peat resulted in (1) a higher fresh and dry plant weight; (2) a lower susceptibility for the fungal pathogen *Botrytis cinerea* on both leaves and fruits; and (3) changes in the rhizosphere microbiology such as an increase of bacterial diversity and a shift in composition of the rhizosphere microbiota. Extra inorganic plant nutrition and lime added to the peat reduced these effects of biochar on the strawberry plants. In [11] the authors reported that the hydrophysical properties of peat based growing media changed with the addition of various biochars up to 70% v/v. Lettuce (*Lactuca sativa*) grown in the peat/biochar mixtures showed substantially higher yields than with peat alone. In addition, this study confirmed the importance of biochar production conditions on the product properties. In [20] Dumroese et al. found that peat moss, amended with various ratios of pellets comprised of equal proportions of biochar and wood flour, generally had chemical and physical properties suitable for service as a rooting medium during nursery production of plants. A mixture of 75% v/v peat and 25% v/v pellets enhanced hydraulic conductivity and water availability at low (< -10 kPa) matric potentials. In [21] it was demonstrated in rooting media used for container production of greenhouse crops that biochars from wood gasification were able to buffer peat acidity, eliminating the need for liming agents. Gasification biochars reduced shrinkage of peat acting as a stable skeleton, and reduced the ammonium/nitrate ratio in the peat after a fertilization event. Lastly, biochars added stable and high levels of potassium to rooting media. In [22], the performance of tomato crop green-waste pyrolysis biochar as a rooting medium for hydroponic tomato production was compared with an existing, commercially acceptable rooting medium, pine sawdust. No peat was used in these rooting media. The EC of rooting media containing, or consisting entirely of biochar was reduced by rinsing with water before use, believing there is potential to capture and recycle nutrients flushed during this process. In terms of growth, yield, or fruit quality no differences were found among biochar and pine sawdust rooting media. In another attempt to completely replace peat [23] as well as vermiculite, a dried anaerobic digestate remaining after the fermentation of potato processing wastes, was mixed with three biochars produced from either wood pellets, pelletized wheat straw or field pennycress press cake. All three biochars were acidified and combined in 50%/50% v/v ratios with the digestate before comparing with a 50%/50% v/v sphagnum peat moss/vermiculite control medium containing slow-release chemical fertilizers. A growth increase of tomato was observed for the mix containing the wood pellet biochar.

Horticultural interest in biochar also seems justified in the light of prior results with biochar-like materials such as charred rice husks [24,25] and torrefied Gramineae [26,27]. Torrefaction or carbonization is a process that happens at temperatures (250–400 °C) with the objective of creating an energy-dense, solid biofuel with gas and liquids as by-products. The term biochar usually describes the by-product of pyrolysis and gasification. Pyrolysis is carried out at temperatures of 400 °C to 650 °C, without oxygen, while gasification occurs at temperatures above 600 °C. Both pyrolysis and gasification have as main objective the production of syngas, liquid fuels and chemicals. Charred rice husks (synonymous with "burnt" rice husks) are used in mixes and on their own on a large scale in horticulture in South East Asia, notably Indonesia, as well as in South America [28].

In [24] the study of gasified rice hull biochar (GRHB) on available nutrients in a rooting medium for containers revealed that GRHB provides sufficient P and K to support a production cycle of geranium, but lacked either the correct concentration or balance of micronutrients for healthy growth. In [25] a number of crops were grown on standard commercial rooting medium composed of sphagnum peat moss/perlite (85%/15% v/v) and mixed with 0%, 5%, or 10% v/v GRHB. GRHB provided a source of readily available phosphate and potassium when incorporated at 5% or 10% v/v. The amount of available phosphate and potassium became depleted after a period of up to 6 weeks. Torrefied reed (*Phragmites australis*) was successfully used in an experiment with up to 50% v/v as compared to peat [27]. Torrefaction was used as low oxygen gasification of organic materials at temperatures of 150–400 °C [26]. Both charred rice husks and torrefied reed indicated potential for charred products at dosages far beyond 25% by volume.

A large and growing body of literature has reported no beneficial, adverse or contradicting effects on plant growth when using biochar [29–31]. The negative and neutral effects of biochar application have resulted in increased attention on methods to characterize biochars in general and for specific applications [32–34]. A wide variety of growing media are already being used in horticulture and a set of specific quality parameters and measuring methods has been developed [35]. Evaluating biochar for horticultural applications requires both, a material characterization with rooting media tests and a quantified link with results in field or container experiments. Such data will then fill the gap between biochar engineering and horticultural application results. Evidence of the importance of production factors has been reported for the nature of feedstocks [36,37]; the temperature of production [34,37,38]; the supply of oxygen [34,39]; and the cooling procedure to prevent condensation of toxic substances [14]. Gray et al., 2014, have reported a decreased hydrophobicity and related greater water entry at higher production temperatures [40].

The objective of our paper is a follow up on an earlier study by Fryda and Visser [41], who related the feedstock materials and thermal processes (pyrolysis and gasification) to the properties of the produced biochar. They concluded that biochars of widely different properties can be produced using the same feedstock under different production conditions. The authors further concluded that phytotoxic properties caused by condensation of tar loaded gas on the biochar particles can be avoided by using higher temperatures and early separation of gas from solids. They also reported increased internal particle porosity with gasification temperatures. In the present paper, we aim to first show the influence of feedstock and production parameters on a set of growth influencing properties, and second, to investigate the impact of two ways of disease suppression on rooting media. Our hypotheses are: (1) biochar is a potting soil constituent which can be used in growing medium blends in quantities of 20% v/v without negative growth effects; (2) biochar can induce disease suppression.

Our approach is to use well-defined materials for testing on horticultural properties, like water retention, water uptake rate, phytotoxicity, pH buffer values, nutrient and EC levels, cation exchange capacity (CEC), microbial stability and nitrate immobilization [35]. In addition we use biochars in plant tests to find potentially positive effects of biochar addition on the suppression of powdery mildew and *Fusarium*. Powdery mildew is selected based on possible biochar induced stimulation of plant hormones, which are related to induced plant resistance against biotrofic pathogens [9]. *Fusarium* is selected based on possible biochar enhanced *Fusarium* suppression by beneficial microorganisms [7].

2. Materials and Methods

2.1. Biochars Produed

Eight biochars were used in this study (Table 1). All biochars were produced by ECN (Petten, The Netherlands) in a lab-scale gasifier, at 670 °C gasification temperature, except for one sample (number 5) which was produced at 750 °C (Table 1). The facilities and test procedures are described elsewhere in detail [41]. All biochars were continuously collected from the fluidized gasification bed under stable and continuous conditions. Initial laboratory testing [41] was supplemented with horticultural tests [35]. In close cooperation between biochar producers and horticultural specialists, it was decided to make biochar grains of 3–4 mm (with high internal particle porosity) because of the foreseen mixing with milled peat and aimed at optimal texture and drainage of the biochar/peat mixture.

Table 1. Overview of the biochars produced by ECN and pyrolysis conditions.

Code	Biomass	T	Moisture	Ash	Volatiles	C	HHV
		°C	%, a.r.	%, d.m.	%, d.m.	%, d.m.	MJ/kg
1. Beech/Tomato	80% beech wood + 20% tomato leaves	670	5.4	17.5	12.3	70.2	27.5
2. Wood/Tomato	80% wood chips-1 * + 20% tomato leaves	670	3.3	22.5	13.3	64.2	24.4
3. Wood chips-1	Batch spring 2015 *	670	-	-	-		-
4. Sweet pepper waste	Vegetable residues (Spain)	670	4.5	33.6	14.7	51.7	21.3
5. Sweet pepper waste	Vegetable residues (Spain)	750	4.5	26.5	15.1	58.4	21.2
6. Wood chips-2 **	Batch July 2015 *	670	3.2	10.7	10.2	79.1	29.7
7. Wood chips-3	Batch August 2015 *	670	-	-	-		-
8. Wood chips ***	Beech wood chips	670	2.7	23.8	6.0	70.2	27.1

T = Temperature; C = organic matter calculated as remaining mass without ash and volatiles; HHV = Higher Heating Value as measure of combustion value; a.r. = as received; d.m. = dry matter; * Wood cuttings from forestry Purmerend; ** Used in greenhouse experiment; *** Used in climate chamber experiment.

2.2. Rooting Medium Testing

Rooting media are tested for about twenty properties to evaluate their suitability for use as a rooting medium in horticulture [35,42]. Not all tests are relevant for all rooting media materials used. For biochar as constituent of peat based potting soil mixes, we limited tests to dry bulk density, organic matter content, ash content, specific surface area (SSA), salt content by electro conductivity (EC), acidity (pH), nutrient availability, total nutrient content, cation exchange capacity (CEC), base saturation, water retention and available air curve, toxicity and degradability. Reports by production sciences and agriculturists often use % w/w and air filled space (a ratio of two unknowns) to describe water content, air content, biochar addition and organic matter content. The consequence is that such reporting often cannot be interpreted for rooting media with a dry bulk density different from soil as those common in horticulture. To evaluate the material's suitability in horticulture, results should be reported in % v/v and air filled space (a ratio of the fixed total sample volume). The reason is that plants and bacteria sense and react to the environment on a per volume basis. We believe the observations above merit a place in the biochar assessment procedures already available [33].

2.2.1. Physical Characterization of Biochars

Dry weight and bulk density were determined after drying at 110 °C [43]. The organic matter content was determined by loss on ignition at 500 °C [44]. The specific surface area (SSA), elemental contents and levels of organic compounds were analyzed according to a biochar protocol by a commercial laboratory (Eurofins, Obritzsch-Hilbersdorf, Saxony, Germany). The proximate analysis of the biochar samples was carried out at ECN under EN ISO/IEC 17025 accreditation (Table 1). The proximate analysis includes moisture content of the sample as received and the remaining dry matter dived into the ash (mineral) content, the volatile content and the organic matter content (also referred to as free carbon). Finally, the high heating value (HHV) based on complete combustion of the sample to carbon dioxide and liquid water is given. The actual carbon content as total C (% in dry sample) was determined later (Table 2).

Table 2. Properties of biochars produced from different feedstocks or with different temperatures.

Parameter	Unit	1	2 *	3	4	5	6 *	7	8 *
		Tomato		Wood	Pepper	Pepper	Residual wood		
		Beech	Wood	Beech	650 °C	750 °C	Batch 1 **	Batch 2	Batch 3 ***
pH	-	11	12	11	12	12	9.4	9.9	10
EC	dS·m^{-1}	6.8	13	0.68	9.6	11	0.53	0.61	0.71
NH$_4$	mmol·L^{-1}	<0.1	0.1	<0.1	<0.1	<0.1	<0.1	<0.1	<0.1
K	mmol·L^{-1}	49.1	94.3	3.2	61	84.3	3.6	3.6	4.5
Na	mmol·L^{-1}	0.5	1	0.1	5.1	7.4	0.2	0.2	0.3
Ca	mmol·L^{-1}	0.3	3	0.3	0.3	0.5	0.2	0.3	0.2
Mg	mmol·L^{-1}	0.1	<0.1	<0.1	0.2	0.1	0.2	0.2	0.1
Si	mmol·L^{-1}	0.3	0.3	<0.1	0.1	0.2	<0.1	<0.1	<0.1
NO$_3$	mmol·L^{-1}	<0.1	0.2	<0.1	0.1	0.1	<0.1	0.1	<0.1
Cl	mmol·L^{-1}	23.6	48.5	0.1	47.8	65.2	0.2	0.4	0.6
SO$_4$	mmol·L^{-1}	10.8	21.4	0.3	4.1	4.3	0.4	0.2	0.2
HCO$_3$	mmol·L^{-1}	3.3	7.2	3.6	18.9	19.5	3.2	3.7	4.3
P	mmol·L^{-1}	<0.05	<0.05	<0.05	<0.05	<0.05	0.1	<0.05	<0.05
Fe	μmol·L^{-1}	0.4	0.9	<0.4	<0.4	<0.4	0.5	0.6	0.5
Mn	μmol·L^{-1}	0.2	<0.1	<0.1	<0.1	<0.1	2.2	0.7	0.2
Zn	μmol·L^{-1}	0.1	0.1	<0.1	<0.1	<0.1	0.2	0.1	0.1
B	μmol·L^{-1}	13	9	10	28	35	8	7	6
Cu	μmol·L^{-1}	0.1	5	<0.1	0.2	0.1	<0.1	<0.1	<0.1
Mo	μmol·L^{-1}	0.5	0.9	<0.1	<0.1	<0.1	<0.1	0.10	<0.1
Dry weight	%	96.7	97.1		95.5	95.5	97.8		
Bulk density	kg·m^{-3}	131	113		104	129	102		
SSA	m^2·g^{-1} d.m.	59	81		39	29	119		
Ash	% d.m.	19	28		35	34	13		
total C	% d.m.	77	68		59	59	82		
total H	% d.m.	1.3	1.4		1.2	1.3	1.5		
Total N	% d.m.	0.7	1.0		0.8	0.9	0.8		
C/N ratio	mol·mol^{-1}	128	79		86	76	119		

Nutrient concentrations and pH were determined in a 1:1.5 v/v water-extract. d.m. stands for dry matter. * Wood cuttings from forestry Purmerend; ** Used in greenhouse experiment; *** Used in climate chamber experiment.

2.2.2. Water Holding Capacity

The water retention at different water potentials (expressed as suction forces) was determined with the sand box method, using a suction device which can be set to a series of standard suction forces [43]. The method allows prediction of the water content in field circumstances as well as comparing various rooting materials. First, the samples were nearly saturated (−3 cm). Thereafter, the water potential was decreased to −31.5 and −50 cm water column and the water content was measured when equilibrium was established. The drying steps were followed by a rewetting step in which the sample was again brought to near-saturation in order to determine the effect of the drying on water uptake.

2.2.3. Chemical Characterization of Biochars

Water-extractable nutrients were determined in a water extract with a sample-solution ratio of 1:1.5 v/v [45]. Extraction after dilution is the accepted method to find plant available elements when extraction by suction or pressure is not applicable. The exchangeable cations were extracted with concentrated $BaCl_2$ [46,47]. Water-extractable nutrients and exchangeable cations were analyzed with ICP by a commercial laboratory (Groen Agro Consult, Delft, The Netherlands).

2.2.4. Acid-Neutralizing Capacity

The acid-neutralizing capacity of the biochar was determined by a method using titration with acid with time-stepped acid addition to allow for dissolution and reaction kinetics [48]. The method uses an automated titration unit (Metrohm, Schiedam, The Netherlands). Concentrated acid (HCl) was dosed for five minutes followed by an equilibration period of 45 min without acid dosing. During dosing, acid was added until a pH of five was reached. During equilibration, the pH slowly increased due to buffering of the medium. This procedure was repeated ten times. The titration took therefore nine hours and was performed in duplicate. The total amount of acid dosed during the titration was thought equal to the acid-neutralizing capacity of the rooting medium expressed in $mol \cdot kg^{-1}$. The same procedure was used to determine the base neutralizing capacity of peat by using concentrated base (KOH) instead of acid.

2.2.5. Phytotoxicity Test

Phytotoxicity was tested using water extracts (1:2 volume ratio) from the biochars as received (without any pre-washing), using an established plant response method for rooting material quality [49,50]. Water-extracts were filtered (8 μm paper, Merck, Schiphol-Rijk, The Netherlands). Thereafter, the pH values of the extracts were adjusted to pH 5.5 using concentrated nitric acid, the EC values were adjusted to $2 \ dS \cdot m^{-1}$ by dilution with demi water, and nutrients (NPK) were added to reach the required standard concentrations [50]. The bioassays were carried out using: *Sorghum saccharatum* (L.) Moench (sorghum), *Lepidium sativum* L. (garden cress) and *Sinapis alba* L. (mustard). For each treatment, 4 plates per plant species with 10 seedlings per plate were prepared, i.e., 120 seedlings per treatment. After incubation for 3 days at 25 °C, in darkness, photos of the plates were taken. The number of germinated seedlings was counted and root- and shoot length of seedlings were measured. Statistical analyses on the data were performed in GenStat (ANOVA).

2.2.6. Oxygen Uptake Rate

The oxygen uptake rate (OUR) method determines the maximum oxygen uptake rate measured under conditions ideal for microbial degradation of organic matter [51]. This method results in a value for the maximum possible breakdown rate of the rooting material tested and allows ranking of materials in classes for stability over time. Media were mixed with a nutrient solution ($4.3 \ g \cdot L^{-1}$ NH_4Cl, $5.4 \ g \cdot L^{-1}$ $CaCl_2 \cdot 2H_2O$, $4.3 \ g \cdot L^{-1}$ $MgSO_4 \cdot 7H_2O$, $0.03 \ g \cdot L^{-1}$ $FeCl_3 \cdot 6H_2O$ $5.0 \ g \cdot L^{-1}$ EDDHA 6% iron chelate, $1.4 \ g \cdot L^{-1}$ $MnSO_4$, $1.1 \ g \cdot L^{-1}$ $ZnSO_4$, $4.2 \ g \cdot L^{-1}$ $Na_2B_4O_7$, $0.2 \ g \cdot L^{-1}$ $CuSO_4$, $0.13 \ g \cdot L^{-1}$ Na_2MoO_4, $1 \ mL \cdot L^{-1}$ HCl (36%)) to ensure that microbial activity was not limited by nutrients or moisture. The pH of the medium-nutrient mixture was adjusted to pH 5.5 and a nitrification inhibitor, allylthiourea (ATU), was added. The rooting media were put in a closed vessel and placed on a horizontal shaker (150 rpm) for five days at a temperature of 30 °C. The pressure in the vessels was continuously measured over the course of the incubation period. CO_2 was scrubbed from the gas phase. Therefore, the decrease in pressure was completely attributed to consumption of O_2. The O_2 concentration was plotted as a function of time and the maximum O_2 consumption rate (dO_2/dt) was derived from this graph. The results are expressed as oxygen consumption per unit of time per mass of dry organic matter ($mmol \cdot h^{-1} \cdot kg^{-1}$ DOM).

2.3. Climate Chamber Experiment with Gerbera and Powdery Mildew

In a pot experiment in a climate chamber, the direct (inherent) effect of biochar addition on the induction of plant resistance in young *Gerbera jamesonii* plants was tested. After three weeks challenging time with biochar, the plants were artificially infected with powdery mildew (*Erysiphales*) and monitored on disease development. Powdery mildew is a bio-trophic fungus that feeds on living plants. The biochar used was beech wood derived (Table 1, code 8).

2.3.1. Treatments

Four treatments, with 20 plants each were included in the climate chamber experiment:

A. Standard rooting medium (milled white peat)
B. Standard rooting medium with 20% v/v biochar
C. Standard rooting medium + fungicide (0.1% triflumizool)
D. Standard rooting medium + SAR elicitor (chemically induced disease resistance)

In treatment B, the rooting medium was produced by mixing 3 L of biochar with 2 L of acid peat in order to neutralize the high pH of the biochar. This neutral peat/biochar blend was then mixed with standard peat medium to achieve the proper treatment ratio. The volume of biochar in the final rooting medium was 20% v/v. In treatment C, the chemical fungicide applied was Rocket which is triflumazool at 0.1% w/w (Certis Europe, Utrecht, The Netherlands). In treatment D the chemical elicitor used was INA which is 2,6-dichloroisonicotinic acid (Sigma-Aldrich, St. Louis, MO, USA). INA is a Systemic Acquired Resistance (SAR) elicitor that induces disease resistance against powdery mildew through stimulation of natural defense processes in the plant (i.e., activation of the salicylic acid route).

The experiment was performed in a climate chamber set at temperature 20 °C; relative humidity 85%; light duration 16 h; light level 240 $\mu mol \cdot m^{-2} \cdot s^{-1}$ PAR light. After three weeks, plants were infected with a spore suspension of powdery mildew spores derived from infested *Gerbera* plants (2×10^4 spores per mL).

2.3.2. Assessment and Harvest

Two weeks after the first symptoms of mildew infection appeared the mildew was scored per plant with an index made by Spencer [52]. After five weeks, a plant health assessment was performed and the leaf chlorophyll content was measured with a SPAD 502 Plus meter (Konica Minolta Business Solutions Europe, Langenhagen, Lower Saxony, Germany). After six weeks, plants were harvested and fresh weights were analyzed.

2.4. Greenhouse Experiment with Added Microorganisms and Chrysanthemum

A potting soil mix with biochar, behaving equally well as a pure peat based medium, would already establish biochar as a peat alternative. However, an additional claim on benefits for plant health would be even more convincing. Such claims could be based on either direct effects of biochar itself, suppressing a particular pathogen, or on indirect effects of biochar, such as the promotion of beneficial microorganisms which then suppress pathogen activity. A greenhouse experiment was performed to assess whether the addition of biochar to peat affects plant growth and plant health. More specific, it was tested whether (1) biochar addition enhances the ability of *Chrysanthemum* × *morifolium* cv Euro to suppress diseases caused by the wilting pathogen, *Fusarium oxysporum*; and whether (2) addition of rhizobacteria spp to the rooting medium enhances the colonization of biochar with bacteria followed by suppression of *Fusarium* incidence.

2.4.1. Treatments

Four treatments, with 30 plants per repetition were included in the greenhouse experiment:

A.   Peat
B.   Peat + plant-growth promoting rhizobacteria
C.   Peat + biochar
D.   Peat + plant-growth promoting rhizobacteria

Treatments A, B and D received Baltic white peat (Jiffy Substrates, Dordrecht, The Netherlands), limed to pH 5.0 with 4 kg·m^{-3} calcium magnesium carbonate.

Treatment C received a mix of the same but unlimed peat and biochar-6 in an 85%:15% v/v ratio limed to pH 5.0 with 2 kg·m^{-3} of calcium-magnesium carbonate. The biochar-6 used was delivered in two batches, which were mixed in a 1:1 ratio (Table 1; ECN, Petten, The Netherlands). The mixing ratios above had been established in a preliminary mixing experiment. NPK 12-14-24 fertilizer including trace elements was added to all treatments at 1 kg·m^{-3} (PG-mix, Yara, Vlaardingen, ZH, The Netherlands). The rooting media were prepared two weeks before the start of the experiment. The pH and nutrient content were analyzed shortly before the start of the experiment.

All treatments were infected with the plant-pathogen *Fusarium oxysporum* isolated from chrysanthemums. *Fusarium* was added on day 14 of the experiment in a concentration of 10^4 cfu per pot (cfu = colony forming units per g dry matter). Because there were only minor indications of infection, *Fusarium* was added again in a concentration 10^4 cfu per pot on day 49.

Treatment B and D were treated with a commercial product, Compete Plus which is a mix of organisms that promote plant-growth and/or increase disease suppression (Plant Health Cure B.V, Oisterwijk, The Netherlands). Compete Plus claims to contain >5 × 10^7 cfu of each of 6 *Bacillus* strains (*licheniformis, megaterium, polymyxa, pumilus, subtilis, azotofixans*) as well as 1 × 10^7 cfu of *Trichoderma harzianum* and 1 × 10^6 cfu *Streptomyces griseoviridis* as well as various organic feed supplements. Rhizobacteria were added every two weeks, totaling four applications per pot and starting one day before planting the chrysanthemums. For the first addition, each pot received 100 mL of solution prepared with 1 g·L^{-1} Compete Plus. For the latter additions, each pot received 50 mL of solution prepared with 0.4 g·L^{-1} Compete Plus. The rhizobacteria were poured on top of the rooting medium in order to ensure they spread throughout the whole pot volume.

Each treatment was repeated 6 times in fields of 30 plants, i.e., 180 plants per treatment, 720 in total. The fields were distributed over 6 blocks, each block having all four treatments. Within a block there were 2 sub irrigation tables with each two treatments. Treatments without rhizobacteria (A and C) and with rhizobacteria (B and D) were kept on separate sub irrigation tables and received water from separated storage tanks to prevent spreading of rhizobacteria. For each block, 14 plants were guard plants and observations were restricted to the 16 remaining plants.

2.4.2. Cultivation

The pot-experiment was performed with chrysanthemums which were reproduced by stem cuttings. Ten days after planting the unrooted cuttings on a standard peat medium, the plants had formed a sufficient rooting system and were transplanted into the final pots. Plants were grown in containers of 0.7 L. For the first two weeks, plants were irrigated manually on top of the containers to ensure mixing of the rhizobacteria throughout the containers. From day 14 onwards, plants were irrigated from below by 15 min flooding cycles every one or two days, depending on weather conditions. After 70 days, plants were harvested and dry weight, fresh weight and nutrient content were determined.

2.4.3. Bacterial Analysis

Abundance of bacteria and fungi in the media was checked with semi quantitative PCR technique (qPCR) with primers sets, and 338F and 518R and 5.8s and ITSrev for bacteria and fungi respectively. Additional qPCR was performed to establish the numbers of *Firmicutes* (bacterial phylum to which *Bacillus* species belong). At the end of the greenhouse experiment, three mixed rooting media samples of each treatment were collected by mixing rooting medium from 5 pots and manually removing plant roots. After manual homogenization of the sample by sieving them through 2 mm sieve, 3 subsamples were collected for DNA isolation. DNA was isolated using the commercially available PowerSoil DNA Isolation kit (MoBio Laboratories, Carlsbad, CA, USA) according to manufacturer's protocol. qPCR was performed using SYBR Green chemistry (Promega Inc., Leiden, The Netherlands).

3. Results

3.1. Suitability of Biochars as a Rooting Medium

This section will show the results on the physical, chemical and biological characterization of the various biochars and will interpret these results in terms of suitability of biochar as a rooting medium.

3.1.1. Physical Properties

The biochars produced had low dry bulk densities 100–130 $kg \cdot m^{-3}$ (Table 2). The low dry bulk densities were related to very high total pore space 92%–94% v/v. Low dry bulk density and high total pore space are hallmarks for established potting soil constituents like peat and coir which also combine high water contents with air contents >15% v/v. The ash content of tomato or sweet pepper derived vegetable waste biochar was 34%–35% w/w, reflecting the high content of mineral nutrients of horticultural crops. The wood biochar sample showed only 10% w/w of ash and the wood/vegetable biochar mixes were in between. The volatile matter of the pepper based biochars in the proximate analysis was 14.7%–15.1% w/w which is higher than the amount of volatiles found in the wood biochars (6.0%–10.2% w/w) with the wood/tomato mixes in between (Table 1). The biochars produced showed high organic carbon contents, 59%–82% w/w (Table 2). The low hydrogen and nitrogen contents of <1.5% w/w reflect the low oxygen level in the gasification. Another apparent difference between biochar produced from wood and vegetable waste is the specific surface area, which is 2 to 3 times higher for biochar produced from wood compared to vegetable waste. A high specific surface area facilitates water uptake and the establishment of microbial life on biochar particles. Two possible explanations of the reduced specific surface area found in sweet pepper based biochar are: (1) Clogging of internal pores by volatile tars [34]; (2) Low temperature melting of ash compounds with high levels of calcium and potassium, which is in line with slagging reported for such samples [41].

3.1.2. Water Holding Capacity

The water holding capacity was measured at different suction values to represent the wetting and drying cycles that occur in rooting media during plant growth. Even under saturated conditions (i.e., after watering), rooting media need to contain a sufficient volume of air (>15%) to prevent anaerobic conditions [53]. Peat has a water holding capacity ranging from 74% to 35% v/v at suction forces of −10 cm respectively −50 cm (Figure 1a). The water holding capacity of the biochar (produced from wood chips) is much lower and ranges from 56% to 25% v/v (Figure 1b). Biochar consists of relatively coarse particles of 3–4 mm. The pores in between these particles are too large to retain moisture. Therefore, the 100% v/v biochar sample has a poor water holding capacity. However, once the biochar is mixed with peat, such pores are filled and the water holding capacity is similar to the water holding capacity of the peat alone (Figures 1 and 2), just as has been reported on perlite-peat mixes [54]. Under nearly saturated conditions, the peat and peat-biochar mixture contain at least 20% v/v air meaning that the rooting media contain a sufficient amount of oxygen. For both biochar and peat, less than 10% v/v of the material volume is occupied by solid particles showing the porous nature of both materials. To conclude, a peat/biochar mixture containing 15% v/v biochar has an almost similar water holding capacity as peat and is in that aspect suitable for application as a rooting medium in horticulture.

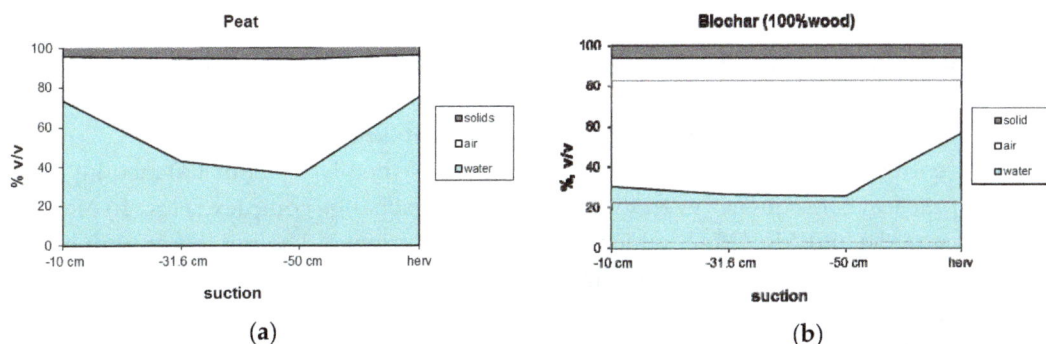

(a)   (b)

Figure 1. Water content of peat (**a**) and 100% v/v biochar from wood chips (**b**) in % v/v as a function of suction strength in cm water column. The water holding capacity of 100% v/v wood biochar of 3–4 mm grains is low compared to peat and too low to ensure sufficient water supply to plants. The higher water-content after re-saturation compared to the initial situation is partly explained by re-arrangement of the biochar particles leading to an increase in the presence of small pores.

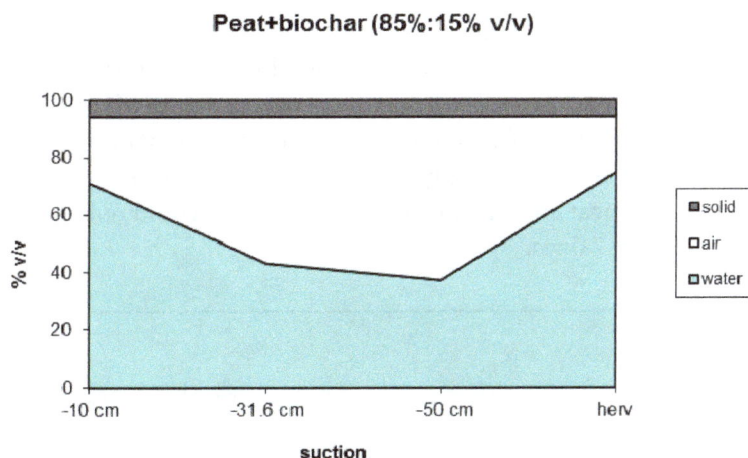

Figure 2. Water content in % v/v of a peat/biochar mixture (85%:15% v/v) as a function of suction strength in cm water column. The biochar particles are 3–4 mm and are produced from wood chips. The water holding capacity of the peat/biochar mixture is similar to the water holding capacity of peat because the large pores in between the biochar particles are now filled with peat. Based on the water holding capacity, the peat-biochar mixture is suitable for application in horticulture.

3.1.3. Chemical Composition and Nutrient Values of Biochar

The high pH values (9.4–12) and high ECs (6.8–13 dS·m^{-1}) of the biochars based on vegetable waste, are problematic for application in horticulture since a pH of about 5.5 and an EC value below 1 dS·m^{-1} are required for growing pot plants (Table 2). The results also show that the EC of the biochar is about 20 times higher when using vegetable waste as feedstock compared to wood materials. Two of the tested biochars were produced from feedstock that contained 80% v/v wood and only 20% v/v tomato leaves. The EC of these biochars is still considerably higher than the biochars produced from 100% wood. The addition of small amounts of nutrient-rich waste materials to the feedstock thus has a strong negative effect on biochar quality. Biochar produced from wood (beech or residual wood) has a low EC value (<0.6 dS·m^{-1}), low Na and SO$_4$ concentrations (<0.5 mmol·L^{-1}) and somewhat lower pH values (9.4–11). The EC value of the wood based biochar remains below the maximum EC value for rooting media. However, the pH value is still much higher than the desired pH range (pH 5.0–6.0) meaning that the biochar should be mixed with other acidic media or additives to produce a mixture with the desired pH.

Biochars do not deliver substantial amounts of plant nutrients as measured with the 1:1.5 extraction, except for potassium. The potassium levels however, especially in the vegetable residues, are up to ten times higher than the desired 3–5 mmol·L^{-1} and may induce shortages of other elements by oversupply. Biochars based on vegetable residues also contain unwanted elements such as Na, Cl, SO$_4$ and HCO$_3$ in quantities high enough to reduce growth at dosages of over 10% v/v.

The concentrations of exchangeable cations are about 10 meq·L^{-1} (about 100 meq·kg^{-1}) which is of limited practical consequence even though the cation exchange complex is low in Na and rich in K (33%), Ca (33%) and Mg (28%), which are all plant nutrients. The potassium will be readily exchanged for calcium, which is routinely handled by potting soil producers preparing mixes, i.e., of no consequence for the grower.

Because of the very high C/N ratio of biochar (65–110), mineralization of biochar will at first lead to N immobilization rather than N mineralization. The high C/N ratio also indicates a material as stable as peat.

3.1.4. Acid-Neutralizing Capacity

In order to produce a biochar/peat mixture with a neutral pH (pH 5), common for horticultural cultivation, the acid-neutralizing capacity of the biochar and the base-neutralizing capacity of the peat were determined. For this test, biochar made from wood chips was used. The results of the acid titration of biochar (Figure 3) show that biochar has a high acid-neutralizing capacity and that 258 mmol H$^+$ per kg dry matter are needed to bring the pH of biochar to pH 5.0 (Table 3). The titration curve shows that multiple acid dosages were needed to reduce the pH to the desired level due to the rather slow buffering of the biochar (Figure 3). Similar slow buffering processes occurred when biochar was mixed with acid peat and the final pH of the mixture could only be established after an equilibration period of at least 7 days.

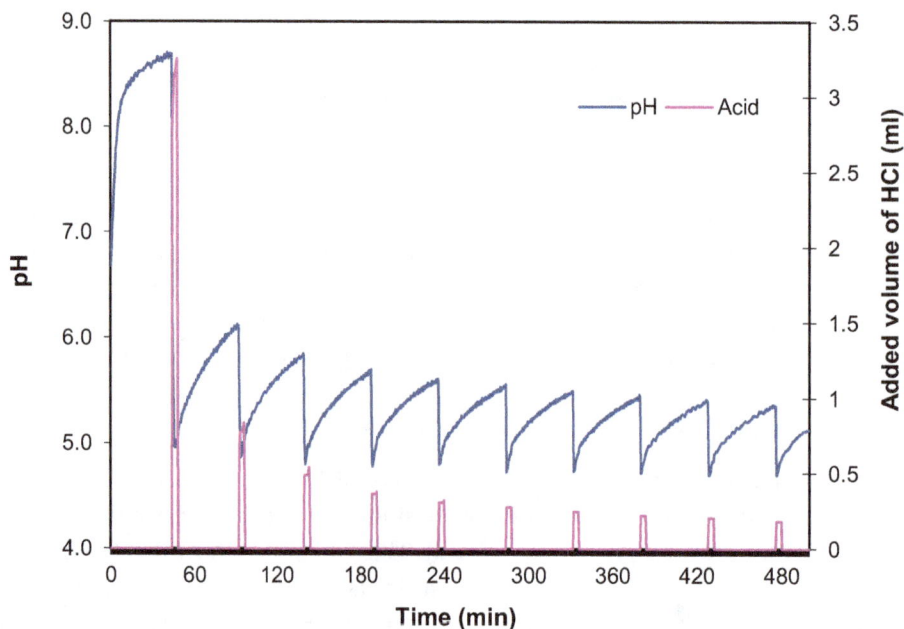

Figure 3. Acid neutralizing capacity of biochar from wood chips as determined by titration. The blue line is the pH; the pink spikes are the acid dosage.

Table 3. Acid buffering capacity of biochar and base buffering capacity of peat, determined by titration [1], expressed per unit weight of dry matter (d.m.) and per unit of volume.

Medium	Acid Buffering Capacity		Base Buffering Capacity	
	mmol·kg^{-1} d.m.	mol·m^{-3}	mmol·kg^{-1} d.m.	mol·m^{-3}
Biochar (wood chips)	258	26	-	-
Peat [2]	-	-	90	9

[1] Titration with a titrino using concentrated HCl or KOH; [2] Acid milled white peat, no fertilizer or calcium-carbonate added.

The acid neutralizing capacity of biochar (258 mmol·kg^{-1} dry matter) is 2.9 times higher compared to the base neutralizing capacity of peat (90 mmol·kg^{-1} dry matter) (Table 3). For horticultural purposes mixing ratios are based on a volume basis using the bulk density of both materials, biochar (102 kg·m^{-3}) and peat (91 kg·m^{-3}), which changes the ratio very slightly to 3.2. This means that mixing 24 kg of biochar and 76 kg of acid peat will give a mixture with a neutral pH (Table 4).

Table 4. Bulk density and volumetric contribution of solid particles and pores in peat, biochar and peat-biochar mixture [1].

Substrate	Bulk Density kg·m^{-3}	Solid Fraction % v/v	Total Pore Space % v/v
Biochar (wood chips)	100 ± 2.6	6.2 ± 0.17	93.8 ± 0.17
Peat [2]	91 ± 0.87	5.8 ± 0.06	94.2 ± 0.06
Peat/biochar mixture (85%:15%)	91 ± 1.7	5.8 ± 0.11	94.2 ± 0.11

[1] Data collected for determination of pF curves; [2] Acid milled white peat, no fertilizer or calcium-carbonate added.

It is usual in this stage to compensate for interstitial filling of the mix, a density increase of the mix caused by small particles filling up larger pores. Interstitial filling up to 20% v/v is common in rooting media and increases production costs. In this case, the dry bulk density and total pore space of the separate peat and biochar and the 85%/15% v/v mixed material remained similar (Table 4). It was therefore concluded that interstitial filling was negligible. This was somewhat unexpected as the high water content of a peat/biochar mix 85%/15% v/v in Figure 2 seemed to indicate some degree of interstitial filling.

3.1.5. Phytotoxicity

The results of the phytotoxicity test (Figure 4) provide insight into the presence of toxic compounds that affect germination and early growth of seedlings. The tested biochar materials were produced from three different feedstocks; sweet pepper waste, wood chips and a mix of tomato leaves and wood chips. The biochar produced from sweet pepper waste performed rather well in the phytotoxicity test (Figure 4) as there was no significant difference in germination and early growth of seedlings as expressed by the observed root and shoot length compared with the control treatment. In contrast, the biochars produced from wood and tomato leaves showed significant (above 20%) adverse effects on root and shoot development indicating that this biochar contains water-soluble phytotoxic compounds. The biochar produced from 100% wood chips is not as phytotoxic as the wood/tomato leaf biochar, but still a significant reduction in root length of Sorghum was found, indicating that even the biochar from 100% wood chips is not completely free of toxic compounds. The presence of phytotoxic compounds is a point of concern. Prior research showed that toxicity is caused by condensation of volatile tar components in the cooler parts of the product [41]. Early separation of solid product and fumes is supposed to prevent this type of problem so further production optimization is called for.

root length

(a)

shoot length

(b)

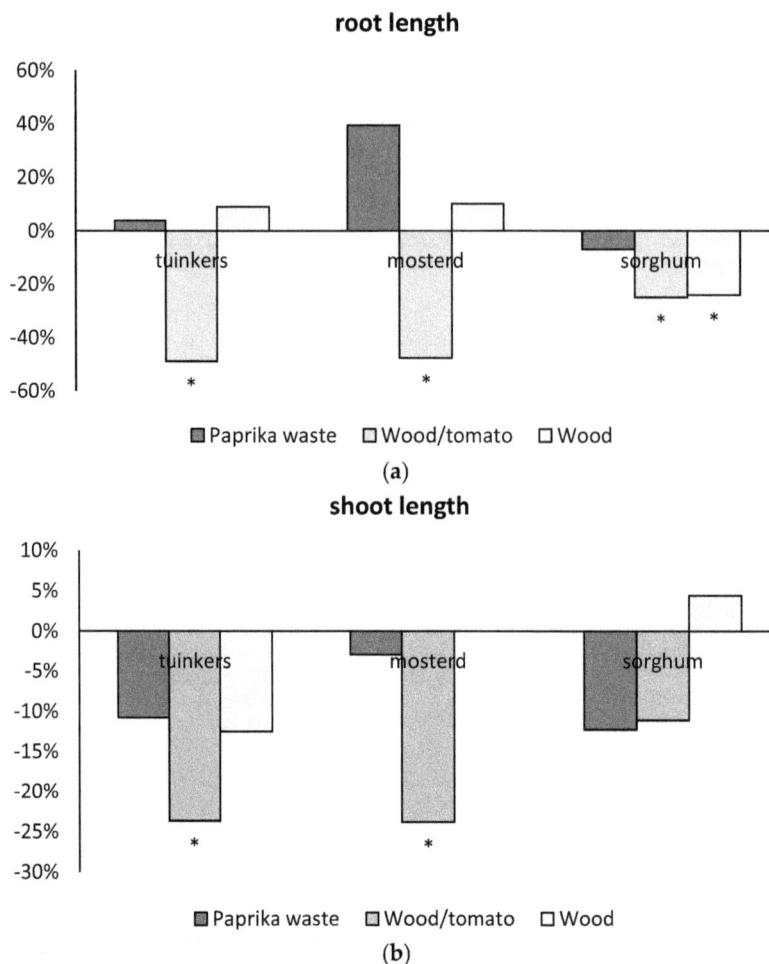

Figure 4. Average difference in root (**a**) and shoot length (**b**) between treatments and the control in the phytotoxicity test for three biochars produced from sweet pepper waste, wood and tomato leaves, and wood chips only. The asterisks denote a statistically significant reduction in root or shoot length at $p = 0.05$.

3.1.6. Stability

The oxygen uptake rate (OUR) is a measure for bacterial degradation under ideal conditions (i.e., ideal pH, EC and sufficient amounts of nutrients present). The result is reported as mmol oxygen used by bacteria per hour per kilogram of dry organic matter (DOM). The OUR of biochar is very low (2.5 mmol·h^{-1}·kg^{-1} DOM) and almost as low as the OUR of peat (Table 5). This is an indication the biochar is very stable. A low OUR is beneficial for application in horticulture because it means that degradation of the material, and consequently shrinkage of the medium, will not occur. A consequence of the low OUR is that biochar is a poor source of carbon for micro-organisms. Thus, if one aims to stimulate microbial activity in potting soil, an additional carbon source like fresh wood fibre may be needed.

Table 5. Oxygen uptake rate (OUR) of biochar and peat in mmol·h^{-1}·kg^{-1} dry organic matter.

Material	OUR Value
Biochar Woodchips	2.5
Peat	1.9
Compost (for reference)	5–10

3.2. Effect of Biochar on Plant Growth and Powdery Mildew Infection of Gerbera

A pot experiment was performed with *Gerbera* plants cultivated on rooting media with and without beech wood chip biochar in a climate controlled room (20 °C, RH 80%, light:dark period, 16 h:8 h). Plants were infected with powdery mildew to assess the effect of biochar addition on the ability to induce plant resistance and thereby suppress powdery mildew infection.

3.2.1. Effect of Biochar Addition on Rooting Medium Quality

The pH in the rooting medium with 20% v/v biochar was higher than the pH in the 100% v/v peat medium (Table 6), which shows that the amount of acidic peat used to neutralize biochar was not yet sufficient. This is consistent with the result of the neutralization test, which showed the test's duration is not long enough to cover the reaction time of biochar.

Table 6. Nutrient concentrations at the end of the cultivation period, including pH and EC of peat and peat with 20% v/v biochar [1].

Parameter	Unit	Medium	Medium
		Peat	Peat + Biochar
pH	-	5.3	6.4
EC	$dS \cdot m^{-1}$	0.65	0.54
NH_4	$mmol \cdot L^{-1}$	<0.1	<0.1
K	$mmol \cdot L^{-1}$	2.3	2.4
Na	$mmol \cdot L^{-1}$	0.3	0.4
Ca	$mmol \cdot L^{-1}$	0.7	0.4
Mg	$mmol \cdot L^{-1}$	0.5	0.3
Si	$mmol \cdot L^{-1}$	<0.1	<0.1
NO_3	$mmol \cdot L^{-1}$	2.7	0.6
Cl	$mmol \cdot L^{-1}$	0.1	0.1
SO_4	$mmol \cdot L^{-1}$	0.8	1.1
HCO_3	$mmol \cdot L^{-1}$	<0.1	<0.1
P	$mmol \cdot L^{-1}$	0.65	0.65
Fe	$\mu mol \cdot L^{-1}$	4.7	2.8
Mn	$\mu mol \cdot L^{-1}$	0.4	0.2
Zn	$\mu mol \cdot L^{-1}$	0.8	0.8
B	$\mu mol \cdot L^{-1}$	6	8
Cu	$\mu mol \cdot L^{-1}$	0.2	0.4
Mo	$\mu mol \cdot L^{-1}$	0.1	0.2

[1] Measured in a 1:1.5 v/v soil/water extract. Biochar was produced from beech wood.

At the end of the pot experiment, the nitrate concentration was much lower in the rooting medium with biochar compared to the standard peat medium. This indicates nitrogen immobilization by microbial life involved in the degradation of biochar. Degradation of biochar was expected because of the high C:N ratio of biochar (>65, Table 2) which is much higher than the C:N ratio needed for bacterial growth. The amount of nitrogen immobilization is acceptable for practical purposes as it can easily be compensated by addition of extra nitrogen in the form of calcium-nitrate. Even though plant yield and leaf chlorophyll content were not visibly affected, the amount of nitrogen immobilization must be known to potting soil producers in order to allow for exact compensation.

3.2.2. Effect of Biochar Addition on Plant Growth and Disease Suppression

The fresh weight of *Gerbera* plants was similar for plants grown on 100% v/v peat than on the peat-biochar 80%–20% v/v after six weeks of cultivation (Figure 5). Thus, biochar addition did not lead to any phytotoxic or other growth-reducing effects. Fresh weight of the plants receiving SAR elicitor, a product to chemically induce disease resistance, was lower compared to other treatments.

This product is known to cause a reduction in yield in *Gerbera* plants but is used as a positive control on induction of plant hormones.

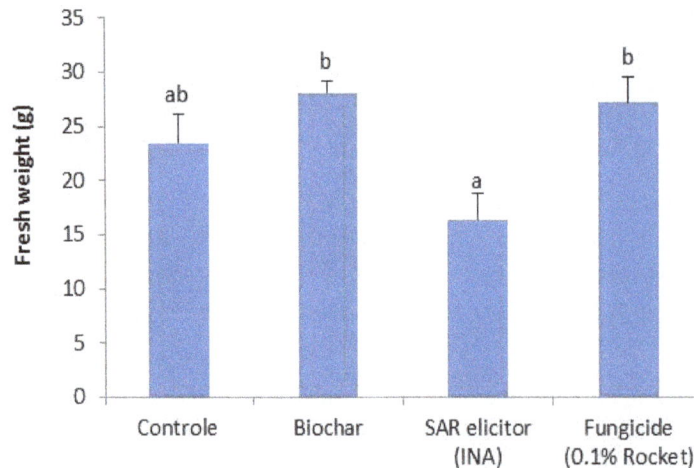

Figure 5. Average fresh weight of *Gerbera* after six weeks. No significant effects of biochar addition were found compared to the untreated plants (Tukey's test, $p < 0.05$). Treatments which do not share the same letters a or b differ significantly (Tukey's test, $p < 0.05$).

Nineteen days after inoculation with mildew, plants were assessed for symptoms of powdery mildew infection (Figure 6). *Gerberas* cultivated on peat and peat/biochar showed symptoms of powdery mildew infection and the area of leaf spots was similar for both treatments. Thus, biochar addition does not affect the level of severity of powdery mildew infection. Generally, the severity of infection with powdery mildew was rather low since less than 5% of the plant surface was affected. Other tests with gerbera plants showed that infection is more efficient at lower relative humidity (<85%). As mildew symptoms were not reduced, the plant material was not further analyzed on increased levels of plant hormones related with systemic acquired resistance. Therefore a systemic effect on disease resistance against biotrophic pathogens, e.g., powdery mildew, was neither proven nor disproven for the wood biochar tested.

Overall, it can be concluded that cultivation of *Gerbera* on media containing 20% v/v of biochar was successful and gave similar yields as compared to the standard peat medium.

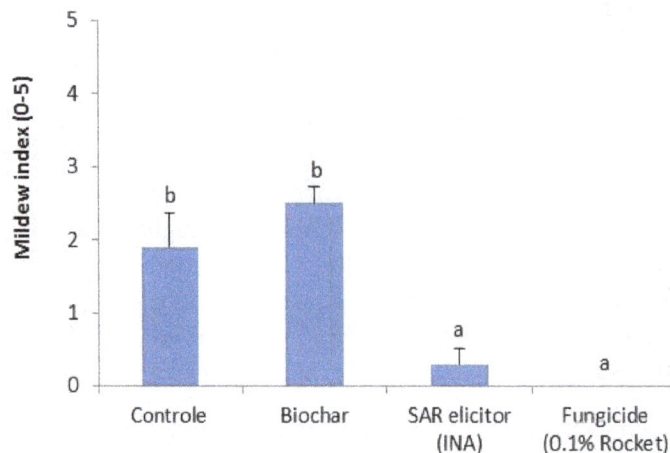

Figure 6. Symptoms of powdery mildew infection (mildew index) on *Gerbera* plants, 19 days after inoculation. Symptoms were quantified on a scale between 0 and 5 (0 = 0, 1 = 0.1%–2%, 2 = 2%–5%, 3 = 5%–20%, 4 = 20%–40%, 5 = >40%). Treatments which do not share the same letters a or b differ significantly (Tukey's test, $p < 0.05$).

3.3. Effect of Biochar and Added Microorganisms on Plant Growth of Chrysanthemum

A pot experiment was performed with *Chrysanthemum* plants cultivated on rooting media with and without biochar in a greenhouse setting. Plants were infected with *Fusarium* and the rooting media were pre-treated with a mix of beneficial microorganisms to assess the effect of biochar addition on the ability of beneficial microorganisms to suppress *Fusarium* infection.

3.3.1. Effect of Biochar Addition on Rooting Medium Quality

EC, pH and elemental composition including nitrate of the mix after cultivation resembled that of the reference, peat only, with the exception of potassium, which is 25% higher in the biochar mixtures because of the high level in the biochar itself (Table 2). This means the addition of 20% v/v of biochar did not require adjustments in the fertilization scheme other than the prior neutralization of alkalinity.

At the start of the experiment, the dissolved organic carbon (DOC) concentration was measured in peat (1392 mg·kg^{-1}) and in the peat/biochar 80%/20% v/v mix (723 mg·kg^{-1}). The level of DOC in peat is caused by naturally occurring organic acids. Addition of the biochar leads to a marked decrease in the DOC concentration, indicating adsorption of organic acids on biochar. This mechanism might be used to load biochar with digestible compounds to support microorganisms.

3.3.2. Effect of Biochar Addition on Plant Growth and Added Microorganisms

No effect of treatment on the plant fresh- and dry weight was observed and no differences in nutrient content of the plants were observed. This is consistent with the observation that rooting medium solution pH, EC and nutrient availability are similar in treatments with and without biochar. Thus, for the tested biochar, biochar addition had no consequences for nutrient uptake or plant growth. No effect of *Fusarium* incidence was found meaning that the effect of biochar addition on disease suppression could not be tested.

Numbers of bacteria, based on the copy numbers of 16S gene per gram of rooting medium, did not differ significantly between treatments (Figure 7).

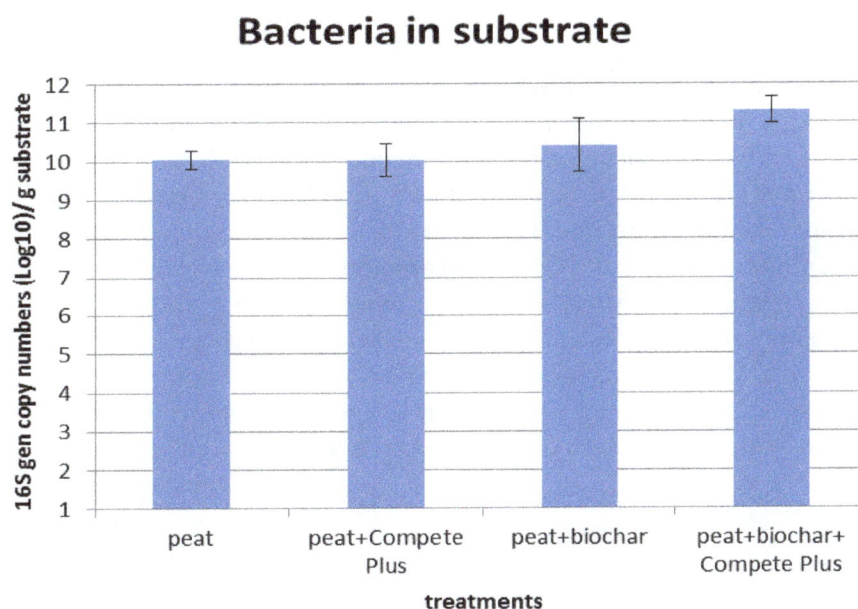

Figure 7. Total Bacteria numbers (on basis of log10 of copy numbers of 16S gene per g dry medium).

Hypothetically, addition of Compete Plus should result in higher number of *Firmicutes* in the studied rooting media with Compete Plus, as *Firmicutes*, i.e., *Bacillus* species, are abundant in Compete Plus. However, the numbers of *Firmicutes* found are not particularly high (Figure 8). We added in total

160 mg of Compete Plus per 700 mL container. With a specified level of about 3×10^8 *Bacilli* per gram Compete Plus, this is about 7×10^5 cfu with a medium of 100 kg·m^{-3}. The actual numbers were over 100 times lower for all treatments (Figure 8). It is difficult to explain why the numbers of this phylum are so low and presumably have decreased after addition.

Firmicutes in substrate

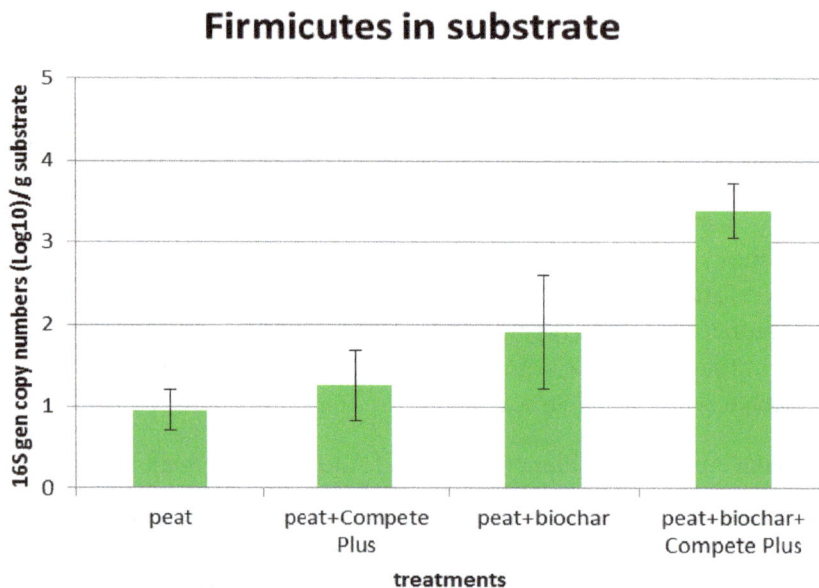

Figure 8. *Firmicutes* numbers (on basis of log10 of copy numbers of 16S gene per g dry medium).

Even so, in the peat+biochar+Compete Plus treatment, the numbers of *Firmicutes* showed a statistically significant higher level compared with the other treatments. This could, arguably, indicate that biochar slows the decline of *Firmicutes* numbers, which remains open for further research.

3.4. Suitability of Feedstocks for Application as Rooting Medium

The biochars based on vegetable waste had to be rejected for further experimentation based on both, a too high level of nutrient elements and a too high alkalinity. Both nutrient level (EC) and alkalinity (pH and buffering capacity) restrict the amounts of vegetable biochar which can be mixed with unlimed acid peat to levels below 10% v/v, which are too low for commercial applications. EC and pH restrictions are valid for rooting media mixes used for protected horticulture but not so much for open field horticulture or agriculture. Open field applications usually use a much smaller volume ratio of biochar to mix with soil, which reduces the effects of salinity and alkalinity. Furthermore, soil can buffer salinity and pH effects much more efficiently than common rooting media. Finally, the nutrient input is a direct substitute for nutrients which open field farmers otherwise have to buy and which represent a large operational cost factor. For protected horticulture growers, nutritional costs are a relatively small factor which plays no role in choosing a rooting medium.

4. Discussion

4.1. Feedstock Related Effects of Biochar

Biochars produced from beech wood or mixed residual wood have a higher specific surface area compared to biochars produced from tomato or sweet pepper waste. This is expected to increase its water retention and to increase the cation exchange capacity (CEC). Increased water content is a positive property if mixtures of >35% v/v biochar are going to be used as these might become too dry on their own. A higher CEC is not valued in professional horticulture as interference of the CEC with the nutrient solutions used, changes the ratios of the elements in the nutrient solutions to undesirable values. The values for CEC measured here are low enough to allow them to be disregarded in mixtures with <50% v/v of biochar. CEC is thought to be proportional to surface area of the biochar but also the result of negatively charged carboxyl (COO^-) groups. The amount of carboxyl groups depends initially on the feedstock, in which composted leaves are known to have more of these groups per unit weight than woody products.

Biochar produced from source materials containing nutrient-rich biomass (i.e., tomato leaves, sweet pepper waste) appears unsuitable for use as potting soil because of its high salt content and high pH. Even biochar produced from feedstock with only 20% v/v tomato leaves and 80% v/v wood chips is too saline to be used as an ingredient for potting soil. We interpret this as evidence that the cations in the feedstock oxidize into oxides (Na_2O, K_2O, CaO and MgO) during biochar production. As these oxides are readily soluble in water and highly alkaline this explains why feedstock influences EC and pH. It also explains how feedstock can be quantitatively tested prior to use on salinity and alkalinity, notably by measuring the content of alkali metals Na, K, Ca and Mg. Even so there are reports where high EC feedstocks are used for biochar for horticultural purposes [23] in which case the EC was lowered by either prior washing with water or dilution with low EC biochar.

4.2. Production Setting Related Effects of Biochar

The temperature of the production process increases salinity in accordance with the mass loss i.e., the salt ions concentrate in the remaining solids. Conversely, to produce higher amounts of biochar from a mass unit of dry feedstock, the process temperature and oxygen levels should be reduced. Higher process temperatures (>500 °C) and higher process oxygen levels, including the oxygen in the process water vapor, will increase the degree of oxidation of feedstock cations which will increase the pH and pH buffer capacity of the material. The influence of temperature and oxidation level on alkalinity was observed before [34], as was a smaller but still relevant relation of ash content with pH [41]. Generally if nutrient poor feedstock is used the biochar qualities are closer to those of peat.

The amount of carboxyl groups in the end product is influenced by the process parameters' temperature and oxygen level. Higher temperatures and lower oxygen levels reduce the amount of carboxylic groups left in the biochar and therefore presumably reduce the cation exchange capacity.

The surface area of the feedstock is reduced by the level of clogging of smaller pores. Such surface area reduction supposedly by clogging is reported for condensation of volatile tar [34] and may also have been caused by slagging [41]. Slagging arguably is worse in materials with higher amounts of potassium and calcium salts which act as viscosity reductants for the slag.

Condensation of plant toxic volatile tar products is possible over the total temperature range of 200–900 °C [36,55,56], as long as the volatiles are not immediately transported away from the biochar mass, especially during the cooling of the mass.

The particle size of the biochar is a production variable which affects physical and chemical properties of the materials as well as the properties of rooting media which are mixed with a biochar. (1) Physical properties affected are density, specific surface area (SSA) and water retention, which all increase with smaller particle size; and air content, which decreases with smaller particle size [28,56,57]; (2) Chemical properties affected are pH, cation exchange capacity (CEC) and surface adsorption, which all increase with smaller particle size [4,34,58]; (3) Rooting media properties require that the biochar particle size is tuned to an intended application such as coarse particles >4 mm to replace peat fractions or bark; 3–4 mm particles to replace milled peat; or powdered <1 mm particles to replace lime [58]. This demonstrates that even a biochar from one specific feedstock, produced at one set of specific production parameters still can produce disappointing results if the particle size does not match the application. On the other hand, particle size manipulation before the heating process is one of the easier ways to diversify biochar product applications.

4.3. The Survival Rate of Beneficial Microorganisms as Affected by Biochar

In two pot experiments, plants were grown on mixtures of peat and biochar (beech wood respectively residual wood). In both experiments, nutrient availability and plant growth were similar in the biochar treatment compared to the control treatment without biochar with the exemption of a slight decrease in nitrate availability. There were no signs of phytotoxic or growth reducing effects of biochar. Thus, the pot experiments confirm that peat/biochar media, with substantial biochar content (20% v/v), can be used in horticulture without yield reduction and without adjustment of the fertilization scheme, other than a preventive increase in nitrate to counteract nitrogen immobilization.

Results of the pot experiment with *Gerbera* show that the severity of powdery mildew infection is similar for *Gerberas* cultivated on rooting medium with 20% v/v biochar and the control treatment without biochar. Thus, biochar addition does neither decrease nor increase severity of powdery mildew infection in *Gerberas* and a systemic mode of action on disease resistance is not proven for the biochar used. In the experiments with chrysanthemums, no *Fusarium* incidence was found meaning that the effect of biochar addition on disease suppression could not be tested. More research is needed in order to elucidate if, and if so under which conditions, biochar additions can enhance disease suppression of diseases.

The addition of biochar did not increase the overall numbers of micro-organisms in the potting soil. A reason the level of micro-organisms did not respond to biochar addition could be the low level of degradation (OUR < 3), showing biochar particles to be a poor environment for micro-organisms needing carbon for growth. Possibly, colonization of biochar can be enhanced by loading the biochar with biodegradable organic acids, which would make biochar a more attractive environment for microbes. This is made more plausible by the observation that 20% v/v biochar can halve the amount of free organic carbon in a peat sample, presumably by absorbing the organic acids. Thus, it may be possible to pre-treat biochar with small organic molecules which can then feed a population of micro-organisms establishing themselves on the biochar particles. This will be the topic for further work.

Our results showed that biochar offered some temporary or more permanent protection to *Firmicutes*, the bacteria which include the plant health supporting *Bacilli*. Whether the level of *Firmicutes* is high enough to influence disease suppression remains to be seen and merits further research.

4.4. Horticultural Perspectives of Biochar

Biochar dry bulk density and total pore space (100 kg·m^{-3} and 93% v/v) are well within the range required for horticultural rooting media.

Peat/biochar mixtures, containing 20% v/v biochar, still have properties (EC, pH, water holding capacity, nutrient availability) similar to standard peat and therefore can be qualified as a proper peat alternative. The water retention of biochars (45% v/v at −10 cm) is about 20% v/v lower than for peat but by no means too dry for optimum growth. The water holding capacity can be improved by using relatively wet rooting materials like peat or coir pith but the water holding capacity of biochar can also be improved by mixing with a finer grade of biochar. The wettability of biochar can be increased by increasing the specific surface area by choosing a feedstock low in nutrients and/or increasing process temperature [34]. It may be of practical interest to note that biochars with higher specific surface areas will reduce the effectivity of some crop protection agents by adsorption [30].

Examples in literature show it is possible to grow plants just as successfully as on standard peat mixtures with 50%–100% v/v rice husk and reed based biochar [24,27]. It is, therefore, expected that the wood based biochars used in this work may be used in ratios up to 50% v/v if the pH is properly neutralized.

Because the pH of most biochars is too high, those biochars have to be acidified to obtain an optimal potting mixture. Most rooting media in their natural i.e., unlimed and unfertilized state are neutral to slightly alkaline (pH 6–7.5). The only exception is peat which can be harvested at pHs as low as pH 3.0–4.0. This leads to this paradox: in order to use biochar as a peat alternative one needs unlimed and acid peat. The mixing ratio differs depending on the biochar as well as on the peat used but expected maximum biochar levels range from 20%–50% v/v. The pH buffer method described is effective but the method needs to be adapted by an increase of the 16 hours measuring period to allow for the slow reaction of biochar with added acids.

The very low oxygen uptake rate (OUR) of biochar is a positive property for horticultural media since it indicates that biochar is poorly degradable, reducing the need for additional nitrate and reducing the risks of shrinkage of the rooting medium due to microbial degradation of biochar.

However, the presence of some phytotoxic compounds in biochar from wood chips in these experiments, despite that the problem was reported as technically solvable [41], remains a point of concern that needs further attention in process design.

5. Conclusions

The first hypothesis was: (1) biochar is a potting soil constituent which can be used in growing medium mixes in quantities of 20% v/v without negative growth effects. This hypothesis is proven for mixtures up to 20% v/v of wood based biochar and may hold true for mixtures up to 50% v/v, provided pH effects can be compensated. Biochar based on vegetable waste had a high EC and a high alkalinity. These characteristics limit the amount of biochar that can be mixed with unlimited peat, to levels below 10%, which are too low for commercial applications; The second hypothesis was: (2) biochar can induce disease suppression. This hypothesis was not proven. In a climate chamber experiment we found no difference in the powdery mildew infection between peat and peat/biochar mixtures. A greenhouse experiment with *Chrysanthemum* and addition of the pathogen *Fusarium* showed no effects of *Fusarium* on both the peat and the peat/biochar mixtures. It was however found that some added and potentially beneficial microorganisms remained more numerous in a mix of peat with 20% v/v biochar than in 100% v/v of peat.

Acknowledgments: We thank the ministry of economic affairs which funded this research with a TO_2 grant, including the costs to publish in open access.

Author Contributions: Rian Visser and Lyda Fryda produced and partly analyzed the biochars. Chris Blok and Jantineke Hofland-Zijlstra conceived and designed the experiments; Inge Regelink and Barbara Eveleens performed the experiments. Inge Regelink and Marta Streminska analyzed the data; Chris Blok and Caroline van der Salm wrote the main body of the paper.

Conflicts of Interest: The authors declare no conflict of interest.

References

1. Lehmann, J.; Joseph, S. *Biochar for Environmental Management: Science and Technology*, 2nd ed.; Earthscan: London, UK, 2015.

2. Laird, D.A. The charcoal vision: A win-win-win scenario for simultaneously producing bioenergy, permanently sequestering carbon, while improving soil and water quality. *Agron. J.* **2008**, *100*, 178–181. [CrossRef]

3. Atkinson, C.; Fitzgerald, J.; Hipps, N. Potential mechanisms for achieving agricultural benefits from biochar application to temperate soils: A review. *Plant Soil* **2010**, *337*, 1–18. [CrossRef]

4. Singh, B.; Singh, B.P.; Cowie, A.L. Characterisation and evaluation of biochars for their application as a soil amendment. *Soil Res.* **2010**, *48*, 516–525. [CrossRef]

5. Glaser, B.; Wiedner, K.; Seelig, S.; Schmidt, H.-P.; Gerber, H. Biochar organic fertilizers from natural resources as substitute for mineral fertilizers. *Agron. Sustain. Dev.* **2014**, *35*, 667–678. [CrossRef]

6. Olmo, M.; Villar, R.; Salazar, P.; Alburquerque, J. Changes in soil nutrient availability explain biochar's impact on wheat root development. *Plant Soil* **2016**, *399*, 333–343. [CrossRef]

7. Akhter, A.; Hage-Ahmed, K.; Soja, G.; Steinkellner, S. Compost and biochar alter mycorrhization, tomato root exudation, and development of *Fusarium oxysporum* f. Sp. *lycopersici*. *Front. Plant Sci.* **2015**, *6*. [CrossRef] [PubMed]

8. Nielsen, S.; Minchin, T.; Kimber, S.; van Zwieten, L.; Gilbert, J.; Munroe, P.; Joseph, S.; Thomas, T. Comparative analysis of the microbial communities in agricultural soil amended with enhanced biochars or traditional fertilisers. *Agric. Ecosyst. Environ.* **2014**, *191*, 73–82. [CrossRef]

9. Elad, Y.; David, D.R.; Harel, Y.M.; Borenshtein, M.; Kalifa, H.B.; Silber, A.; Graber, E.R. Induction of systemic resistance in plants by biochar, a soil-applied carbon sequestering agent. *Phytopathology* **2010**, *100*, 913–921. [CrossRef] [PubMed]

10. Nemati, M.R.; Simard, F.; Fortin, J.-P.; Beaudoin, J. Potential use of biochar in growing media. *Vadose Zone J.* **2015**, *14*. [CrossRef]

11. Nieto, A.; Gascó, G.; Paz-Ferreiro, J.; Fernández, J.M.; Plaza, C.; Méndez, A. The effect of pruning waste and biochar addition on brown peat based growing media properties. *Sci. Hortic.* **2016**, *199*, 142–148. [CrossRef]

12. Steiner, C.; Harttung, T. Biochar as growing media additive and peat substitute. *Solid Earth Discuss.* **2014**, *6*, 1023–1035. [CrossRef]

13. Verhagen, J.; van den Akker, J.; Blok, C.; Diemont, H.; Joosten, H.; Schouten, M.; Schrijver, R.; Verweij, P.; Wösten, H. *Climate Change. Scientific Assessment and Policy Analysis: Peatlands and Carbon Flows: Outlook and Importance for the Netherlands*; Wab 500102 027; Netherlands Environmental Assessment Agency PBL: Bilthoven, The Netherlands, 2009.

14. Graber, E.R.; Harel, Y.M.; Kolton, M.; Cytryn, E.; Silber, A.; David, D.R.; Tsechansky, L.; Borenshtein, M.; Elad, Y. Biochar impact on development and productivity of pepper and tomato grown in fertigated soilless media. *Plant Soil* **2010**, *337*, 481–496. [CrossRef]

15. Jaiswal, A.; Frenkel, O.; Elad, Y.; Lew, B.; Graber, E. Non-monotonic influence of biochar dose on bean seedling growth and susceptibility to rhizoctonia solani: The "shifted rmax-effect". *Plant Soil* **2015**, *395*, 125–140. [CrossRef]

16. Elad, Y.; Cytryn, E.; Meller Harel, Y.; Lew, B.; Graber, E.R. The biochar effect: Plant resistance to biotic stresses. *Phytopathol. Mediterr.* **2012**, *50*, 335–349.

17. De Tender, C.A.; Debode, J.; Vandecasteele, B.; D'Hose, T.; Cremelie, P.; Haegeman, A.; Ruttink, T.; Dawyndt, P.; Maes, M. Biological, physicochemical and plant health responses in lettuce and strawberry in soil or peat amended with biochar. *Appl. Soil Ecol.* **2016**, *107*, 1–12. [CrossRef]

18. Altland, J.E.; Locke, J.C. Biochar affects macronutrient leaching from a soilless substrate. *HortScience* **2012**, *47*, 1136–1140.

19. Kloss, S.; Zehetner, F.; Wimmer, B.; Buecker, J.; Rempt, F.; Soja, G. Biochar application to temperate soils: Effects on soil fertility and crop growth under greenhouse conditions. *J. Plant Nutr. Soil Sci.* **2014**, *177*, 3–15. [CrossRef]

20. Dumroese, R.K.; Heiskanen, J.; Englund, K.; Tervahauta, A. Pelleted biochar: Chemical and physical properties show potential use as a substrate in container nurseries. *Biomass Bioenergy* **2011**, *35*, 2018–2027. [CrossRef]

21. Bedussi, F.; Zaccheo, P.; Crippa, L. Pattern of pore water nutrients in planted and non-planted soilless substrates as affected by the addition of biochars from wood gasification. *Biol. Fertil. Soils* **2015**, *51*, 625–635. [CrossRef]

22. Dunlop, S.J.; Arbestain, M.C.; Bishop, P.A.; Wargent, J.J. Closing the loop: Use of biochar produced from tomato crop green waste as a substrate for soilless, hydroponic tomato production. *Hortscience* **2015**, *50*, 1572–1581.

23. Vaughn, S.F.; Eller, F.J.; Evangelista, R.L.; Moser, B.R.; Lee, E.; Wagner, R.E.; Peterson, S.C. Evaluation of biochar-anaerobic potato digestate mixtures as renewable components of horticultural potting media. *Ind. Crops Prod.* **2015**, *65*, 467–471. [CrossRef]

24. Altland, J.E.; Locke, J.C. Gasified rice hull biochar is a source of phosphorus and potassium for container-grown plants. *J. Environ. Hortic.* **2013**, *31*, 138–144.

25. Locke, J.C.; Altland, J.E.; Ford, C.W. Gasified rice hull biochar affects nutrition and growth of horticultural crops in container substrates. *J. Environ. Hortic.* **2013**, *31*, 195–202.

26. Trifonova, R.; Postma, J.; Ketelaars, J.J.M.H.; van Elsas, J.D. Torrefied grass fibres as a substitute for peat in potting soil. Presented at ORBIT Conference, Wageningen, The Netherlands, 13–15 October 2008.

27. Blok, C.; Rijpsma, E.; Ketelaars, J.J.M.H. New growing media and value added organic waste processing. *ISHS Acta Hortic.* **2016**, *1112*, 269–280. [CrossRef]

28. Quintero, M.F.; Ortega, D.; Valenzuela, J.L.; Guzman, M. Variation of hydro-physical properties of burnt rice husk used for carnation crops: Improvement of fertigation criteria. *Sci. Hortic.* **2013**, *154*, 82–87. [CrossRef]

29. Singh, B.; Macdonald, L.M.; Kookana, R.S.; van Zwieten, L.; Butler, G.; Joseph, S.; Weatherley, A.; Kaudal, B.B.; Regan, A.; Cattle, J.; et al. Opportunities and constraints for biochar technology in australian agriculture: Looking beyond carbon sequestration. *Soil Res.* **2014**, *52*, 739–750. [CrossRef]

30. Graber, E.; Tsechansky, L.; Gerstl, Z.; Lew, B. High surface area biochar negatively impacts herbicide efficacy. *Plant Soil* **2012**, *353*, 95–106. [CrossRef]

31. Abiven, S.; Schmidt, M.W.I.; Lehmann, J. Biochar by design. *Nat. Geosci.* **2014**, *7*, 326–327. [CrossRef]

32. Schimmelpfennig, S.; Glaser, B. One step forward toward characterization: Some important material properties to distinguish biochars. *J. Environ. Qual.* **2012**, *41*, 1001–1013. [CrossRef] [PubMed]

33. International Biochar Initiative. *Standardized Product Definition and Product Testing Guidelines for Biochar That Is Used in Soil*; International Biochar Initiative: Westerville, OH, USA, 2012.

34. Mukherjee, A.; Zimmerman, A.R.; Harris, W. Surface chemistry variations among a series of laboratory-produced biochars. *Geoderma* **2011**, *163*, 247–255. [CrossRef]

35. Blok, C.; Kreij, C.D.; Baas, R.; Wever, G. Analytical Methods Used in Soilless Cultivation. In *Soilless Culture : Theory and Practice*; Elsevier: Amsterdam, The Netherlands, 2008.

36. Kloss, S.; Zehetner, F.; Dellantonio, A.; Hamid, R.; Ottner, F.; Liedtke, V.; Schwanninger, M.; Gerzabek, M.H.; Soja, G. Characterization of slow pyrolysis biochars: Effects of feedstocks and pyrolysis temperature on biochar properties. *J. Environ. Qual.* **2012**, *41*, 990–1000. [CrossRef] [PubMed]

37. Morales, V.L.; Pérez-Reche, F.J.; Hapca, S.M.; Hanley, K.L.; Lehmann, J.; Zhang, W. Reverse engineering of biochar. *Bioresour. Technol.* **2015**, *183*, 163–174. [CrossRef] [PubMed]

38. Mimmo, T.; Panzacchi, P.; Baratieri, M.; Davies, C.A.; Tonon, G. Effect of pyrolysis temperature on miscanthus (*Miscanthus × giganteus*) biochar physical, chemical and functional properties. *Biomass Bioenergy* **2014**, *62*, 149–157. [CrossRef]

39. Keiluweit, M.; Nico, P.S.; Johnson, M.G.; Kleber, M. Dynamic molecular structure of plant biomass-derived black carbon (biochar). *Environ. Sci. Technol.* **2010**, *44*, 1247–1253. [CrossRef] [PubMed]

40. Gray, M.; Johnson, M.; Dragila, M.I.; Kleber, M. Water uptake in biochars: The roles of porosity and hydrophobicity. *Biomass Bioenergy* **2014**, *61*, 196–205. [CrossRef]

41. Fryda, L.; Visser, R. Biochar for soil improvement: Evaluation of biochar from gasification and slow pyrolysis. *Agriculture* **2015**, *5*, 1076–1115. [CrossRef]

42. Urrestarazu Gavilán, M. *Tratado de Cultivo Sin Suelo*; Mundi-Prensa: Madrid, Spain, 2004.

43. CEN, European Committee for Standardization. *Soil Improvers and Growing Media—Determination of Physical Properties—Dry Bulk Density, Air Volume, Water Volume, Shrinkage Value and Total Pore Space*; prEN 13041; Technical Committee 223: Berlin, Germany, 2006.

44. CEN, European Committee for Standardization. *Soil Improvers and Growing Media—Determination of Organic Content and Ash*; prEN 13039; Technical Committee 223: Berlin, Germany, 2007.

45. CEN, European Committee for Standardization. *Soil Improvers and Growing Media—Determination of Electrical Conductivity*; prEN 13038; Technical Committee 223: Berlin, Germany, 2008.

46. Verhagen, J.B.G.M. Cec and the saturation of the adsorption complex of coir dust. *Acta Hortic.* **1999**, *481*, 151–155. [CrossRef]

47. ISO/DIS 11260. *Soil Quality—Determination of Effective Cation Exchange Capacity and Base Saturation Level Using Barium Chloride Solution*; Technical Committee 190: Berlin, Germany, 2011.

48. Blok, C.; Kaarsemaker, R. Ph in rockwool propagation blocks: A method to measure the ph buffer capacity of rockwool and other mineral wool media. *Acta Hortic.* **2013**, *1013*, 65–72. [CrossRef]

49. Blok, C.; Wever, G.; Persoone, G. A practical and low cost microbiotest to assess the phytotoxic potential of growing media and soil. *Acta Hortic.* **2008**, *779*, 367–374. [CrossRef]

50. CEN, European Committee for Standardization. *Soil Improvers and Growing Media—Determination of Plant Response—Part 2: Petri Dish Test*; EN 16086; Technical Committee 223: Berlin, Germany, 2008.

51. CEN, European Committee for Standardization. *Pren_00223089 2009 Determination of Aerobic Biological actIvity—Part 1: Oxygen Uptake Rate (Our)*; EN 16087; Technical Committee 223: Berlin, Germany, 2009.

52. Spencer, D.M. Standardized methods for the evaluation of fungicides to control cucumber powdery mildew. In *Crop Protection Agents—Their Biological Evaluation*; McFarlane, N.R., Ed.; Academic Press: London, UK, 1977; pp. 455–464.

53.   Verhagen, J.B.G.M. Oxygen diffusion in relation to physical characteristics of growing media. *Acta Hortic.* **2013**, *1013*, 313–318. [CrossRef]

54.   Londra, P.A. Simultaneous determination of water retention curve and unsaturated hydraulic conductivity of substrates using a steady-state laboratory method. *HortScience* **2010**, *45*, 1106–1112.

55.   Trifonova, R.; Postma, J.; Verstappen, F.W.A.; Bouwmeester, H.J.; Ketelaars, J.J.; Van Elsas, J.-D. Removal of phytotoxic compounds from torrefied grass fibres by plant-beneficial microorganisms. *FEMS Microbiol. Ecol.* **2008**, *66*, 158–166. [CrossRef] [PubMed]

56.   Giuffrida, F.; Consoli, S. Reusing perlite substrates in soilless cultivation: Analysis of particle size, hydraulic properties, and solarization effects. *J. Irrig. Drain. Eng.* **2016**, *142*. [CrossRef]

57.   Dede, O.H.; Dede, G.; Ozdemir, S.; Abad, M. Physicochemical characterization of hazelnut husk residues with different decomposition degrees for soilless growing media preparation. *J. Plant Nutr.* **2011**, *34*, 1973–1984. [CrossRef]

58.   Jeffery, S.; Verheijen, F.G.A.; van der Velde, M.; Bastos, A.C. A quantitative review of the effects of biochar application to soils on crop productivity using meta-analysis. *Agric. Ecosyst. Environ.* **2011**, *144*, 175–187. [CrossRef]

14

Development and Testing of Cool-Season Grass Species, Varieties and Hybrids for Biomass Feedstock Production in Western North America

Steven R. Larson [1,*], Calvin H. Pearson [2], Kevin B. Jensen [1], Thomas A. Jones [1], Ivan W. Mott [1], Matthew D. Robbins [1], Jack E. Staub [1] and Blair L. Waldron [1]

[1] USDA-ARS, Forage and Range Research, Utah State University, Logan, UT 84322, USA; Kevin.Jensen@ars.usda.gov (K.B.J.); Thomas.Jones@ars.usda.gov (T.A.J.); Ivan.Mott@ars.usda.gov (I.W.M.); Matthew.Robbins@ars.usda.gov (M.D.R.); Jack.Staub@ars.usda.gov (J.E.S.); Blair.Waldron@ars.usda.gov (B.L.W.)

[2] Agriculture Experiment Station, Department of Soil and Crop Sciences, Colorado State University, Fruita, CO 81521, USA; Calvin.Pearson@colostate.edu

*   Correspondence: Steve.Larson@ars.usda.gov

Academic Editors: John W. Forster and Smith Kevin F. Smith

Abstract: Breeding of native cool-season grasses has the potential to improve forage production and expand the range of bioenergy feedstocks throughout western North America. Basin wildrye (*Leymus cinereus*) and creeping wildrye (*Leymus triticoides*) rank among the tallest and most rhizomatous grasses of this region, respectively. The objectives of this study were to develop interspecific creeping wildrye (CWR) × basin wildrye (BWR) hybrids and evaluate their biomass yield relative to tetraploid 'Trailhead', octoploid 'Magnar' and interploidy-hybrid 'Continental' BWR cultivars in comparison with other perennial grasses across diverse single-harvest dryland range sites and a two-harvest irrigated production system. Two half-sib hybrid populations were produced by harvesting seed from the tetraploid self-incompatible Acc:641.T CWR genet, which was clonally propagated by rhizomes into isolated hybridization blocks with two tetraploid BWR pollen parents: Acc:636 and 'Trailhead'. Full-sib hybrid seed was also produced from a controlled cross of tetraploid 'Rio' CWR and 'Trailhead' BWR plants. In space-planted range plots, the 'Rio' CWR × 'Trailhead' BWR and Acc:641.T CWR × Acc:636 BWR hybrids displayed high-parent heterosis with 75% and 36% yield advantages, respectively, but the Acc:641.T CWR × 'Trailhead' BWR hybrid yielded significantly less than its BWR high-parent in this evaluation. Half-sib CWR × BWR hybrids of Acc:636 and 'Trailhead' both yielded as good as or better than available BWR cultivars, with yields similar to switchgrass (*Panicum virgatum*), in the irrigated sward plots. These results elucidate opportunity to harness genetic variation among native grass species for the development of forage and bioenergy feedstocks in western North America.

Keywords: arid ecosystems; cold-desert; high-elevation; interspecific hybrids; self-incompatibility

1. Introduction

Development of biofuel feedstocks in the United States has been focused on switchgrass (*Panicum virgatum*) as a model crop system in part because of its performance in herbaceous crop screening trials conducted across Alabama, Iowa, Indiana, New York, North Dakota, Ohio and Virginia; and also because decision makers recognized the strategic importance to demonstrate the feasibility of developing a cellulosic biofuel crop with limited funding [1–4]. Switchgrass is a warm-season rhizomatous perennial grass endemic to southeastern North America encompassing parts of Cuba, southeastern Canada, northeastern Mexico, and most the United States east of the

Rocky Mountains [5] including all of the states where it was tested during the initial phase of biofuel feedstock screening in the United States [1–4]. However, with the exception of western North Dakota, located near the geographic center of North America, most of the western United States was excluded from the initial herbaceous crop screening process [1–4].   Studies have shown that cool-season perennial Triticeae grasses including crested wheatgrass (*Agropyron desertorum*), intermediate wheatgrass (*Thinopyrum intermedium*), mammoth wildrye (*Leymus racemosus*), tall wheatgrass (*Thinopyrum ponticum*), and native western wheatgrass (*Pascopyrum smithii*) may be useful for forage and bioenergy feedstock production across the prairie provinces of Canada and the central United States including Alberta, Kansas, Manitoba, North Dakota, Saskatchewan, and South Dakota [1,4,6–10]. These cool-season perennial Triticeae grasses are also well adapted to the high-elevation cold-desert Great Basin region of the western United States including large regions of Nevada and Utah [11–13]. Intermediate wheatgrass and tall wheatgrass produced substantially more biomass than warm-season grasses, including five varieties of switchgrass, in a five-harvest irrigated production system in northern Utah [11]. Basin wildrye (*Leymus cinereus*) performed relatively well in low-irrigation single-harvest management systems in the cold-desert environments of the Idaho, Nevada, and Utah [12,14], where this species is native. However, efforts to improve biomass productivity of these cool-season perennial grasses lag behind switchgrass or miscanthus.

The development and testing of intraspecific and interspecific hybrids has been a focal point of breeding and genetic research in bioenergy grasses [15], including switchgrass and miscanthus. One of the most productive forms of miscanthus (*Miscanthus × giganteus*) is a sterile hybrid between *Miscanthus sacchriflorus* and *Miscanthus sinensis* that is clonally propagated by rhizomes [1,16–19]. Some hybrids of *M. sacchriflorus* and *M. sinensis* display up to 30%–35% F_1 high-parent heterosis without selection for specific combining ability [20], and there has been considerable effort to identify, develop, and test new miscanthus hybrids [19,21–24]. Switchgrass hybrids also show evidence of heterosis. Full-sib hybrid families, population hybrids, and advanced generation synthetic hybrid populations of upland and lowland switchgrass ecotypes showed evidence of mid-parent heterosis in spaced-plant plots [25] and high-parent heterosis in sward plots [26]. Extensive testing of reciprocal crosses within and between the upland and lowland ecotypes detected no evidence of high-parent heterosis and mid-parent heterosis was limited to a small fraction of the hybrids [27]. Nevertheless, the identification of complementary gene pools is expected to help produce useful hybrids and hybrid-derived populations for switchgrass biomass production [5,28,29]. As a native plant species, there is a wealth of regionally adapted genetic resources available for the development of switchgrass varieties and hybrids in North America [1,5,29,30].   Likewise, development and testing of native cool-season grasses has the potential to diversify cropping systems and expand the expected range of adaptation of bioenergy feedstocks throughout western North America.

Basin wildrye (BWR) is considered one of the largest and most conspicuous native bunchgrasses in western North America, with aerial stems in excess of 2 m [31] and a deep fibrous root system [12,32]. Although the native range of BWR is quite large, its distribution is restricted to sites where water and soil accumulate, which includes road sides and field margins [33]. Three genetically distinct races, including allotetraploid ($2n = 4x = 28$) and auto-duplicated octoploid ($2n = 8x = 56$) cytotypes, have been identified and named based on their corresponding distributions across the Columbia, Rocky Mountain, and Great Basin ecogeographic regions [33]. The octoploid (8X) cytotypes have significantly larger leaves, longer culms, and greater crown circumference compared to tetraploid (4X) cytotypes, and there is evidence of climatic adaptations within BWR [34]. However, it is not known how these differences related to biomass productivity under cultivation. The two standard BWR varieties—octoploid (8X) 'Magnar' and tetraploid (4X) 'Trailhead'—were collected directly from wild natural populations in southeastern British Columbia (Columbia race) and southcentral Montana (Rocky Mountain race), respectively [33], without subsequent selection. A new synthetic BWR variety, 'Continental', was selected from a hybrid population of chromosome-doubled (4X + 4X) 'Trailhead' and 8X 'Magnar' [35]. 'Continental' was released based on its superior stand establishment in range seeding evaluations [35], but the biomass-related traits of 'Continental' have not been compared to 'Trailhead', 'Magnar', or any other grass. Although BWR shows potentially useful biomass yields in a single-harvest management systems [12,14], none of the BWR cultivars were bred for biomass yield and they may lack the defoliation tolerance needed for more intensive multiple-harvest production systems [14].

Creeping wildrye (*Leymus triticoides*) is closely related to BWR and has a similar range of distribution throughout western North America, but it is usually found in different habitats such as saline meadows and harsh alkaline sites in [33]. Creeping wildrye (CWR) is also different from BWR in that it has extensive rhizomes, typically grows much shorter than BWR, and it had significantly lower biomass yield, by about 50%, in comparisons with BWR [14,31]. The only available CWR cultivar, 'Rio', was collected from a natural population in Kings Valley of California [36] and has been evaluated as a forage and biomass crop in the western San Joaquin Valley using saline-sodic drainage water for irrigation [37,38].

Interspecific hybrids of CWR and BWR have been developed and tested for biomass yield and other agronomic traits [14,31]. In a single-harvest management system, two CWR × BWR single-cross hybrid genets showed indications of mid-parent heterosis for dry matter yield (DMY), with both hybrids showing substantially better yields than the lower-yielding CWR parent, and one of the single-cross hybrid genets showed significantly ($p < 0.001$) greater yields than the higher-yield BWR parent [14]. It was also postulated that increased rhizomatousness in the CWR × BWR hybrids may provide a mechanism of defoliation tolerance and regrowth not present in BWR [14,39], which may in turn facilitate management in multiple-harvest production systems. However, these previous studies [14,39] were based on evaluations of clonal propagules from two single-cross hybrid genets. The difficulty of controlling pollination and producing substantial quantities of hybrid seed has limited the testing of CWR × BWR hybrids on the field-scale level [14], which has also been a challenge with switchgrass and other allogamous perennial grasses [26]. Thus, for this study, we have proposed a strategy to produce larger volumes of hybrid seed for different CWR × BWR hybrids by growing rhizome propagules from one self-incompatible CWR genet, Acc:641.T, in isolated field hybridization plots with different varieties or natural populations of BWR as the only available pollen source. Using this approach, we anticipated several possible outcomes: (1) distinct half-sib hybrid seed populations resulting from crosses of Acc:641.T CWR pollinated by different BWR populations; (2) no seed set due to the lack of compatible pollen; and (3) seeds resulting from self-fertilization of the Acc:641.T CWR genet.

The overall purpose of the research reported herein was to develop and test BWR and CWR × BWR hybrids for efficient biomass feedstock production in western North America. This study addresses several questions:

- Is it possible to utilize the gametophytic self-incompatibility mechanism of perennial grasses combined with the highly rhizomatous nature of CWR to mass produce half-sib hybrid seed for different 4X CWR × 4X BWR or 4X CWR × 8X BWR crosses on a field scale level?
- How do the relative biomass yields of CWR × BWR half-sib hybrids and higher-yielding BWR parent varieties compare to relative yields of previously tested CWR × BWR hybrid genets and their BWR parent varieties [14], in single-harvest dryland range production systems?
- What is the relative performance of CWR × BWR hybrids, interploidy BWR hybrid, standard BWR varieties, and other bioenergy candidate species in terms of DMY in a two-harvest irrigated production system of western North America?
- Do the CWR × BWR hybrids show evidence of biomass heterosis?

Specific experimental objectives that were formulated to address these questions were to (1) compare the yield, average seed size, and percent germination of seed harvested from one 4X CWR genet (Acc:641.T) grown in combination with different 4X or 8X BWR populations in isolated hybridization plots; (2) examine the genetic identity of seed harvested from the 4X Acc:641.T

CWR genet, in different hybridization plots, relative to CWR and BWR parental genotypes, using DNA markers; (3) compare biomass accumulation potential of half-sib CWR × BWR hybrids and the interploidy BWR hybrid to the high-parent BWR reference populations in dryland range environments, and; (4) compare early-season, late-season, and average yearly biomass yields of half-sib CWR × BWR hybrids, the interploidy BWR hybrid, two standard BWR varieties, intermediate wheatgrass, switchgrass, and tall wheatgrass in a two-harvest irrigated management system designed for efficient biomass production in this region.

2. Results

2.1. Yield and Quality of Half-Sib Hybrid Seed Production

Yield, average seed weights, and percent germination of seed harvested from hybridization plots were significantly influenced ($p < 0.001$) by the presence of different BWR pollen populations (Table 1). Seed yield and average seed weights also showed significant variation ($p < 0.001$) over years. Seed yield in the second year was significantly higher ($p < 0.05$) than other years (Table 1) but seed weights were significantly greater ($p < 0.001$) in the first year compared to other years (results not shown). Seed yields in the fourth year were significantly lower than the second year ($p < 0.001$) or third year ($p < 0.05$) and seed weights in the fourth year were significantly lower than all other years ($p < 0.001$). The average seed yields in hybridization plots containing 4X BWR pollen parents (5.9 g·m^{-2}) was significantly greater ($p < 0.001$) than the average seed yields in hybridization plots containing 4X BWR pollen parents (1.0 g·m^{-2}). The average seed weights in hybridization plots containing 4X BWR pollen parents (2.68 mg) was significantly greater ($p < 0.001$) than the average seed yields in hybridization plots containing 4X BWR pollen parents (1.74 mg). Finally, the percent germination in hybridization plots containing 4X BWR pollen parents (56.3%) was significantly greater ($p < 0.001$) than the average seed yields in hybridization plots containing 4X BWR pollen parents (11.0%).

Table 1. Yield, average seed weight, and percent germination of seed harvested from the tetraploid (4X) Acc:641.T creeping wildrye genet grown in hybridization plots containing octoploid (8X) or tetraploid (4X) basin wildrye (BWR) pollen-parent populations. Significant differences among groups of entry means are indicated by lettered ranks, within table columns, based on least significant differences ($p < 0.05$).

BWR Pollen-Parent Population	Year 1 Yield (g·m^{-2})	Year 2 Yield (g·m^{-2})	Year 3 Yield (g·m^{-2})	Year 4 Yield (g·m^{-2})	Avg. Yield (g·m^{-2})	Seed Weight (mg)	Percent Germination
4X Acc:636 BWR	3.8 a	10.2 a	4.5 b	1.9 ab	5.1 b	2.70 a	54.1 a
8X 'Continental' BWR	0.6 b	1.8 b	1.0 c	0.2 c	0.9 c	1.72 b	12.5 b
8X 'Magnar' BWR	0.4 b	2.1 b	1.3 c	0.5 bc	1.1 c	1.75 b	9.6 b
4X 'Trailhead' BWR	3.7 a	12.1 a	7.7 a	3.1 a	6.6 a	2.67 a	58.5 a
Standard error	0.6	2.5	0.6	0.5	0.6	0.05	2.8
Average	2.1	6.6	3.6	1.4	3.4	2.21	33.7

2.2. Genetic Identity and Genetic Diversity of Seed Harvested from CWR × BWR Hybridization Plots

Principle coordinates analysis of DNA profiles from individual plants elucidated genetic diversity within and between three major groups comprised of CWR, BWR, and apparent half-sib hybrids of CWR and BWR (Figure 1). Moreover, genetically distinct subgroups corresponding to the 4X Acc:636, 4X 'Trailhead', and 8X 'Magnar' populations were also detectable within the highly-diverse BWR group (Figure 1). The 8X 'Continental' BWR interploidy hybrid population showed considerable overlap with its 8X 'Magnar' BWR parent population, but it did not overlap with its other 4X 'Trailhead' BWR parent population (Figure 1). Most of the progeny sampled from the hybridization plots appeared to be hybrids of CWR and BWR, similar to previously described TC1 and TC2 single-cross hybrid genets of the 4X Acc:641 CWR × 4X Acc:636 BWR cross (Figure 1). In fact, the TC1 and TC2 hybrids were indistinguishable from progeny sampled from hybridization plots containing 4X Acc:636 BWR (Figure 1), which all have a very similar genetic background except that the TC1 and TC2 single-cross hybrids originated from 4X Acc:641 CWR plants that were not genetically identical to the 4X Acc:641.T CWR genet.

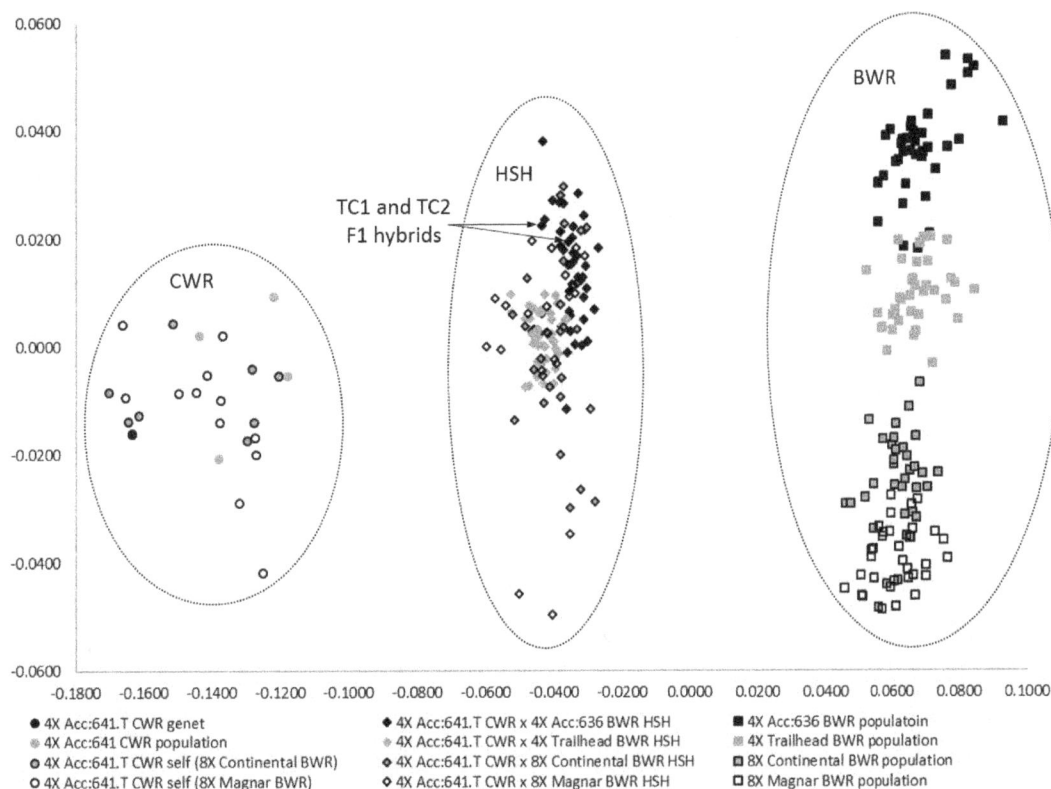

Figure 1. Principle coordinates analysis of genetic similarity coefficients among DNA genotypes from 314 individual plants of creeping wildrye (CWR), basin wildrye (BWR), and half-sib hybrid (HSH) populations harvested from the 4X Acc:641.T CWR genet in hybridization plots containing tetraploid (4X) or octoploid (8X) pollen-parent populations of 4X Acc:636 BWR, 4X 'Trailhead' BWR, 8X 'Continental' BWR, and 8X 'Magnar' BWR. Population identifiers include presumed self-progeny of the 4X Acc:641.T CWR genet (from hybridization plots containing 8X 'Continental' BWR or 8X 'Magnar' BWR). The identity of two 4x Acc:641 CWR × 4X Acc:636 BWR single-cross hybrids, TC1 and TC2, are also identified.

Not all of the progeny harvested from hybridization plots containing 8X BWR pollen sources appeared to be CWR × BWR hybrids. Eight of the 33 progeny sampled from hybridization plots containing the 8X 'Continental' BWR pollen source grouped with other 4X Acc:641 CWR plants, indicating that approximately 24% of these progeny resulted from self-pollination of the 4X Acc:641.T CWR genet (Figure 1). Likewise, 12 of the 38 progeny sampled from hybridization plots containing the 8X 'Magnar' BWR pollen source grouped with other 4X Acc:641 CWR plants, indicating that approximately 32% of these progeny resulted from self-pollination of the 4X Acc:641.T CWR genet (Figure 1). All of the remaining progeny sampled from hybridization plots containing 4X or 8X BWR pollen parents were CWR × BWR hybrids, but many of these apparent hybrids did not separate into genetically distinct groups corresponding to the BWR pollen parents (Figure 1).

Bayesian cluster analysis (Figure 2) provided further insights into the ancestry of hybrid progeny. Four nearly-pure ancestry groups corresponding to 4X Acc:641 CWR, 4X Acc:636 BWR, 4X 'Trailhead' BWR, and 8X 'Magnar' were identified using an a priori model of K = 4 Bayesian groups (Figure 2). As might be expected, plant samples from the 8X 'Continental' interploidy BWR hybrid showed mixed ancestry coefficients ranging from about 0.25 to 0.75 'Trailhead' BWR and 0.75 to 0.25 'Magnar' BWR. As expected, all of the other progeny sampled from the 4X Acc:641.T CWR genet in hybridization plots containing BWR pollen sources, had nearly pure CWR ancestry or mixed ancestry coefficients of about 0.65 CWR and 0.35 BWR (Figure 2). All of the progeny sampled from hybridization plots containing the 4X 'Trailhead' BWR pollen source appeared to be half-sib hybrids of 4X Acc:641.T CWR genet × 4X 'Trailhead' BWR (Figure 2). Likewise, most of the progeny sampled from hybridization plots containing the 4X Acc636 BWR pollen source appeared to be half-sib hybrids of 4X Acc:641.T CWR genet × 4X Acc636 BWR except that a few (at least six) appeared to be half-sib hybrids of 4X Acc:641.T CWR genet × 4X 'Trailhead' BWR (Figure 2). Conversely, most of the progeny sampled from hybridization plots containing 8X BWR pollen sources, 'Magnar' or 'Continental', appeared to be half-sib hybrids of 4X BWR (either 'Trailhead' or Acc:636) or self-progeny of the 4X Acc:641.T CWR genet (Figure 2). Only three progeny from hybridization plots containing 8X 'Magnar' BWR as the pollen source actually contain 'Magnar' ancestry (Figure 2). Likewise, only three of the sampled progeny from hybridization plots containing 8X 'Continental' BWR as the pollen source actually contained portions of 4X Acc:641 CWR, 'Trailhead' BWR, and 'Magnar' ancestry expected from a "three-way hybrid" of the 4X Acc:641.T CWR genet and the 8X 'Continental' interploidy BWR hybrid (Figure 2). Interestingly, most of the apparent hybrids of 4X Acc:641.T CWR genet × 8X BWR or self-pollinated 4X Acc:641.T CWR progeny occurred in the second evaluation year.

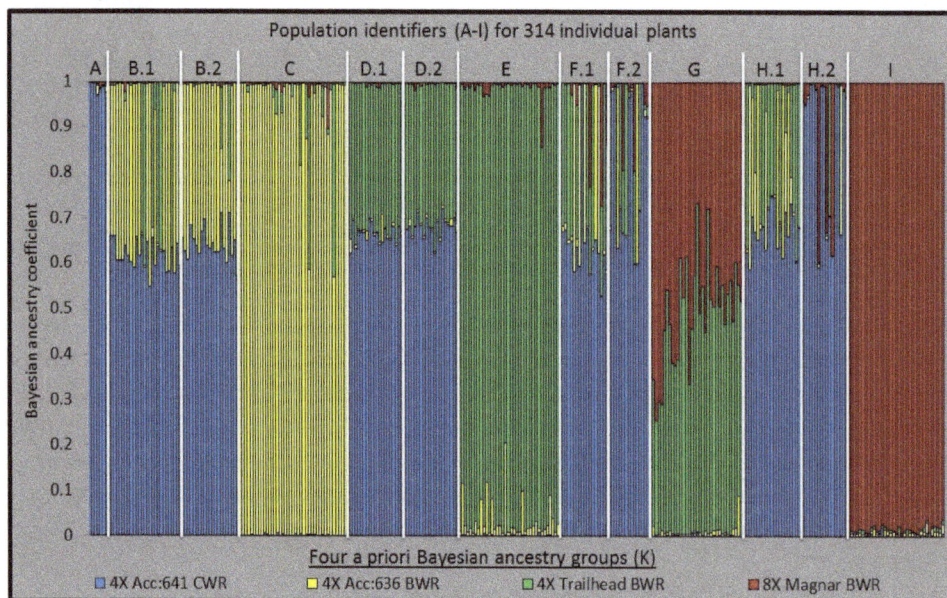

Figure 2. Bayesian cluster analysis of DNA genotypes from 314 individual plants of creeping wildrye (CWR), basin wildrye (BWR), and half-sib hybrid (HSH) populations harvested from the 4X Acc:641.T CWR genet in hybridization plots (HPs) containing tetraploid (4X) or octoploid (8X) pollen-parent populations of 4X Acc:636 BWR, 4X 'Trailhead' BWR, 8X 'Continental' BWR, and 8X 'Magnar' BWR. Population identifiers include (**A**) 4X Acc:641 CWR (including the 4X Acc:641.T CWR genet); (**B**) progeny of the 4X Acc:641.T CWR genet in HPs containing 4X Acc:646 BWR harvested in years 1 and 2; (**C**) 4X Acc:646 BWR; (**D**) progeny of the 4X Acc:641.T CWR genet in HPs containing 4X 'Trailhead' BWR harvested in years 1 and 2; (**E**) 4X 'Trailhead' BWR; (**F**) progeny of the 4X Acc:641.T CWR genet in HPs containing 8X 'Continental' BWR harvested in years 1 and 2; (**G**) 8X 'Continental' BWR; (**H**) progeny of the 4X Acc:641.T CWR genet in HPs containing 8X 'Magnar' BWR harvested in years 1 and 2; and (**I**) 8X 'Magnar' BWR.

The average genetic similarity coefficients (S ± SE) varied from lowest to highest within populations of 4X 'Trailhead' BWR (0.641 ± 0.005), 8X 'Continental' BWR (0.676 ± 0.005), 8X 'Magnar'

BWR (0.707 ± 0.006), 4X Acc:636 BWR (0.710 ± 0.005), 4X Acc:641.T CWR × 4X 'Trailhead' BWR half-sib hybrid (0.732 ± 0.004), 4X Acc:641.T CWR × 4X Acc:636 BWR half-sib hybrid (0.742 ± 0.004), and 4X Acc:641.T CWR (0.746 ± 0.012). The average genetic similarity coefficients within the two half-sib hybrid families were both significantly higher ($p < 0.001$) than their respective BWR pollen-parent populations. The average genetic similarity coefficient with the 8X 'Continental' BWR synthetic interploidy hybrid was significantly ($p < 0.001$) higher than its 4X 'Trailhead' BWR parent and significantly ($p < 0.001$) lower than its 8X 'Magnar' BWR parent.

2.3. Biomass Yields in Dryland Range Environments

Dry matter yield of two half-sib hybrid populations was evaluated relative to four BWR reference populations in spaced-plant plots over two years and two dryland range environments: Providence and Tintic (Table S1). A full-sib hybrid population of 4X 'Rio' CWR × 4X 'Trailhead' BWR was also evaluated at the Providence site (Table 2) but there was insufficient seed to test this hybrid population at Tintic or any of the other sites. Variation in DMY was significantly influenced ($p < 0.001$) by main effects of population, location, and year. There was a significant interaction of year × location ($p < 0.001$), but there were no significant interactions of population × year or population × location. Values of DMY for each population were compared by location, in part because the full-sib hybrid population was not planted at the Tintic location (Table 2). The 4X 'Rio' CWR × 4X 'Trailhead' BWR full-sib hybrid population showed significantly ($p < 0.05$) greater DMY, compared to any other BWR or CWR × BWR population, at the Providence location (Table 2), which was true in the first and second year. The 4X Acc:641.T CWR × 4X Acc:636 BWR half-sib hybrid population showed significantly ($p < 0.05$) greater DMY than its 4X Acc:636 BWR parent population, averaged over both years and both locations (Table 2). Conversely, DMY of the 4X 'Trailhead' CWR × 4X Acc:636 BWR half-sib hybrid population was significantly ($p < 0.05$) lower than its 4X 'Trailhead' BWR parent population (Table 2). Dry matter yield of the 4X Acc:641.T CWR × 4X Acc:636 BWR half-sib hybrid population was not significantly different ($p < 0.05$) from the three release varieties of BWR at Providence but it was significantly lower ($p < 0.05$) than 'Continental' and 'Trailhead' at the Tintic location. Moreover, DMY of the synthetic-hybrid 8X 'Continental' BWR population was equal to or significantly ($p < 0.05$) greater than either of its parental populations, 'Trailhead' or 'Magnar', at the Tintic location and ranked highest based on the overall averages (Table 2).

Table 2. Dry matter yield (Mg·ha^{-1}) for tetraploid (4X) and octoploid (8X) populations of basin wildrye (BWR), creeping wildrye (CWR) × BWR half-sib hybrids (HSH), and CWR × BWR full-sib hybrids (FSH) in spaced-plant dryland range plots evaluated over two years and two locations (Providence and Tintic) in Utah. Significant differences among groups of entry means are indicated by lettered ranks, within table columns, based on least significant difference.

Population	Providence	Tintic	Avg.
4X Acc:636 BWR	0.9 c	0.8 de	0.9 d
4X 'Trailhead' BWR	1.6 b	1.3 ab	1.5 b
8X 'Continental' BWR	1.5 b	1.6 a	1.5 b
8X 'Magnar' BWR	1.4 b	1.2 bc	1.3 bc
4X Acc:641.T CWR × 4X Acc:636 HSH	1.4 b	1.0 cd	1.2 c
4X Acc:641.T CWR × 4X 'Trailhead' HSH	0.9 c	0.7 e	0.8 d
4X 'Rio' CWR × 'Trailhead' BWR FSH	2.8 a	—	2.8 a
Standard error	0.1	0.1	0.1
Average	1.5	1.3	1.4

For comparison, the relative dry matter yields among two 4X Acc:641 CWR × 4X Acc:636 BWR single-cross hybrid genets (TC1 and TC2) and their parent populations (4X Acc:641 CWR and 4X Acc:636 BWR) from other experiments conducted over four years at Hyde Park and Tetonia (Table S1), were summarized here (Table 3). Moreover, field evaluations at the Hyde Park and Tetonia sites also included other important reference populations such as 'Rio' CWR, 'Trailhead' BWR, 'Continental' BWR, 'Magnar' BWR, 'Alkar' tall wheatgrass, 'Mustang' Altai wildrye (*Leymus angustus*), and four switchgrass cultivars ('Alamo', 'Dacotah', 'Falcon' and 'Sunburst'). 'Falcon' was the only switchgrass cultivar that successfully established and persisted at both locations, Hyde Park and Tetonia, thus it was the only one reported (Table 3). Dry matter yield variation among CWR × BWR s hybrid genets and other reference populations, at Hyde Park and Tetonia, was significantly influenced ($p < 0.001$) by population and there was a significant interaction of population × location ($p < 0.01$). Yields of the 4X Acc:641 CWR × 4X Acc:636 TC1 and TC2 single-cross hybrid genets were significantly higher than their 4X Acc:641 CWR parent population and yields of the TC1 hybrid were significantly greater than the 4X Acc:636 BWR high-parent. However, yields of the TC1 single-cross hybrid genet were not significantly different than the other BWR cultivars (Table 3). Moreover, it was also important to note that the 4X Acc:641.T CWR genet and 4X Acc:641 CWR reference population displayed substantially lower yields compared to all other populations including the 4X 'Rio' CWR cultivar (Table 3). These CWR reference populations were not included in other field experiments reported herein, in part because their aggressive rhizomes are difficult to manage and also because initial hybrid field evaluations, conducted at Hyde Park and Tetonia, showed that BWR is the higher-yielding reference parent (Table 3).

Table 3. Dry matter yield (Mg·ha^{-1}) for tetraploid (4X) and octoploid (8X) forms of basin wildrye (BWR), creeping wildrye (CWR), CWR × BWR single-cross hybrids (SCH), Altai wildrye (AWR), switchgrass (SG), and tall wheagrass (TWG) in dryland range plots evaluated over four years and two locations (Hyde Park, Utah and Tetonia, Idaho). Differences among groups of entry means are indicated by lettered ranks, within table columns, based on least significant difference.

Population	Hyde Park	Tetonia	Avg.
4X Acc:636 BWR	5.1 d	4.8 ab	5.7 cd
4X 'Trailhead' BWR	7.8 ab	4.6 ab	7.1 ab
8X 'Continental' BWR	7.1 ab	5.9 a	7.5 ab
8X 'Magnar' BWR	6.7 bc	6.1 a	7.1 ab
4X Acc:641 CWR	2.5 e	1.5 de	2.3 f
4X Acc:641 CWR.T genet	1.9 e	1.0 e	1.7 f
4X 'Rio' CWR	4.5 d	4.2 abc	4.8 de
4X Acc:641 CWR × 4X Acc:636 BWR TC1 SCH	7.2 ab	6.0 a	7.8 a
4X Acc:641 CWR × 4X Acc:636 BWR TC2 SCH	5.6 cd	4.1 bc	5.4 cd
'Alkar' TWG	8.2 a	4.5 abc	6.4 bc
'Falcon' SG	4.6 d	2.1 de	3.8 e
'Mustang' AWR	4.8 d	2.9 cd	4.0 e
Standard error	0.5	0.6	0.4
Average	5.5	4.0	5.3

2.4. Biomass Yields in an Irrigated Production System

Dry matter yield of two CWR × BWR half-sib hybrid populations was evaluated relative to other perennial grasses, including three released varieties of BWR, and two species mixtures, in a two-harvest irrigated production system at the Western Colorado Agriculture Experiment Station, Fruita location (Table S1).

Dry matter yield showed significant effects ($p < 0.001$) for population, year, harvest, and relative regrowth (difference between late- and early-season harvests). All two-way interactions and the three-way interactions among these four fixed effects were also significant ($p < 0.001$). Combined overall years and harvests, 'Alkar' tall wheatgrass produced significantly ($p < 0.05$) more dry matter than any other population (Table 4), but it also had significantly lower relative regrowth values indicating that it showed the greatest DMY decline from early-season to late-season harvests. Moreover, DMY of 'Alkar' tall wheatgrass was not significantly greater than the switchgrass variety mixture in the third and fourth evaluation years. Switchgrass consistently produced significantly ($p < 0.05$) more dry matter in the late-season harvest, overall years, compared to any of the other perennial grasses evaluated in this study (Table 4). Switchgrass also had significantly greater relative regrowth (Table 4), meaning that it produced relatively low DMY in the early-season harvest and relatively high DMY in the late-season harvest compared to the cool-season grasses. The overall DMY of the 'Trailhead' BWR variety was significantly greater ($p < 0.05$) than the 'Continental' and 'Magnar' BWR populations, and it was also significantly greater ($p < 0.05$) than the 4X Acc:641.T CWR × 4X Acc:636 BWR half-sib hybrid population. However, the overall DMY of the 'Trailhead' BWR variety was not significantly greater than the 4X Acc:641.T CWR × 4X 'Trailhead' BWR half-sib hybrid population. The 4X Acc:641.T CWR × 4X 'Trailhead' BWR half-sib hybrid population showed significantly greater ($p < 0.05$) relative regrowth than its 'Trailhead' BWR parent population, which in this case indicated that the hybrid showed less DMY decline between the early- and late-season harvests (Table 4).

Table 4. Dry matter yield (Mg·ha^{-1}) for basin wildrye (BWR), creeping wildrye (CWR) × BWR half-sib hybrids (HSH), Altai wildrye (AWR), intermediate wheatgrass (IWG), switchgrass (SG), and tall wheatgrass (TWG) in a two-harvest, irrigated production experiment at the Western Colorado Agriculture Experiment Station. Significant differences among groups of entry means are indicated by lettered ranks, within table columns, based on least significant differences.

Population	Year 1	Year 2	Year 3	Year 4	Early Cut	Late Cut	%Δ [1]	Avg. Total
4X 'Trailhead' BWR	9.9 c–e	20.2 b	9.4 b	17.9 bc	9.4 b	4.9 de	−47 d	14.3 b
8X 'Continental' BWR	9.0 ef	18.3 bc	7.5 cd	16.5 c	7.8 cd	5 de	−36 c	12.8 c–e
8X 'Magnar' BWR	9.2 d–f	18.4 bc	7.5 cd	13.7 d	7.6 cd	4.7 e	−38 c	12.2 e
4X Acc:641.T CWR × 4X Acc:636 BWR HSH	10.0 c–e	17.3 cd	7.2 d	15.8 cd	7.8 cd	4.8 de	−37 c	12.6 de
4X Acc:641.T CWR × 4X 'Trailhead' BWR HSH	10.3 cd	17.6 cd	7.9 cd	17.5 c	8.0 c	5.4 cd	−33 c	13.3 b–e
'Mustang' AWR	10.4 c	18.0 b–d	9.7 b	16.8 c	7.2 d	6.5 b	−5 b	13.7 b–d
'Rush' IWG	12.7 b	18.3 bc	8.5 bc	16.5 c	9.1 b	4.9 de	−46 d	14.0 bc
'Oahe' IWG	12.9 b	17.6 cd	7.9 cd	17.7 bc	8.9 b	5.1 c–e	−42 d	14.0 bc
'Blackwell' and 'Dacotah' SG	8.5 f	15.6 d	10.9 a	22.0 a	6.4 e	7.9 a	+23 a	14.3 b
'Alkar' TWG	14.8 a	23.0 a	11.1 a	20.1 ab	11.6 a	5.7 c	−51 e	17.3 a
Standar error	0.4	0.8	0.5	1.0	0.3	0.3	0.3	0.4
average	10.8	18.4	8.8	17.4	8.4	5.5	−35	13.9

[1] Percent change (%Δ) calculated as (early cut yield—late cut yield)·100/(early cut yield).

2.5. Summarized Analysis of Biomass Yield from Dryland and Irrigated Testing Sites

A summarized analysis of overall years and testing sites was performed by nesting populations within groups including the 8X BWR, 4X BWR, 4X CWR, 4X CWR × 4X BWR hybrids (Table 5). Significant variation ($p < 0.0001$) was detected for fixed effects of locations, years, population within group, and group. However, these results (Table 5) should be viewed cautiously because not all populations were tested at all sites. In fact, the only populations that were tested across all sites were 'Trailhead' BWR, 'Continental' BWR, 'Magnar' BWR, and the 4X Acc:641 CWR × 4X Acc:636 BWR hybrids (as single-cross or half-sib hybrids). This meta-analysis shows that BWR and 4X CWR × 4X BWR hybrids displayed significantly more dry matter yield than CWR at dryland sites and suggests that this could be extrapolated to other sites, even though CWR was only tested at two of the four dryland testing sites (Table 5). This meta-analysis also shows that 8X BWR performed better than 4X BWR in dryland environments, whereas 4X BWR performed better than 8X BWR in the irrigated environment (Table 5). Estimated yields of the 4X CWR × 4X BWR hybrids including all half-sib hybrids, full-sib hybrids, and single-cross hybrids were significantly less than 8X BWR in dryland environments, significantly less than 4X BWR in the irrigated environment, but not significantly different from 4X or 8X BWR overall (Table 5). The estimated yield of switchgrass in dryland environments was based only on the best of four populations, where three populations failed to flourish, at only two of the four sites. Thus, estimated yield of switchgrass in dryland environments should also be viewed cautiously.

Table 5. Dry matter yield estimates for tetraploid (4X) and octoploid (8X) forms of basin wildrye (BWR), creeping wildrye (CWR), CWR × BWR hybrids, Altai wildyre (AWR), intermediate wheatgrass (IWG), switchgrass (SG), and tall wheatgrass (TWG) across four dryland ranges and one irrigated testing environment. Significant differences among groups of entry means are indicated by lettered ranks based on least significant difference.

Population	Dryland Range	Irrigated	Overall
4X BWR	4.0 b	14.3 b	5.6 b
8X BWR	4.4 a	12.5 d	5.5 b
4X CWR	0.6 d	—	2.7 c
4X CWR × 4X BWR	4.2 ab	13.0 cd	5.8 b
AWR	2.0 c	13.7 bc	5.2 b
IWG	—	14.0 b	6.1 b
SG	1.6 c	14.3 b	5.0 b
TWG	4.7 ab	17.3 a	8.3 a
Standard error	0.2	0.5	0.3
Average	3.1	14.1	5.5

3. Discussion

3.1. Development and Testing of CWR × BWR Hybrids

Our study demonstrated useful methods of producing hybrid seed for allogamous plants on a field scale, which has been a considerable challenge in perennial grass breeding [15,26]. Our approach worked effectively for crosses of 4X CWR and 4X BWR, but it did not work in crosses of 4X CWR and 8X BWR. One possible reason for low production of 4X CWR × 8X BWR hybrids may be delayed flowering of the 8X BWR populations. However, other intrinsic fertility barriers such as abnormal endosperm development are expected from interploidy crosses of 4X CWR and 8X BWR, 4X and 8X BWR, or 4X and 8X switchgrass interploidy hybrids [40,41]. This is why chromosome doubling of 4X 'Trailhead' BWR was required to make the 8X 'Continental' interploidy hybrid of 4X 'Trailhead' and 8X 'Magnar' BWR cultivars. Thus, reproductive incompatibility mechanisms such as the endosperm genic balance number requirement [41] explain the prevalence of relatively light seed and poor germination of seed obtained from the 4X CWR × 8X BWR hybridization plots (Table 1). Most of the seed from the 4X CWR × 8X BWR hybridization plots that did germinate was derived from self-pollination or pollination by 4X BWR from other hybridization plots. Conversely, all of the seed sampled from 4X CWR × 4X BWR hybridization plots was of hybrid origin with relatively high seed weights and germination rates. The average estimated seed yields of the 4X Acc:641.T CWR genet within hybridization plots containing 4X BWR (Table 1) were about five-fold lower than the reported seed yield averages of about 336 kg·ha^{-1} (33.6 g·m^{-2}) from 'Rio' CWR [42]. However, it is impossible to say whether our yields were limited by inherent limitations of this technique, such as gamete incompatibility, or other factors such as differences in the productivity of CWR genotypes or environments.

The half-sib 4X CWR × 4X BWR hybrid populations are similar to that of semi-hybrids of grass or alfalfa [43–46] in the sense that the hybrids are comprised of genetically heterogeneous individuals, but different in that virtually all of the seeds harvested from the 4X CWR seed parent were hybrids. The semi-hybrids, in contrast, contain an equal mixture hybrid and parental populations [43–46]. With prolific production of rhizomes and tillers, clonal production of the CWR seed parent is feasible. This approach of half-sib hybrid seed production, using one clonally propagated seed parent, creates opportunity to select individual genets that have good combining ability and also provides a rapid way to introduce novel genes into hybrid populations. Although similar approaches of producing full-sib single-cross hybrids has been proposed by clonal propagation of two self-incompatible genets, the difficulty of clonal propagation limits the application of this approach in perennial ryegrass (*Lolium perenne*) or switchgrass [15].

As a group, 4X CWR × 4X BWR hybrids showed yields that were at least equal to their 4X BWR parental populations, overall sites, and significantly higher than CWR yields in dryland environments (Table 5). Moreover, 4X BWR populations displayed substantially and statistically greater yields than 4X CWR in dryland environments (Tables 3 and 5). Thus, we cautiously assume that our 4X CWR × 4X BWR hybrid populations show indication of mid-parent heterosis if their yields are equivalent to their 4X BWR parent populations and show indication of high-parent heterosis if their yields are significantly greater than their 4X BWR parent populations in dryland range growing environments. The 4X Acc:641.T CWR × 4X Acc:636 BWR half-sib hybrid population showed a significant 33% advantage over its 4X Acc:636 BWR parent population at the Providence and Tintic testing sites (Table 2), indicating possible high-parent heterosis of this half-sib hybrid. One of the 4X Acc:641 CWR × 4X Acc:636 BWR single-cross hybrid genets also displayed a significant 36% yield advantage over the 4X Acc:636 BWR population at the Hyde Park and Tetonia testing sites (Table 3), which may indicate high-parent heterosis, but we did not have the actual 4X Acc:636 parent plant that was needed as a reference for this single-cross hybrid. Nevertheless, the 4X Acc:636 BWR parent population and both 4X Acc:641 CWR × 4X Acc:636 BWR single-cross hybrids all showed at least two-to three-fold greater yields than the 4X Acc:641.T CWR genet or 4X Acc:641 CWR population at the Hyde Park and Tetonia testing sites (Table 3). Thus, we conclude that yields of 4X Acc:641.T CWR × 4X Acc:636 BWR hybrids show indications of high-parent heterosis at the Providence and Tintic testing sites (Table 2) and that the 4X Acc:641 CWR × 4X Acc:636 BWR single-cross hybrids showed indications of mid-parent and possible high-parent heterosis at the Hyde Park and Tetonia testing sites (Table 3). Yields of the 4X 'Rio' CWR × 4X 'Trailhead' BWR full-sib hybrid population were 75% greater than its 4X 'Trailhead' BWR parent population at the Providence site (Table 2) and the 4X 'Trailhead' BWR population showed yields that were nearly 50% greater than the 4X 'Rio' CWR at the Hyde Park and Tetonia sites (Table 3). Therefore, the 4X 'Rio' CWR × 4X 'Trailhead' BWR full-sib hybrid population also showed indications of high-parent heterosis by outperforming its 4X 'Trailhead' BWR parent population by a 75% margin. Although yields of 4X CWR × 4X BWR hybrid populations were slightly but not significantly greater than 4X BWR populations' overall dryland environments (Table 5), the 4X Acc:641 CWR.T × 4X 'Trailhead' BWR half-sib hybrid performed much worse than its 4X 'Trailhead' BWR population at the Providence and Tintic dryland testing sites (Table 2), indicating that performance of this hybrid may be more similar to its presumed low-parent, 4X Acc:641 CWR.T.. As a group, yields of the 4X CWR × 4X BWR hybrids were significantly less than estimated yields of the 4X BWR populations in the irrigated testing site (Table 5), but yields of the 4X Acc:641.T CWR × 4X 'Trailhead' BWR half-sib hybrid were significantly less than its 4X 'Trailhead' BWR parent population in this two-harvest irrigated management system (Table 4). Thus, we conclude that some but not all 4X CWR × 4X BWR hybrids show indications of high-parent heterosis in dryland environments. Although performance of 4X CWR × 4X BWR half-sib hybrids was comparable to 4X BWR in the two-harvest irrigated production system, there were no indications of high-parent heterosis in this type of management system. Thus, it seems that heterosis in the 4X CWR × 4X BWR hybrids showed stronger indications of expression in the less competitive conditions of space-planted plots, which was also true for population hybrids of perennial ryegrass [46].

Dryland biomass yields of the 4X 'Rio' CWR × 4X 'Trailhead' BWR single-cross hybrid population were at least two-fold greater than biomass yields of either of the two half-sib hybrid populations produced by the 4X Acc:641.T CWR genet. One factor that may explain higher yields of the 4X 'Rio' CWR × 4X 'Trailhead' BWR hybrid is that the 4X 'Rio' CWR showed about two- to three-fold greater yields than 4X Acc:641 CWR population or 4X Acc:641.T CWR genet, respectively, in dryland yield trials conducted at Hyde Park and Tetonia. Also, the 4X 'Trailhead' BWR showed significantly greater yields than the other 4X BWR parent, Acc:636, used in this study (Tables 2 and 3). Delayed flowering may be another factor that may have contributed to relatively strong heterosis expressed in the 4X 'Rio' CWR × 4X 'Trailhead' BWR hybrid. It has been shown that delayed flowering can be a strong driver of biomass yield in switchgrass hybrids [27]. In our observations, 'Rio' CWR and the 4X 'Rio'

CWR × 4X 'Trailhead' BWR hybrid flowered about one week later than the 4X BWR, 4X Acc:641 CWR, or 4X Acc:641 CWR × 4X BWR populations. Biomass yields up to 11.5 Mg·ha^{-1} per year with saline irrigation water and repeated harvesting have been reported for 'Rio' CWR in the San Joaquin Valley of California [37]. Thus, an important future goal will be to generate half-sib hybrid seed for the cross of 4X 'Rio' CWR × 4X 'Trailhead' BWR and test these hybrids in multiple dryland and irrigated environments of California, Colorado, Nevada, and Utah where BWR and CWR may be useful as biomass feedstocks [12,37]. In any case, comparisons of hybrids made using two different CWR genets—4X Acc:641.T and one 4X 'Rio' plant—indicate that the identification and selection of superior CWR seed-parents, as well as superior BWR pollen parents, may be the fastest and most promising approach to develop higher-yielding CWR × BWR hybrids and native grass bioenergy feedstocks for western North America.

3.2. Dryland Yield Potential of Perennial Grasses in Cold-Desert Environments

Average yearly dryland biomass yields ranging from 2.2 to 9.6 Mg·ha^{-1} for the 4X Acc:636 BWR population were previously reported based on testing over four years and two locations near Hyde Park, Utah and Tetonia, Idaho [14] but otherwise we are not aware of any other published reports on non-irrigated biomass yields of BWR in its native range environments. The average yearly dryland biomass yields of the 4x Acc:636, 4X 'Trailhead', 8X 'Continental', and 8X 'Magnar' BWR populations ranged from 0.5 to 2.1 Mg·ha^{-1} at Providence and Tintic (Table 2), whereas the average yearly biomass yields of these same four populations ranged from 2.2 to 13.3 Mg·ha^{-1} at the Hyde Park and Tetonia testing sites (Table 3). Thus, dryland biomass yields from experiments conducted at the Providence and Tintic sites were lower than previously described experiments conducted at the Hyde Park and Tetonia testing sites [14]. One possible explanation for these differences is that plant densities in the Hyde Park and Tetonia sites were twice as high as the Providence and Tintic sites. The plants at Providence and Tintic were, however, simply less vigorous. Under these conditions, we would expect BWR plants to grow larger, not smaller, at the lower plant densities such as those used at Providence and Tintic (1 plant·m^{-2}). Although the Tetonia and Tintic testing sites are located in dissimilar environments (Table S1), the Hyde Park and Providence sites were located at nearly identical elevations in the same valley, only 12.8 km apart, with the same soil types (Nibley silty clay loam). The average yearly biomass yields at Hyde Park and Tetonia varied nearly three-fold, between 3.1 and 8.6 Mg·ha^{-1}, over four harvest years (Table 3), indicating that seasonal variations in climate can have dramatic effects on perennial grass productivity especially under low-input management. However, other factors probably contributed to differences between evaluations conducted at the Providence and Tintic sites (Table 2) versus the Hyde Park and Tetonia sites (Table 3). Nevertheless, the 4X 'Trailhead' BWR population ranked relatively high and the 4X Acc:636 BWR population ranked relatively low across all dryland environments (Tables 2 and 3). Moreover, 8X BWR populations performed significantly better across all dryland range testing sites (Table 5). These results demonstrate significant genetic variation for BWR biomass yields within its native growing environments. Wide variation in the average yields among sites, and significant genotype by environment interactions, observed across our dryland testing sites also demonstrate the importance of testing plant materials over multiple locations and years.

The average total biomass yields of all four BWR populations, two CWR × BWR single-cross hybrid genets, and tall wheatgrass were all significantly greater than the best of four switchgrass varieties tested in non-irrigated experiments conducted at Hyde Park and Tetonia (Table 3). This was not surprising considering that these relatively high elevation, cold-desert testing sites are not located within the native range of switchgrass [1,5]. However, there did appear to be some variation for adaptation to these environments, among the four switchgrass varieties tested, and it was interesting that the best adapted variety, Falcon, was an upland ecotype originating from relatively high-elevations (1517 m) in New Mexico on the western range of switchgrass distribution [5]. Falcon did not perform as well as Dacotah in high-latitude regions of Europe [47], but it did perform better than Dacotah in our high-elevation cold-desert testing sites (Table S1). Additional screening and breeding research may help identify and develop switchgrass populations that perform better in cold high-latitude or high-elevation growing environments [48], such as those found in western North America.

Although 'Alkar' tall wheatgrass was clearly the best yielding population in the irrigated testing environment, it is interesting that overall dryland biomass yields of BWR cultivars and the 4X Acc:641 CWR × 4X Acc:636 BWR TC2 single-cross hybrid genet were equal to or significantly greater than 'Alkar' tall wheatgrass (Table 3). Tall wheatgrass was introduced and widely naturalized throughout North America [13], and has proven to be a high yielding cool-season grass in other non-irrigated field evaluations in studies conducted in western North Dakota [6] and Kansas [9].

3.3. Irrigated Yield Potential of Perennial Grasses in Cold-Desert Environments

The early-season biomass yields and average yearly biomass yields of perennial grasses including 'Trailhead' BWR, 'Alkar' tall wheatgrass, and switchgrass evaluated in our two-harvest irrigated production experiment in western Colorado can be compared to the total annual yields in a single-harvest irrigation system in western Nevada [12]. In the single-harvest evaluation conducted in western Nevada, DMY values of 'Trailhead' BWR and 'Alkar' tall wheatgrass averaged 7.2 and 7.3 $Mg \cdot ha^{-1}$ under low-water treatments (71 cm water annually) or 8.7 and 9.9 $Mg \cdot ha^{-1}$ in the high irrigation (120 cm water annually) treatment, respectively [12]. These values from single-harvest irrigated field trial [12] were similar but slightly lower than the average early-season harvest yields of 'Trailhead' BWR and 'Alkar' tall wheatgrass from western Colorado (Table 4). Since all of the cool-season grasses evaluated in the two-harvest irrigated production experiment were in the anthesis or post-anthesis stage of development on the early-harvest dates, we believe that they would have reached maximum or near-maximum annual yields in a single-harvest system. However, this was not true for switchgrass, which was not yet flowering and did not have a chance to reach maximum biomass accumulation values on the early-harvest dates in this two-harvest experiment (Table 4). Nevertheless, the switchgrass yields from both the early- and late-harvests in western Colorado (Table 4) were similar to the observed DMY of 7.8 $Mg \cdot ha^{-1}$ in the high-water treatment in western Nevada and substantially greater than 2.8 $Mg \cdot ha^{-1}$ observed in the low-water treatment [12]. Although the timing of our early-season harvest was optimized for cool-season grasses, it provided time for all of the perennial grasses, including switchgrass, to regrow and produce significantly more yearly total biomass (Table 4) than the single-harvest irrigated production system used in western Nevada [12]. The overall biomass yields showed nearly three-fold variation, from 5 to 15 $Mg \cdot ha^{-1}$, over three years in Nevada [12] and over two-fold variation over four years in Colorado (Table 4).

The average yearly total yields of 'Alkar' tall wheatgrass, Rush intermediate wheatgrass, and switchgrass from our two-harvest irrigated evaluation in western Colorado (Table 4) can be compared to the average yearly total yields from another multiple-harvest irrigated study, conducted in northern Utah, containing these same species [11]. The average yearly total yields of 'Alkar' tall wheatgrass and Rush intermediate wheatgrass from Colorado (Table 4) were substantially lower than the maximum DMY values of 27.6 and 26.8 $Mg \cdot ha^{-1}$ observed for these same varieties, respectively, at the optimum water level of a five-harvest irrigated production evaluation in northern Utah [11]. Yet, the average yearly total DMY of one switchgrass variety mixture tested in the two-harvest irrigated production system (Table 4) was very similar to the average yearly total of about 14.0 $Mg \cdot ha^{-1}$ observed for the four switchgrass varieties tested in the five-harvest irrigated production system [11]. Thus, with adequate irrigation, switchgrass was competitive with cool-season grasses in one-harvest [12] and two-harvest management systems (Table 4), but substantially greater yields can be attained from repeated harvesting of cool-season grasses including tall wheatgrass and intermediate wheatgrass [11]. Comparisons of these three studies suggest that something, perhaps cool-nights, may be impeding regrowth of warm-season grasses with repeated harvests in these high-elevation cold-desert environments. The early-season DMY values of tall wheatgrass and intermediate wheatgrass were greater than switchgrass in multiple-harvest studies conducted in Utah [11] and Colorado (Table 4), whereas the late-season yields of switchgrass were always greater than cool-season grasses in both studies, over all irrigation levels [11]. Presumably this difference in yield phenology is related to the C4 photosynthesis system of switchgrass and the low-temperature growth potential of cool-season grasses.

Although early-season yields of 'Trailhead' BWR were substantially greater than switchgrass, the average total yields were not significantly different in the two-harvest irrigated experiment in Colorado (Table 4) or either of the two irrigation treatment levels tested in Nevada [12]. The early-season and average total biomass yields of 'Alkar' tall wheatgrass were both significantly greater than switchgrass in the two-harvest irrigated experiment in Colorado (Table 4) and the low-water treatment in the one-harvest experiment conducted in Nevada [12]. The average total biomass yields of 'Alkar' tall wheatgrass was nearly two-fold higher than the best of four switchgrass varieties at the highest water levels and more than three-fold higher at the lowest water levels in the five-harvest irrigated experiment in Utah [11]. Thus, results from irrigated production systems in Colorado (Table 4), Nevada [12] and Utah [11] indicate that cool-season grasses such as BWR, intermediate wheatgrass, and tall wheatgrass could be better than switchgrass when irrigation water supplies are limited in the cold-desert environments of western North America.

Significant variance among the relative regrowth of cool-season grasses (Table 4) may reflect dissimilarities in phenology or defoliation tolerance. Tall wheatgrass displayed significantly greater yield declines from early- to late-season harvests, compared to other grasses (Table 4), which might be related to its relatively late flowering times. Yield declines of the 4X 'Trailhead' BWR population were significantly greater than the 4X Acc:641.T CWR × 4X 'Trailhead' BWR hybrid. The most obvious explanation for this is that CWR and the interspecific hybrids are much more rhizomatous and produce more tillers than BWR, which may provide mechanisms for better defoliation tolerance and regrowth potential [14,39]. Expression of rhizomatousness in the CWR × BWR hybrids is largely controlled by a combination of two recessive genes: one dominant gene, and one partially-dominant gene with major effects [14,39]. We are not aware of any other differences in phenology or physiology that would likely explain differences in the regrowth potential between BWR and the CWR × BWR hybrids. However, it was somewhat surprising that most of the BWR and CWR × BWR plant materials displayed significantly less yield declines, from early- to late-season harvests, compared to tall wheatgrass or intermediate wheatgrass. Native range grasses, such as BWR, are usually not included in multiple-harvest production systems, such as the five-harvest irrigated production trial conducted in northern Utah [11], in part because they do not have the forage quality of conventional pasture grasses and also because it has been generally presumed that they do not have significant regrowth potential. In fact, to our knowledge, this is the first documented experiment involving seasonal regrowth of BWR in an irrigated production system. Thus, improved regrowth of the 4X Acc:641.T CWR × 4X 'Trailhead' BWR hybrid could be a potentially useful attribute of the CWR × BWR hybrids [14,39].

Although yields of switchgrass from the two-harvest irrigated production study in Colorado (Table 4) were good, these yields have been surpassed in other environments. Average annual yields of upland and lowland switchgrass ecotypes ranged from 4 to 21 Mg·ha^{-1} in single-harvest production experiments conducted across various latitudes of central North America, from 36 to 46° N, including Kansas, Nebraska, Oklahoma, and Wisconsin where switchgrass is well adapted [28]. Switchgrass yields up to 26 and 33 Mg·ha^{-1} have been reported in one- and two-harvest production systems in the southeastern parts of North America [1].

4. Materials and Methods

4.1. Plant Materials Used for Making CWR × BWR Hybrids

The term "population" was used herein as a generic term for individuals sampled from genetically heterogeneous cultivars, hybrids, or natural germplasm accessions. Plant materials used for making and testing experimental CWR × BWR hybrids included two tetraploid (4X) BWR populations, two octoploid (8X) BWR populations, and two 4X CWR populations. The BWR populations included 8X 'Continental', 4X 'Trailhead', and 8X 'Magnar' in addition to one natural 4X population Acc:636 collected near Lethbridge, Alberta sometime prior to 1974 by Sylvester Smoliak. 'Trailhead' originated from a natural population near Roundup, Montana and was released in 1991 [49]. 'Magnar' originated from a natural population in south-eastern British Columbia and was released sometime before 1995 [50]. The 4X CWR populations included a natural germplasm accession Acc:641 originally collected near Jamieson, Oregon in 1975 by Kay Asay and the cultivar 'Rio', which was originally collected from Kings Valley, California and released in 1991 [36].

The term "genet" was used herein to identify and reference clonally propagated individuals such as the two single-cross hybrids—TC1 and TC2—of 4X Acc:641 CWR × 4X Acc:636 BWR and the 4X Acc:641.T CWR plant used to make the full-sib pseudo-backcross families [51] and identify chromosome regions controlling biomass yield [14]. Both BWR and CWR are considered highly self-incompatible [52] and previous studies showed that the 4X Acc:641.T CWR genet was receptive to pollen from the TC1 and TC2 single-cross hybrids with less than 4% selfing using paper bags to cover its spike-inflorescences [51]. However, the quantities of seed produced by these techniques is insufficient for extensive testing in seeded field trials. Thus, we hypothesized that it would be possible to produce half-sib hybrid seed by growing the highly rhizomatous 4X Acc:641.T CWR genet in isolated "hybridization plots" with different 4X or 8X BWR populations, described above, as the only available pollen source. For purposes of testing other sources of CWR in hybrids with BWR, a new full-sib hybrid population was developed using paper bags to cover the spike-inflorescences of one 4X 'Rio' CWR plant and manually shaking pollen from extruded anthers of spikes from one 4X 'Trailhead' BWR genet.

4.2. Description of Locations Used to Develop and Test CWR × BWR Hybrids

All testing sites (Table S1) were located on properties of the Utah State University, Utah Agriculture Experiment Station; the Colorado State University, Western Colorado Research Center; or the University of Idaho, Eastern Idaho Agriculture Experiment Station. The average precipitation and temperature data was obtained from the Prism Climate Group [53] using geographic coordinates (Table S1) and 30-year normals (1981–2010) data setting at 800 m resolution. Precipitation data was summarized as seasonal averages for winter (December, January and February), spring (March, April and May), summer (June, July and August), and fall (September, October and November). The average, minimum, and maximum number of freeze-free days above 0 °C were determined using data available from the nearest station of similar elevation available from the Utah Climate Center [54].

4.3. Evaluation of Half-Sib Hybrid Seed Production

This study included an experiment designed to compare the yield and quality of seed from the 4X Acc:641.T CWR genet pollinated using the two 4X (Acc:636 and 'Trailhead') and the two 8X (Continental and 'Magnar') BWR populations as different fixed effects. The 4X Acc:641.T CWR genet was grown in isolated hybridization plots with each of the four different BWR pollen parent populations in a randomized complete block (RCB) design with four different hybridization plots (four different BWR pollen parents) within each block and three replicated blocks for a total of 12 hybridization plots. Beginning in December of 2007, rhizome propagules from the 4X Acc:641.T CWR genet and seedlings of the four BWR pollen parents were grown in a sufficient number of racked soil containers (4-cm diameter) to produce 972 clones of the 4X Acc:641.T CWR genet and 408 individual seedlings from each of the four BWR pollen parents. These plant materials were transplanted to a testing site near Richmond, Utah in the spring of 2008. Plants were irrigated after planting, but not irrigated during subsequent years of seed production. Each hybridization plot had three rows of the 4X Acc:641.T CWR genet flanked by two rows of the BWR pollen parent on each side. Rows were 13 m long with 0.91 m (36 inch) spacing between rows with plants spaced 0.5 m apart within rows. The hybridization plots were cultivated between rows and fertilized once per year with approximately

56 kg N ha^{-1} in the fall (October) and 34 kg N ha^{-1} in the spring (May) in the form of urea. Thus, the harvested CWR plot size was approximately 37 m^2, not including space used by the BWR pollen parents. The 12 hybridization plots were tandemly arranged in seven parallel rows, with 13 m space between test plots, for a total length of 323 m and total width of 5.5 m. The randomization scheme was restricted such that hybridization plots containing the 4X or 8X BWR pollen parents were alternated.

The total seed yield, average seed weight, and seed germination rates were measured using seed harvested from the 4X Acc:641.T CWR genet in each of the 12 hybridization plots over the course of four years, from 2009 to 2012, using a plot combine (Wintersteiger Inc., Salt Lake City, Utah, USA). Additional fine threshing and cleaning was performed manually before the seed yield (g·m^{-2}) was determined for the harvested plot area (37 m^2). Three sets of 100 seed were counted and weighed from each plot, each year, to determine the average seed weight (mg·seed^{-1}) and germination rates (%). Germination rates were determined by treating each 100-seed sample with tetramethythiuram disulfide, and then spreading each sample on blue germination blotter paper (Anchor Paper, St. Paul, Minnesota, USA) moistened with distilled water in 11 × 11 × 3.5-cm Cont 156C germination boxes (Hoffman Manufacturing, Inc., Jefferson, Oregon, USA). Using standard seed testing procedures for CWR and BWR, seeds were imbibed for 3 d at 25 °C, stratified at 4 °C for 14 d in dark chambers, and then tested for germination in dark plant-growth chambers maintained on a diurnal cycle of 16 h at 15 °C and 8 h at 25 °C. Germinated seeds were counted each week and until the final 28 d count.

4.4. Genetic Testing of Half-Sib Hybrid Seed

The genetic identity of seed produced by the 4X Acc:641.T CWR genet, from the first and second years of seed production, was tested by comparing it with representative genotypes of the CWR and BWR parents using a principle coordinate analysis of genetic distances based on comparisons of DNA profiles. Fresh leaf samples were lypholized and milled using a MM300 (Retsch Inc., Newtown, Pennsylvania, USA) mixer mill for DNA extraction using the DNeasy 96 Plant Kit (Qiagen, Germantown, Maryland, USA). Multi-locus DNA profiles of each DNA sample were developed using the AFLP technique [55] with modifications for detection using fluorescent labels and capillary electrophoresis. Briefly, the selective *Eco*RI primers were fluorescent labeled with 6-carboxyfluorescein and fractionated by capillary electrophoresis on an ABI3100 instrument (PE Applied Biosystems, Foster City, California, USA) with internal GS-500 size standards (PE Applied Biosystems) for each sample in each channel. The *Eco*RI and *Mse*I preamplification primers both included one selective nucleotide A and C, respectively. The selective amplification primers included two additional selective nucleotides in four different combinations including E36(ACC)//M61(CTG), E37(ACG)//M60(CTC), E38(ACT)//M60(CTC) and E41(AGG)//M47(CAA). Different AFLP markers were identified and scored for the presence or absence of bands (DNA amplicons) based on the relative mobility corresponding to about 1 bp using Genographer open source software. A principle coordinates analysis of DNA profiles was performed using NTSYSpc, Numerical Taxonomy System version 2.21 (Exeter Software, Setauket, New York, USA) based on pairwise comparisons of genetic similarity among individual plants computed using the similarity index (S) formula $S_{XY} = 2N_{XY}/(N_X + N_Y)$, where N_{XY} is the number of shared bands and where N_X and N_Y are the numbers of bands in plants X and Y. Population structure and ancestry coefficients of seed harvested from the 4X Acc:641.T CWR genet, in different hybridization plots, were also compared to BWR and CWR reference samples using Bayesian clustering [56,57] modeled a priori for possible admixture among the four expected base populations of 4X Acc:641 CWR, 4X Acc:636 BWR, 4X 'Trailhead' BWR, and 8X 'Magnar' BWR. All other genotypes were anticipated to be hybrid- or direct-descendants of these four base populations. Finally, statistical comparisons of genetic diversity within populations were based on the similarity index (S), based on unbiased estimates of the variance for S corrected for covariance [58].

4.5. Evaluation of DMY in Dryland Range Environments

In many cases, natural stands of BWR and other caespitose grasses grow with open spaces between plants rather than solid swards in the Great Basin region of western North America. Dry matter yields of experimental half-sib hybrid and full-sib hybrid populations were compared to the four BWR reference populations (Acc:636, 'Continental', 'Magnar', and 'Trailhead') in simulated range plots established using seedlings transplanted from soil containers at Tintic, Utah and Providence, Utah testing sites (Table S1). Germinated seedlings were planted into soil containers (4-cm diameter) in December, 2010 and then transplanted into a cultivated soil with dibbled holes in May, 2010 so that each plot was comprised of seven plants in a row, with 1 m between plants and 1 m between rows for a total plot size of 7 m^2 (1 plant·m^{-2}). Plots were replicated in a randomized complete block (RCB) design with six replications at each testing site. Plots were maintained with light cultivation to remove weeds as needed, with no fertilizer or irrigation. Once established, these plots have plant densities similar to typical natural stands of BWR in its native range environments. The plots were harvested to a height of 8 cm and weighed using a Swift Forage Harvester (Swift Machine and Welding LTD, Swift Current, Saskatchewan, Canada) once each year on 9 August 2011 and 28 August 2012 at the Providence site and 25 August 2011 and 29 August 2012 at the Tintic site. Results were reported on a DMY per unit area basis (Mg·ha^{-1}). Subsamples of harvested plant material were weighed, at the time of harvest, and dried at 60 °C in a forced-air oven to constant weights that were used to convert fresh harvest weights to DMY.

Prior to development of the half-sib hybrids, biomass yields of two single-cross 4X Acc:641 CWR × 4X Acc:636 BWR hybrid genets—TC1 and TC2—were also evaluated relative to the 4X Acc:641.T CWR genet, 4X Acc:641 CWR population, 4X Acc:636 , 'Magnar' BWR, 'Continental' BWR, 'Trailhead' BWR, 'Rio' CWR, 'Mustang' Altai wildrye, 'Alkar' tall wheatgrass, and four varieties of switchgrass ('Alamo', 'Dacotah', 'Falcon', and 'Sunburst') in space-planted dryland range plots over four years (2008–2011) and two locations near Hyde Park, Utah and Tetonia, Idaho (Table S1). The management of the experiments conducted at the Hyde Park and Tetonia sites was similar to the dryland range evaluations at the Providence and Tintic sites, except that plants were spaced 0.5 m within rows [14] and cultivated at least once each year, deeply, to maintain separation between highly rhizomatous plots containing CWR and CWR backcross lines [14].

4.6. Evaluation of Biomass-Related Traits in an Irrigated Production System

Dry matter yield BWR, CWR × BWR half-sib hybrid populations, and other large-statured perennial grasses were evaluated in a two-harvest irrigated production system over four years at the Western Colorado Research Center located in the Grand Valley of the Colorado River near Fruita (Table S1).

A seedbed was prepared for furrow irrigation using equipment commonly used in Grand Valley region, which resulted in a smooth, flat-bed surface with V-shaped furrows approximately 10-cm deep with 76-cm space between the bottom of each furrow. The experiment was planted on 20 October 2011 using a cone plot planter. Two seed rows were sown about 2-cm deep with 30-cm spacing within each bed and about 46-cm spacing between beds. The plot size was 3.05 m wide (four seed beds with a total of eight seed rows) and 4.57 m long for a total area of 13.94 m^2. The soil was a Glenton very fine sandy loam (coarse-loamy, mixed (calcareous), mesic family of Typic Torrifluvents). The plot area was soil sampled prior to planting, which was performed on 20 October 2011. The results of the soil test analysis were: pH 7.7, 0.4 mmhos/cm, 1.4% organic matter, 8 ppm NO_3-N, 13 ppm P, and 62 ppm K.

This experiment included two CWR × BWR half-sib hybrid populations plus selected reference populations of Altai wildrye (*Leymus angustus*); intermediate wheatgrass (*Thinopyrum intermedium*), switchgrass (*Panicum virgatum*), tall wheatgrass (*Thinopyrum intermedium*), and two species mixtures. 'Mustang' Altai wildrye; four BWR populations 4X Acc:636, 8X 'Continental', 8X 'Magnar', and 4X 'Trailhead'; 'Oahe' and 'Rush' intermediate wheatgrass; 'Alkar' tall wheatgrass; and two CWR × BWR half-sib hybrid populations were all seeded at a rate of 16.8 kg·ha^{-1} for each population in each plot.

A switchgrass variety mixture, comprised of cultivars 'Blackwell' and 'Dacotah', was seeded at rates of 3.36 kg·ha^{-1} for each variety with an overall seeding rate of 6.72 kg·ha^{-1}.

Plots were harvested twice each year, over a period of four years, with an automated forage plot harvester [59]. In the first harvest year of 2012, while plants were in the juvenile phase of development, plots were harvested on 24 July and 11 October with an application of 56 kg N ha^{-1} on 27 July. In the second harvest year of 2013, plots were harvested on 20 June and 22 October with an application of 56 kg N ha^{-1} on 24 June. In the third harvest year of 2014, plots were harvested on 19 June and 9 October, with an application of 56 kg N ha^{-1} + 122 kg P$_2$O$_5$ ha^{-1} on 20 June. In the fourth harvest year of 2015, plots were harvested on 19 June and 8 October, with applications of 56 kg N ha^{-1} on 10 April and 56 kg N ha^{-1} + 40 kg P$_2$O$_5$ ha^{-1} + 13 kg K$_2$O ha^{-1} on 21 June. During harvest, a small forage sample was obtained from each plot and oven-dried at 50 °C to a constant weight that was used to convert fresh weight yields to DMY. The experiment was furrow-irrigated each year with irrigation water from the Colorado River delivered through a canal system. No herbicides were applied at any time to the plots during the study.

4.7. Statistical Analysis of Trait Data

Field trait data were analyzed within and among years and locations using the MIXED procedure of SAS version 9.4 with the repeated option to model covariance structure between years (SAS Institute, Cary, North Carolina, USA). Population, years, and locations were assumed to be fixed effects with replications as random effects. Mean comparisons were made among populations using Fisher Protected Least Significant Difference tests at the $p = 0.05$ level of probability.

5. Conclusions

This study demonstrated a useful approach to produce hybrid seed for allogamous perennial grasses, which has been a long-standing problem in perennial grass improvement. Our study also provided new information on the biomass yield accumulation potential of BWR and other cool-season grasses in a two-harvest irrigated production system in a high-elevation, cold-desert environment of the western United States. Performance of BWR and other grasses in a two-harvest irrigated production system demonstrated significant regrowth potential of native grasses such as BWR. Comparisons of relative biomass yields across a wide range of irrigated and dryland environments of western North America, including new testing sites utilized in this study, indicate that cool-season grasses may have significant advantages over switchgrass under non-irrigated or limited-irrigation systems in this region. Although tall wheatgrass was clearly the best yielding population in this irrigated testing environment, BWR and CWR × BWR hybrids performed as good as or better than tall wheatgrass and other introduced species in the dryland range environments. Results the two-harvest irrigated field experiment demonstrated that switchgrass can also do well in a cold-desert environment if irrigation supplies are sufficient throughout the entire growing season. However, development and testing of cool-season perennial grass species, hybrids, and varieties across a wide range of growing environments is needed to expand the range of adaptation of grasses used for low-input bioenergy and forage production in the western United States and other parts of North America. Although biomass yields of BWR and other perennial grasses varied greatly over different environments, we detected significant genetic variation for this trait across widely various conditions. Some hybrids of CWR × BWR exhibited indications of mid- or high-parent heterosis and there were very large differences between the biomass yields obtained using a small sample of only two different CWR genets as seed parents and two BWR populations as pollen parents. Future efforts to develop CWR × BWR hybrids should focus on the identification and utilization of superior CWR genets as half-sib hybrid seed parents in addition to superior BWR pollen parents.

Author Contributions: "S.R.L., C.H.P., K.B.J., and I.W.M. conceived and designed experiments; T.A.J. developed the TC1 and TC2 F_1 hybrids and the interploidy hybrid 'Continental'; S.R.L., M.D.R., and B.L.W. analyzed data; S.R.L. and J.E.S. wrote the paper." All authors read and approved the paper.

Conflicts of Interest: The authors declare no conflict of interest.

References

1. Lewandowski, I.; Scurlock, J.M.O.; Lindvall, E.; Christou, M. The development and current status of perennial rhizomatous grasses as energy crops in the U.S. and Europe. *Biomass Bioenergy* **2003**, *25*, 335–361. [CrossRef]

2. McLaughlin, S.B.; Kszos, L.A. Development of switchgrass (*Panicum virgatum*) as a bioenergy feedstock in the United States. *Biomass Bioenergy* **2005**, *28*, 515–535. [CrossRef]

3. Parrish, D.J.; Fike, J.H. The biology and agronomy of switchgrass for biofuels. *Crit. Rev. Plant Sci.* **2005**, *24*, 423–459. [CrossRef]

4. Wright, L.; Turhollow, A. Switchgrass selection as a "model" bioenergy crop: A history of the process. *Biomass Bioenergy* **2010**, *34*, 851–868. [CrossRef]

5. Casler, M.D.; Vogel, K.P.; Harrison, M. Switchgrass germplasm resources. *Crop Sci.* **2015**, *55*, 2463–2478. [CrossRef]

6. Monono, E.M.; Nyren, P.E.; Berti, M.T.; Pryor, S.W. Variability in biomass yield, chemical composition, and ethanol potential of individual and mixed herbaceous biomass species grown in North Dakota. *Ind. Crops Prod.* **2013**, *41*, 331–339. [CrossRef]

7. Wang, G.J.; Nyren, P.; Xue, Q.W.; Aberle, E.; Eriksmoen, E.; Tjelde, T.; Liebig, M.; Nichols, K.; Nyren, A. Establishment and yield of perennial grass monocultures and binary mixtures for bioenergy in North Dakota. *Agron. J.* **2014**, *106*, 1605–1613. [CrossRef]

8. Lee, D.; Owens, V.N.; Boe, A.; Koo, B.C. Biomass and seed yields of big bluestem, switchgrass, and intermediate wheatgrass in response to manure and harvest timing at two topographic positions. *CB Bioenergy* **2009**, *1*, 171–179. [CrossRef]

9. Harmoney, K.R. Cool-season grass biomass in the southern mixed-grass prairie region of the USA. *Bioenergy Res.* **2015**, *8*, 203–210. [CrossRef]

10. Jefferson, P.G.; McCaughey, W.P.; May, K.; Wosaree, J.; MacFarlane, L.; Wright, S.M.B. Performance of American native grass cultivars in the Canadian Prairie Provinces. *Nativ. Plants J.* **2002**, *3*, 24–33. [CrossRef]

11. Robins, J.G. Cool-season grasses produce more total biomass across the growing season than do warm-season grasses when managed with an applied irrigation gradient. *Biomass Bioenergy* **2010**, *34*, 500–505. [CrossRef]

12. Porensky, L.M.; Davison, J.; Leger, E.A.; Miller, W.W.; Goergen, E.M.; Espeland, E.K.; Carroll-Moore, E.M. Grasses for biofuels: A low water-use alternative for cold desert agriculture? *Biomass Bioenergy* **2014**, *66*, 133–142. [CrossRef]

13. Pearson, C.H.; Larson, S.R.; Keske, C.M.H.; Jensen, K.B. Native grasss for biomass production at high elevations. In *Industrial Crops: Breeding for Bioenergy and Bioproducts*; Cruz, M.V., Dierig, D.A., Eds.; Springer: New York, NY, USA, 2015; pp. 101–132.

14. Larson, S.R.; Jensen, K.B.; Robins, J.G.; Waldron, B.L. Genes and quantitative trait loci controlling biomass yield and forage quality traits in perennial wildrye. *Crop Sci.* **2014**, *54*, 111–126. [CrossRef]

15. Aguirre, A.A.; Studer, B.; Frei, U.; Lubberstedt, T. Prospects for hybrid breeding in bioenergy grasses. *Bioenergy Res.* **2012**, *5*, 10–19. [CrossRef]

16. Heaton, E.A.; Dohleman, F.G.; Miguez, A.F.; Juvik, J.A.; Lozovaya, V.; Widholm, J.; Zabotina, O.A.; Mcisaac, G.F.; David, M.B.; Voigt, T.B.; et al. *Miscanthus*: A promising biomass crop. *Adv. Bot. Res.* **2010**, *56*, 75–137.

17. Linde-Laursen, I. Cytogenetic analysis of *Miscanthus* 'Giganteus', an interspecific hybrid. *Hereditas* **1993**, *119*, 297–300. [CrossRef]

18. Hodkinson, T.R.; Chase, M.W.; Takahashi, C.; Leitch, I.J.; Bennett, M.D.; Renvoize, S.A. The use of DNA sequencing (ITS and *trnl*-f), AFLP, and fluorescent in situ hybridization to study allopolyploid *Miscanthus* (Poaceae). *Am. J. Bot.* **2002**, *89*, 279–286. [CrossRef] [PubMed]

19. Yamada, T. *Miscanthus*. In *Industrial Crops: Breeding for Bioenergy and Bioproducts*; Cruz, M.V., Dierig, D.A., Eds.; Springer: New York, NY, USA, 2015; pp. 43–66.

20. Robson, P.; Jensen, E.; Hawkins, S.; White, S.R.; Kenobi, K.; Clifton-Brown, J.; Donnison, I.; Farrar, K. Accelerating the domestication of a bioenergy crop: Identifying and modelling morphological targets for sustainable yield increase in *Miscanthus*. *J. Exp. Bot.* **2013**, *64*, 4143–4155. [CrossRef] [PubMed]

21. Nishiwaki, A.; Mizuguti, A.; Kuwabara, S.; Toma, Y.; Ishigaki, G.; Miyashita, T.; Yamada, T.; Matuura, H.; Yamaguchi, S.; Rayburn, A.L.; et al. Discovery of natural *Miscanthus* (Poaceae) triploid plants in sympatric populations of *Miscanthus sacchariflorus* and *Miscanthus sinensis* in southern Japan. *Am. J. Bot.* **2011**, *98*, 154–159. [CrossRef] [PubMed]

22. Dwiyanti, M.S.; Rudolph, A.; Swaminathan, K.; Nishiwaki, A.; Shimono, Y.; Kuwabara, S.; Matuura, H.; Nadir, M.; Moose, S.; Stewart, J.R.; et al. Genetic analysis of putative triploid *Miscanthus* hybrids and tetraploid *M. sacchariflorus* collected from sympatric populations of Kushima, Japan. *Bioenergy Res.* **2013**, *6*, 486–493. [CrossRef]

23. Tamura, K.; Sanada, Y.; Shoji, A.; Okumura, K.; Uwatoko, N.; Anzoua, K.G.; Sacks, E.J.; Yamada, T. DNA markers for identifying interspecific hybrids between *Miscanthus sacchariflorus* and *Miscanthus sinensis*. *Grassl. Sci.* **2015**, *61*, 160–166. [CrossRef]

24. Hodkinson, T.R.; Klaas, M.; Jones, M.B.; Prickett, R.; Barth, S. *Miscanthus*: A case study for the utilization of natural genetic variation. *Plant Genet. Resour. C* **2015**, *13*, 219–237. [CrossRef]

25. Martinez-Reyna, J.M.; Vogel, K.P. Heterosis in switchgrass: Spaced plants. *Crop Sci.* **2008**, *48*, 1312–1320. [CrossRef]

26. Vogel, K.P.; Mitchell, K.B. Heterosis in switchgrass: Biomass yield in swards. *Crop Sci.* **2008**, *48*, 2159–2164. [CrossRef]

27. Casler, M.D. Heterosis and reciprocal-cross effects in tetraploid switchgrass. *Crop Sci.* **2014**, *54*, 2063–2069. [CrossRef]

28. Casler, M.D.; Vogel, K.P. Selection for biomass yield in upland, lowland, and hybrid switchgrass. *Crop Sci.* **2014**, *54*, 626–636. [CrossRef]

29. Zalapa, J.E.; Price, D.L.; Kaeppler, S.M.; Tobias, C.M.; Okada, M.; Casler, M.D. Hierarchical classification of switchgrass genotypes using SSR and chloroplast sequences: Ecotypes, ploidies, gene pools, and cultivars. *Theor. Appl. Genet.* **2011**, *122*, 805–817. [CrossRef] [PubMed]

30. Lu, F.; Lipka, A.E.; Glaubitz, J.; Elshire, R.; Cherney, J.H.; Casler, M.D.; Buckler, E.S.; Costich, D.E. Switchgrass genomic diversity, ploidy, and evolution: Novel insights from a network-based SNP discovery protocol. *PLoS Genet.* **2013**, *9*, 139–147. [CrossRef] [PubMed]

31. Larson, S.R.; Wu, X.L.; Jones, T.A.; Jensen, K.B.; Chatterton, N.J.; Waldron, B.L.; Robins, J.G.; Bushman, B.S.; Palazzo, A.J. Comparative mapping of growth habit, plant height, and flowering QTLs in two interspecific families of *Leymus*. *Crop Sci.* **2006**, *46*, 2526–2539. [CrossRef]

32. Reynolds, T.D.; Fraley, L. Root profiles of some native and exotic plant-species in southeastern Idaho. *Environ. Exp. Bot.* **1989**, *29*, 241–248. [CrossRef]

33. Culumber, C.M.; Larson, S.R.; Jensen, K.B.; Jones, T.A. Genetic structure of Eurasian and North American *Leymus* (Triticeae) wildryes assessed by chloroplast DNA sequences and AFLP profiles. *Plant Syst. Evol.* **2011**, *294*, 207–225. [CrossRef]

34. Johnson, R.; Vance-Borland, K. Linking genetic variation in adaptive plant traits to climate in tetraploid and octoploid basin wildrye [*Leymus cinereus* (Scribn. & Merr.) a. Love] in the western U.S. *PLoS ONE* **2016**, *11*, e0148982.

35. Jones, T.A.; Parr, S.D.; Winslow, S.R.; Rosales, M.A. Notice of release of 'Continental' basin wildrye. *Nativ. Plants J.* **2009**, *10*, 57–61. [CrossRef]

36. USDA-NRCS. *Conservation Plant Release Brochure for 'Rio' Beardless Wild Rye (Leymus triticoides Buckley)*; United States Department of Agriculture, Natural Resources Conservation Service, California Plant Materials Center: Lockeford, CA, USA, 2014.

37. Suyama, H.; Benes, S.E.; Robinson, P.H.; Getachew, G.; Grattan, S.R.; Grieve, C.M. Biomass yield and nutritional quality of forage species under long-term irrigation with saline-sodic drainage water: Field evaluation. *Anim. Feed Sci. Technol.* **2007**, *135*, 329–345. [CrossRef]

38. Benes, S.E.; Adhikari, D.D.; Grattan, S.R.; Snyder, R.L. Evapotranspiration potential of forages irrigated with saline-sodic drainage water. *Agric. Water Manag.* **2012**, *105*, 1–7. [CrossRef]

39. Yun, L.; Larson, S.R.; Mott, I.W.; Jensen, K.B.; Staub, J.E. Genetic control of rhizomes and genomic localization of a major-effect growth habit QTL in perennial wildrye. *Mol. Genet. Genom.* **2014**, *289*, 383–397. [CrossRef] [PubMed]

40. Martinez-Reyna, J.M.; Vogel, K.P.; Caha, C.; Lee, D.J. Meiotic stability, chloroplast DNA polymorphisms, and morphological traits of upland × lowland switchgrass reciprocal hybrids. *Crop Sci.* **2001**, *41*, 1579–1583. [CrossRef]

41. Johnston, S.A.; Dennijs, T.P.M.; Peloquin, S.J.; Hanneman, R.E. The significance of genic balance to endosperm development in interspecific crosses. *Theor. Appl. Genet.* **1980**, *57*, 5–9. [CrossRef] [PubMed]

42. USDA-NRCS. *Notice of Release of 'Rio' Beardless Widlrye*; USDA-Natural Resources Conservation Service: Lockeford, CA, USA, 1991.

43. Brummer, E.C. Capturing heterosis in forage crop cultivar development. *Crop Sci.* **1999**, *39*, 943–954. [CrossRef]

44. Knowles, R.P. Comparison of cultivar hybrids and blends with pure cultivars in crested wheatgrass. *Can. J. Plant Sci.* **1979**, *59*, 1019–1023. [CrossRef]

45. Foster, C.A. Interpopulational and intervarietal hybridization in *Lolium perenne* breeding: Heterosis under non-competitive conditions. *J. Agric. Sci.* **1971**, *76*, 107–130. [CrossRef]

46. Foster, C.A. Interpopulational and intervarietal F1 hybrids in *Lolium perenne*: Performance in field sward conditions. *J. Agric. Sci.* **1973**, *80*, 463–477. [CrossRef]

47. Lemeziene, N.; Norkeviciene, E.; Liatukas, Z.; Dabkeviciene, G.; Ceceviciene, J.; Butkute, B. Switchgrass from North Dakota - an adaptable and promising energy crop for northern regions of Europe. *Acta Agric. Scand.* **2015**, *65*, 118–124.

48. Sage, R.F.; Peixoto, M.D.; Friesen, P.; Deen, B. C-4 bioenergy crops for cool climates, with special emphasis on perennial c-4 grasses. *J. Exp. Bot.* **2015**, *66*, 4195–4212. [CrossRef] [PubMed]

49. Cash, S.D.; Majerus, M.E.; Scheetz, J.C.; Holzworth, L.K.; Murphy, C.L.; Wichman, D.M.; Bowman, H.F.; Ditterline, R.L. Registration of 'Trailhead' basin wildrye. *Crop Sci.* **1998**, *38*, 278. [CrossRef]

50. Alderson, J.S.; Sharp, W.C.; Hanson, A.A.; U.S. Department of Agriculture. *Grass Varieties in the United States*; CRC Lewis Publishers: Boca Raton, FL, USA, 1995.

51. Wu, X.L.; Larson, S.R.; Hu, Z.M.; Palazzo, A.J.; Jones, T.A.; Wang, R.R.C.; Jensen, K.B.; Chatterton, N.J. Molecular genetic linkage maps for allotetraploid *Leymus* wildryes (Gramineae:Triticeae). *Genome* **2003**, *46*, 627–646. [CrossRef] [PubMed]

52. Jensen, K.B.; Zhang, Y.F.; Dewey, D.R. Mode of pollination of perennial species of the Triticeae in relation to genomically defined genera. *Can. J. Plant Sci.* **1990**, *70*, 215–225. [CrossRef]

53. PRISM Climate Group. Oregon State University. Available online: http://prism.oregonstate.edu (accessed on 23 December 2016).

54. Utah Climate Center, Utah State University. Available online: https://climate.usurf.usu.edu (accessed on 23 December 2016).

55. Vos, P.; Hogers, R.; Bleeker, M.; Reijans, M.; van de Lee, T.; Hornes, M.; Frijters, A.; Pot, J.; Peleman, J.; Kuiper, M.; et al. AFLP: A new technique for DNA fingerprinting. *Nucleic Acids Res.* **1995**, *23*, 4407–4414. [CrossRef] [PubMed]

56. Falush, D.; Stephens, M.; Pritchard, J.K. Inference of population structure using multilocus genotype data: Dominant markers and null alleles. *Mol. Ecol. Notes* **2007**, *7*, 574–578. [CrossRef] [PubMed]

57. Pritchard, J.K.; Stephens, M.; Donnelly, P. Inference of population structure using multilocus genotype data. *Genetics* **2000**, *155*, 945–959. [PubMed]

58. Leonard, A.C.; Franson, S.E.; Hertzberg, V.S.; Smith, M.K.; Toth, G.P. Hypothesis testing with the similarity index. *Mol. Ecol.* **1999**, *8*, 2105–2114. [CrossRef] [PubMed]

59. Pearson, C.H. An updated, automated commercial swather for harvesting forage plots. *Agron. J.* **2007**, *99*, 1382–1388. [CrossRef]

PERMISSIONS

All chapters in this book were first published in AGRONOMY, by MDPI; hereby published with permission under the Creative Commons Attribution License or equivalent. Every chapter published in this book has been scrutinized by our experts. Their significance has been extensively debated. The topics covered herein carry significant findings which will fuel the growth of the discipline. They may even be implemented as practical applications or may be referred to as a beginning point for another development.

The contributors of this book come from diverse backgrounds, making this book a truly international effort. This book will bring forth new frontiers with its revolutionizing research information and detailed analysis of the nascent developments around the world.

We would like to thank all the contributing authors for lending their expertise to make the book truly unique. They have played a crucial role in the development of this book. Without their invaluable contributions this book wouldn't have been possible. They have made vital efforts to compile up to date information on the varied aspects of this subject to make this book a valuable addition to the collection of many professionals and students.

This book was conceptualized with the vision of imparting up-to-date information and advanced data in this field. To ensure the same, a matchless editorial board was set up. Every individual on the board went through rigorous rounds of assessment to prove their worth. After which they invested a large part of their time researching and compiling the most relevant data for our readers.

The editorial board has been involved in producing this book since its inception. They have spent rigorous hours researching and exploring the diverse topics which have resulted in the successful publishing of this book. They have passed on their knowledge of decades through this book. To expedite this challenging task, the publisher supported the team at every step. A small team of assistant editors was also appointed to further simplify the editing procedure and attain best results for the readers.

Apart from the editorial board, the designing team has also invested a significant amount of their time inunderstanding the subject and creating the most relevant covers. They scrutinized every image to scout for the most suitable representation of the subject and create an appropriate cover for the book.

The publishing team has been an ardent support to the editorial, designing and production team. Their endless efforts to recruit the best for this project, has resulted in the accomplishment of this book. They are a veteran in the field of academics and their pool of knowledge is as vast as their experience in printing. Their expertise and guidance has proved useful at every step. Their uncompromising quality standards have made this book an exceptional effort. Their encouragement from time to time has been an inspiration for everyone.

The publisher and the editorial board hope that this book will prove to be a valuable piece of knowledge forresearchers, students, practitioners and scholars across the globe.

LIST OF CONTRIBUTORS

Sameh H. Youseif, Fayrouz H. Abd El-Megeed and Saleh A. Saleh
National Gene Bank and Genetic Resources, Agricultural Research Center, Giza 12619, Egypt

Satoru Fukagawa and Kenichi Kataoka
Nagasaki Agricultural and Forestry Technical Development Center, Shimabara, Nagasaki 859-1404, Japan

Yasuyuki Ishii
Faculty of Agriculture, University of Miyazaki, Miyazaki 889-2192, Japan

Huihui Zhang
USDA-ARS Water Management Research Unit, San Joaquin Valley Agricultural Sciences Center, 9611 S. Riverbend Ave., Parlier, CA 93648, USA
Curent Affiliation: USDA-ARS Water Management and System Research Unit, 2150 Centre Avenue, Building D, Suite 320, Fort Collins, CO 80526, USA

Dong Wang and Jim L. Gartung
USDA-ARS Water Management Research Unit, San Joaquin Valley Agricultural Sciences Center, 9611 S. Riverbend Ave., Parlier, CA 93648, USA

M. Lukas Seehausen, Nigel V. Gale, Stefana Dranga, Virginia Hudson, Norman Liu, Jane Michener, Emma Thurston, Charlene Williams, Sandy M. Smith and Sean C. Thomas
Faculty of Forestry, University of Toronto, 33 Willcocks Street, Toronto, ON M5S 3B3, Canada

Liliana Avila-Ospina, Gilles Clément and Céline Masclaux-Daubresse
INRA-AgroParis Tech, Institut Jean-Pierre Bourgin, UMR1318, ERL CNRS 3559, Saclay Plant Sciences, Versailles 78000, France

James Bunce
Crop Systems and Global Change Laboratory, USDA-ARS, Beltsville Agricultural Research Center, 10300 Baltimore Avenue, Beltsville, MD 20705-2350, USA

Klaus Sieling, Ulf Böttcher and Henning Kage
Agronomy and Crop Science, Institute of Crop Science and Plant Breeding, Christian-Albrechts-University of Kiel, Hermann-Rodewald-Str. 9, D-24118 Kiel, Germany

Elvire Line Sossa and Guillaume Lucien Amadji
Faculty of Agronomic Sciences, Research Unit Eco-Pedology, University of Abomey-Calavi, Laboratory of Soil Sciences, Cotonou, Benin

Codjo Emile Agbangba
Laboratory of Biomathematics and Forests Estimations, Faculty of Agronomic Sciences, University of Abomey-Calavi, Cotonou, Benin

Sènan Gbèmawonmèdé Gwladys Stéfania Accalogoun and Djidjoho Joseph Hounhouigan
Laboratory of Nutrition and Alimentary Sciences, Faculty of Agronomic Sciences, University of Abomey-Calavi, Cotonou, Benin

Kossi Euloge Agbossou
Laboratory of Hydraulic and Water Control, Faculty of Agronomic Sciences, University of Abomey-Calavi, 01 BP 526, Cotonou, Benin

Emma J. Bennett and Carol Wagstaff
Department of Food and Nutritional Sciences and Centre for Food Security, University of Reading, Whiteknights, Reading, Berkshire RG6 6AP, UK

Christopher J. Brignell
School of Mathematical Sciences, University of Nottingham, University Park, Nottingham NG7 2RD, UK

Pierre W. C. Carion, Samantha M. Cook and Peter J. Eastmond
Rothamsted Research, Harpenden, Hertfordshire AL5 2JQ, UK

Graham R. Teakle
Warwick Crop Centre, University of Warwick, Wellesbourne CV35 9EF, UK

John P. Hammond
School of Agriculture, Policy and Development and Centre for Food Security, University of Reading, Earley Gate, Whiteknights Road, Reading RG6 6AR, UK

Clare Love
Rothamsted Research, Harpenden, Hertfordshire AL5 2JQ, UK

Graham J. King
Southern Cross Plant Science, Southern Cross University, Lismore, NSW 2480, Australia

Jeremy A. Roberts
Office of the Vice-Chancellor, 18 Portland Villas, University of Plymouth, Plymouth, Devon PL4 8AA, UK

Malin C. Broberg and Håkan Pleijel
Department of Biological and Environmental Sciences, University of Gothenburg, SE-40530 Göteborg, Sweden

Petra Högy
Institute of Landscape and Plant Ecology, University of Hohenheim, Ökologiezentrum 2, August-von-Hartmann Str. 3, D-70599 Stuttgart, Germany

Julien Verzeaux, Elodie Nivelle, David Roger, Frédéric Dubois and Thierry Tetu
Ecologie et Dynamique des Systèmes Anthropisés (EDYSAN, FRE 3498 CNRS UPJV), Laboratoire d'Agroécologie, Ecophysiologie et Biologie intégrative, Université de Picardie Jules Verne, 33 rue St Leu, Amiens CEDEX 80039, France

Bertrand Hirel
Adaptation des Plantes à leur Environnement, Unité Mixte de Recherche 1318, Institut Jean-Pierre Bourgin, Institut National de la Recherche Agronomique, Centre de Versailles-Grignon, R.D. 10, Versailles CEDEX F-78026, France

Amanda J. Ashworth
United States Department of Agriculture, Agricultural Research Service, Poultry Production and Product Safety Research Unit, Fayetteville, AR 72927, USA

Fred L. Allen, Kara S. Warwick and Gary E. Bates
Department of Plant Sciences, University of Tennessee, Knoxville, TN 37996, USA

Patrick D. Keyser
Department of Forestry, Wildlife & Fisheries, University of Tennessee, Knoxville, TN 37996, USA

Don D. Tyler
Department of Biosystems Engineering and Soil Science, University of Tennessee, Knoxville, TN 37996, USA

Paris L. Lambdin
Department of Entomology and Plant Pathology, University of Tennessee, Knoxville, TN 37996, USA

Dan H. Pote
United States Department of Agriculture, Agricultural Research Service, Dale Bumpers Small Farms Research Center, 6883 S. Hwy 23, Booneville, AR 72927, USA

Chris Blok, Caroline van der Salm, Jantineke Hofland-Zijlstra, Marta Streminska and Barbara Eveleens
Wageningen Plant Research, Glasshouse Horticulture, Violierenweg 1, 2665 MV Bleiswijk, The Netherlands

Inge Regelink
Wageningen Environmental Research, Wageningen University & Research, Droevendaalsesteeg 3, 6708 PB Wageningen, The Netherlands

Lydia Fryda and Rianne Visser
Energy Research Centre of the Netherlands (ECN), 1755 ZG Petten, The Netherlands

Steven R. Larson, Kevin B. Jensen, Thomas A. Jones, Ivan W. Mott, Matthew D. Robbins, Jack E. Staub and Blair L. Waldron
USDA-ARS, Forage and Range Research, Utah State University, Logan, UT 84322, USA

Calvin H. Pearson
Agriculture Experiment Station, Department of Soil and Crop Sciences, Colorado State University, Fruita, CO 81521, USA

Index